New Challenges in Marine Aquaculture Research

New Challenges in Marine Aquaculture Research

Guest Editors

Ana Catarina Matias
Carlos Andrade

MDPI

Basel • Beijing • Wuhan • Barcelona • Belgrade • Novi Sad • Cluj • Manchester

Guest Editors
Ana Catarina Matias
Aquaculture Research Station
Portuguese Institute of the Sea and Atmosphere
Olhão
Portugal

Carlos Andrade
Marine and Environmental Sciences
Agência Regional para o Desenvolvimento da Investigação Tecnologia e Inovação
Funchal
Portugal

Editorial Office
MDPI AG
Grosspeteranlage 5
4052 Basel, Switzerland

This is a reprint of the Special Issue, published open access by the journal *Journal of Marine Science and Engineering* (ISSN 2077-1312), freely accessible at: https://www.mdpi.com/journal/jmse/special_issues/M37G7483V8.

For citation purposes, cite each article independently as indicated on the article page online and as indicated below:

Lastname, A.A.; Lastname, B.B. Article Title. *Journal Name* **Year**, *Volume Number*, Page Range.

ISBN 978-3-7258-3805-9 (Hbk)
ISBN 978-3-7258-3806-6 (PDF)
https://doi.org/10.3390/books978-3-7258-3806-6

Contents

About the Editors

Ana Catarina Matias

Ana Catarina Parreira Peres Matias Luna Araujo earned both her degree and Ph.D. in Biochemistry, with a specialization in Cellular and Molecular Biology, from the University of Algarve, Portugal. She is currently a contracted researcher at the Aquaculture Research Station (EPPO) of the Portuguese Institute of the Sea and Atmosphere (IPMA). Her comprehensive background and dedication make her a valuable asset in the realm of biochemical research and aquaculture development. To date, her work has resulted in seventeen publications, one book chapter, and three magazine articles. Her research interests include fish physiology, metabolism, and nutrition.

Carlos Andrade

Dr. Carlos Andrade is a Principal Researcher with extensive experience across all major aquaculture taxa and maritime planning. His focus lies in sustainable, low-trophic aquaculture species and innovative production systems. He leads a research unit affiliated with the University of Madeira and the Regional Agency for Development of Research, Technology and Innovation, fosters collaborative projects uniting academia, government, and the private sector, and holds active roles with international research and conservation organizations, including the IUCN—Ecosystem-based Aquaculture Group. He has published widely in peer-reviewed journals.

Preface

The COVID-19 pandemic posed significant challenges to scientific research, forcing research institutes to respond rapidly to an urgent global humanitarian crisis. Just as the pandemic began to ease, conflict in Europe emerged, further exacerbating concerns about future uncertainties. This unpredictable situation triggered a global energy and food crisis, highlighting the critical need for increased investment in food systems to ensure secure, affordable, healthy, and nutritious food for a growing population. Aquatic products have gained recognition as key components of these systems, with production in 2020 exceeding 60 percent above the average from the 1990s, largely driven by the expansion of aquaculture.

Despite notable progress in marine aquaculture, researchers still face significant challenges, such as species diversification, improving animal growth efficiency, mitigating climate change impacts, developing alternatives to fishmeal and oil, safeguarding natural resources, and minimizing the industry's environmental footprint. To meet the rising global demand for aquatic foods, marine aquaculture research remains crucial and must adopt an integrated approach to further advance and strengthen the sector.

Ana Catarina Matias and Carlos Andrade
Guest Editors

Journal of
Marine Science and Engineering

Editorial

New Challenges in Marine Aquaculture Research

Ana Catarina Matias [1,*] and Carlos Andrade [2,*]

1 EPPO—Aquaculture Research Station, IPMA—Institute of the Sea and Atmosphere, Av. do Parque Natural da Ria Formosa s/n, 8700-194 Olhão, Portugal
2 MARE—Marine and Environmental Sciences Centre/ARNET—Aquatic Research Network, Agência Regional para o Desenvolvimento da Investigação Tecnologia e Inovação (ARDITI), Caminho da Penteada Edif Madeira Tecnopolo, 9020-105 Funchal, Portugal
* Correspondence: ana.matias@ipma.pt (A.C.M.); carlos.a.andrade@madeira.gov.pt (C.A.)

Marine aquaculture has witnessed remarkable growth in recent decades, contributing significantly to global food security and economic development. As the world population continues to expand and dietary preferences shift towards healthier, protein-rich options, the demand for seafood, particularly from aquaculture, is expected to rise significantly [1,2]. However, the sector faces numerous hurdles to ensure its continued success and sustainability. One of the most pressing challenges is the increasing pressure on natural ecosystems. The expansion of aquaculture operations can lead to environmental impacts such as water quality and habitat degradation, and competition with wild fish populations [3]. Moreover, the industry partly relies on wild-caught fish for feed ingredients [4]. Climate change poses another significant threat to marine aquaculture. Rising sea temperatures, ocean acidification, and changes in ocean currents can have detrimental effects on aquaculture species, leading to diseases, increased mortality rates, reduced growth, and altered reproductive cycles [5,6]. In addition, the marine aquaculture industry also faces economic and social pressures. Market fluctuations and regulatory hurdles can significantly impact the profitability and sustainability of operations, whereas ensuring fair labor practices and the well-being of workers in the aquaculture sector remains a critical challenge [7]. To address these complex challenges and ensure the long-term sustainability of marine aquaculture, innovative and collaborative research and development are essential. This Special Issue of the *Journal of Marine Science and Engineering* presents a comprehensive collection of research papers spanning a wide range of topics. It includes studies focused on environmental, ecological, and sustainability aspects (contributions 1 to 5) as well as innovative approaches to feed and growth performance optimization (contributions 6 to 13). It addresses critical challenges such as species diversification, animal growth efficiency, and the impact of climate change, while also focusing on alternative feed sources, natural resource conservation, and minimizing environmental impacts. This Special Issue offers valuable insights and solutions for promoting a more sustainable and productive future in the field. The research topics and key findings featured in this Special Issue include the following.

Hengjie et al. (Contribution 1) presented a systematic review examining the effects of ocean acidification (OA) on seaweed aquaculture, analyzing studies from 2001 to 2022. The review explored how seaweeds acclimate to increasing CO_2, focusing on the effects of OA on photosynthesis and nutrient uptake, while identifying knowledge gaps in mitigation strategies. The findings revealed that while certain seaweed species may experience enhanced productivity, others could suffer reduced growth and economic losses if critical biological systems are compromised. The review highlighted the need for future research to focus on actionable mitigation strategies and to assess the economic impacts of OA on seaweed farming.

Received: 5 October 2024
Accepted: 18 January 2025
Published: 10 February 2025

Citation: Matias, A.C.; Andrade, C. New Challenges in Marine Aquaculture Research. *J. Mar. Sci. Eng.* **2025**, *13*, 324. https://doi.org/10.3390/jmse13020324

Deen et al. (Contribution 2) investigated the effects of coastal shellfish aquaculture on benthic–pelagic systems in Onagawa Bay, Japan, through monthly field surveys conducted in 2023. The study demonstrated that shellfish farms significantly influence both the water column and the seafloor by consuming phytoplankton and depositing organic matter as biodeposits, which alter sediment characteristics and the local benthic community. The surveys revealed that polychaetes dominated the benthic macrofauna (86.3%). Additionally, biomass and biodiversity were higher near the aquaculture sites, while chlorophyll-a concentrations and sediment organic matter were lower. However, the combination of global-warming-induced water column stratification and excess organic input may result in hypoxic conditions, presenting new challenges for the sustainability of aquaculture.

Slobodskova et al. (Contribution 3) addressed the increasing mass mortality of cultured seaside scallops (*Mizuhopecten yessoensis*) in the Sea of Japan, proposing a rapid diagnostic method utilizing DNA damage and oxidative stress markers for the early detection of health threats. The study revealed a correlation between elevated DNA damage, malondialdehyde accumulation in tissues, and the mortality of older scallops. This predictive approach offers aquaculture farms a valuable tool for mitigating losses and improving scallop farming management practices.

Lee et al. (Contribution 4) explored the potential of utilizing offshore wind farms (OWFs) for scallop aquaculture, focusing on the growth of Zhikong scallops (*Chlamys farreri*) and bay scallops (*Argopecten irradians*) in South Korea. The results indicated that scallop growth, in terms of shell size and weight, was significantly lower at OWFs compared to traditional farming sites, though consistent growth patterns were observed. Additionally, species-specific differences in mortality rates were noted, with temperature identified as a key factor affecting scallop growth. This study offers valuable insights into the feasibility of farming marine bivalves, such as scallops, in underutilized OWF areas, providing guidance for future aquaculture practices in these environments.

Castilla-Gavilán et al. (Contribution 5) examined the carbon footprint of various land-based marine aquaculture systems, focusing on strategies to reduce emissions and enhance sustainability. Key approaches include polyculture, the use of low-trophic-level species, and improvements in waste and energy management. Systems such as Recirculating Aquaculture Systems (RASs), Biofloc Technology (BFT), and Integrated Multi-Trophic Aquaculture (IMTA) were identified as offering significant environmental benefits, with IMTA considered the most eco-friendly option, though its carbon footprint remains challenging to quantify. The study emphasized the need for further research to optimize these methods and improve their sustainability.

Basto et al. (Contribution 6) explored the effects of replacing fishmeal (FM) with defatted *Tenebrio molitor* larvae meal (dTM) on the stress response of European sea bass. Three diets were tested: a control diet with FM, and two others where 50% and 100% of FM was replaced by dTM. After a 16-week feeding period, the fish were exposed to an acute stressor (1 min air exposure). The study found that seabass exhibited a clear stress response, shown by increased hemoglobin, hematocrit, and various plasma metabolites, regardless of diet. While partial FM substitution did not affect stress responses, full FM replacement led to higher liver antioxidant enzyme activity, particularly in total peroxidase and superoxide dismutase. Despite this increase in antioxidant activity, insect meal remains a promising sustainable alternative in aquafeeds.

Claessens et al. (Contribution 7) evaluated the use of mussel meal (*Perna perna*) as a feed additive for whiteleg shrimp (*Litopenaeus vannamei*) to improve growth and cold resistance. Five diets, containing varying levels of mussel meal inclusion (0–4%), were tested over an 8-week period. Shrimp fed diets with 1–2% mussel meal exhibited significantly better growth performance and a lower feed conversion ratio compared to other groups, while no differences in cold resistance were noted. The optimal inclusion level for enhancing growth was identified as 1.73–2.00% based on quadratic regression models.

Vale Pereira et al. (Contribution 8) compared alternative feed formulations for rainbow trout, demonstrating improved growth performance and higher consumer acceptance for diets without processed animal proteins. Long-term predictions suggest that some of these alternative diets could also be economically viable for trout farming, offering a sustainable option for the industry.

Rajendran et al. (Contribution 9) explored the anti-inflammatory properties of the acetone extract from the lichen *Parmotrema austrosinense* using a zebrafish model. The study demonstrated that *P. austrosinense* possesses significant anti-inflammatory potential, making it one of the most comprehensive in vivo analyses using zebrafish for modeling inflammation.

Soares et al. (Contribution 10) developed FEEDNETICS, a nutrient-based model designed to aid fish farming decisions by simulating individual fish growth and scaling the results to population levels. The model was validated across key aquaculture species, including gilthead seabream, European seabass, Atlantic salmon, rainbow trout, and Nile tilapia. FEEDNETICS helps optimize aquaculture by converting data into actionable insights, supporting experimental design, and assessing the nutritional and environmental impacts on fish farming systems.

Hotos et al. (Contribution 11) assessed the salinity tolerance of two copepod species, *Tigriopus* sp. and *Tisbe holothuriae*, from the Messolonghi Lagoon in Greece. Both species demonstrated strong resilience across a wide salinity range. *Tetraselmis suecica* was identified as the most effective feed for *Tigriopus nauplii*, while *Rhodomonas salina* and *Dunaliella salina* were optimal for *T. holothuriae*. However, high salinity levels negatively impacted reproduction. These copepods show significant potential for use in ecological research and as live feed in marine hatcheries.

Luís et al. (Contribution 12) investigated the effects of different microalgal diets and dietary proportions on the larval development, growth, and survival of the sea urchin *Sphaerechinus granularis* at the premetamorphosis stage. Three diets were tested: *Dunaliella tertiolecta*, *Rhodomonas marina*, and a combination of both, with varying cell densities (low, medium, and high dietary proportions). The results showed that lower cell densities and a combined microalgae diet positively influenced larval survival. This approach is expected to enhance larval production and provide valuable insights for future research in sea urchin aquaculture.

Matias et al. (Contribution 13) evaluated the impact of long-term day/night temperature oscillations on gilthead seabream (*Sparus aurata*) juveniles. Fish were exposed to two thermal regimes: a constant temperature of approximately 19 °C and a daily temperature cycle oscillating between 19 °C and 13 °C for a 67-day period. The results showed that temperature fluctuations negatively impacted fish growth efficiency, reduced fatty acid levels in tissues, and altered blood parameters. In contrast, maintaining a constant temperature of ~19 °C optimized both fish growth and health. Therefore, a constant temperature is recommended for gilthead seabream production, although the choice of heating energy source should align with operational conditions and business strategy.

This Special Issue showcases groundbreaking research that sheds light on the pressing challenges confronting marine aquaculture and highlights the innovative solutions being developed to address them. By focusing on species diversification, improving production efficiency, mitigating climate change impacts, exploring alternative feed sources, and adopting sustainable practices, the marine aquaculture industry can continue to grow and thrive while minimizing its environmental footprint. These topics remain critical areas for future research, which is the focus of *New Challenges in Marine Aquaculture Research—2nd Edition*.

Author Contributions: Conceptualization, A.C.M.; investigation, A.C.M. and C.A.; resources, A.C.M. and C.A.; writing—original draft preparation, C.A.; writing—review and editing, A.C.M. and C.A. All authors have read and agreed to the published version of the manuscript.

Funding: This research received no external funding.

Acknowledgments: As Guest Editors of the Special Issue *New Challenges in Marine Aquaculture*, we would like to express our sincere gratitude to the authors for sharing their valuable research contributions, to the reviewers for their through and insightful evaluations, and to the editorial team for their dedication. Each played a crucial role in making this Special Issue a success.

Conflicts of Interest: The authors declare no conflicts of interest.

List of Contributions:

1. Hengjie, T.; Das, S.K.; Zainee, N.F.A.; Yana, R.; Rozaimi, M. Ocean Acidification and Aquacultured Seaweeds: Progress and Knowledge Gaps. *J. Mar. Sci. Eng.* **2023**, *11*, 78. https://doi.org/10.3390/jmse11010078.
2. Deen, A.; Kitajima, S.; Sato-Okoshi, W.; Fujii, T. Seasonal Variability in the Influence of Coastal Aquaculture Operation on Benthic–Pelagic Coupling Processes in Shallow Aquatic Ecosystems. *J. Mar. Sci. Eng.* **2024**, *12*, 1293. https://doi.org/10.3390/jmse12081293.
3. Slobodskova, V.V.; Dovzhenko, N.V.; Kukla, S.P.; Chelomin, V.P.; Mazur, A.A. Express Diagnosis and Prediction of Remote Mass Mortality of Scallop Mizuhopecten yessoensis in Mariculture Farms Using Biomarkers. *J. Mar. Sci. Eng.* **2024**, *12*, 1151. https://doi.org/10.3390/jmse12071151.
4. Lee, D.-W.; Oh, S.-Y.; Park, J.J.C.; Jung, Y.-H.; Kim, H.-J.; Choi, D.M.; Choi, Y.-U.; Han, J. Offshore Wind Farms in South Korea: A Potential Site for Scallop Culture. *J. Mar. Sci. Eng.* **2023**, *11*, 1988. https://doi.org/10.3390/jmse11101988.
5. Castilla-Gavilán, M.; Guerra-García, J.M.; Hachero-Cruzado, I.; Herrera, M. Understanding Carbon Footprint in Sustainable Land-Based Marine Aquaculture: Exploring Production Techniques. *J. Mar. Sci. Eng.* **2024**, *12*, 1192. https://doi.org/10.3390/jmse12071192.
6. Basto, A.; Peixoto, D.; Machado, M.; Costas, B.; Murta, D.; Valente, L.M.P. Physiological Response of European Sea Bass (*Dicentrarchus labrax*) Juveniles to an Acute Stress Challenge: The Impact of Partial and Total Dietary Fishmeal Replacement by an Insect Meal. *J. Mar. Sci. Eng.* **2024**, *12*, 815. https://doi.org/10.3390/jmse12050815.
7. Claessens, S.; Aragão, C.; Hoffling, F.B.; Pinheiro, I.; Fracalossi, D.M.; Vieira, F.N. Mussel Meal as a Promotor of Growth Performance for the Whiteleg Shrimp (*Litopenaeus vannamei*). *J. Mar. Sci. Eng.* **2023**, *11*, 1670. https://doi.org/10.3390/jmse11091670.
8. Vale Pereira, G.d.; Conceição, L.E.C.; Soares, F.; Petereit, J.; Buck, B.H.; Johansen, J.; Dias, J.; Faccenda, F. Alternative Feed Formulations Impact Growth Performance, Flesh Quality and Consumer Acceptance of Rainbow Trout (*Oncorhynchus mykiss*). *J. Mar. Sci. Eng.* **2023**, *11*, 1135. https://doi.org/10.3390/jmse11061135.
9. Rajendran, K.; Karuppiah, P.; Ponnusamy, P.; Shaik, M.R.; Khan, M.; Oh, T.H.; Shaik, B. Anti-Inflammatory Activity of Mycobiont Extract of Parmotrema austrosinense (Zahlbr.) Hale in a Zebrafish Model. *J. Mar. Sci. Eng.* **2023**, *11*, 1081. https://doi.org/10.3390/jmse11051081.
10. Soares, F.M.R.C.; Nobre, A.M.D.; Raposo, A.I.G.; Mendes, R.C.P.; Engrola, S.A.D.; Rema, P.J.A.P.; Conceição, L.E.C.; Silva, T.S. Development and Application of a Mechanistic Nutrient-Based Model for Precision Fish Farming. *J. Mar. Sci. Eng.* **2023**, *11*, 472. https://doi.org/10.3390/jmse11030472.
11. Hotos, G.N.; Kourelea, E.; Fotodimas, I. Salinity Tolerance and the Effect of Salinity and Algal Feed on the Demographics of Cultured Harpacticoid Copepods Tisbe holothuriae and Tigriopus sp. from the Messolonghi Lagoon (W. Greece). *J. Mar. Sci. Eng.* **2022**, *10*, 1663. https://doi.org/10.3390/jmse10111663.
12. Luís, R.; José, R.; Castro, J.; Andrade, C. A Preliminary Assessment of Microalgal Diets for Echinopluteus Larvae Culture of the Sea Urchin Sphaerechinus granularis (Lamarck, 1816) (Echinoidea: Toxopneustidae). *J. Mar. Sci. Eng.* **2023**, *11*, 1870. https://doi.org/10.3390/jmse11101870.

13. Matias, A.C.; Araújo, R.L.; Ribeiro, L.; Bandarra, N.M.; Gonçalves, A.; Pousão-Ferreira, P. Effect of Long-Term Day/Night Temperature Oscillations on the Overall Performance of Gilthead Seabream (*Sparus aurata*) Juveniles. *J. Mar. Sci. Eng.* **2023**, *11*, 1687. https://doi.org/10.3390/jmse11091687.

References

1. DNV. *Ocean ´s Future for 2050–Marine Aquaculture Forecast, 2021 Edition*; DNV AS: Hovik, Norway, 2021; p. 47.
2. FAO. *The State of World Fisheries and Aquaculture*; Towards Blue Transformation; FAO: Rome, Italy, 2022. [CrossRef]
3. FAO. *Environmental Impact Assessment and Monitoring in Aquaculture*; FAO Fisheries and Aquaculture Technical Paper; FAO: Rome, Italy, 2009; Volume 527, 57p.
4. Troell, M.; Naylor, R.L.; Metian, M.; Beveridge, M.; Tyedmers, P.H.; Folke, C.; Arrow, K.J.; Barrett, S.; Crépin, A.-S.; Ehrlich, P.R.; et al. Does aquaculture add resilience to the global food system. *Proc. Natl. Acad. Sci. USA* **2014**, *111*, 13257–13263. [CrossRef] [PubMed]
5. Clements, J.C.; Chopin, T. Ocean acidification and marine aquaculture in North America: Potential impacts and mitigation strategies. *Rev. Aquacult.* **2017**, *9*, 326–341. [CrossRef]
6. Maulu, S.; Hasimuna, O.J.; Haambiya, L.H.; Monde, C.; Musuka, C.G.; Makorwa, T.H.; Munganga, B.P.; Phiri, K.J.; Nsekanabo, J.D. Climate Change Effects on Aquaculture Production: Sustainability Implications, Mitigation, and Adaptations. *Front. Sustain. Food Syst.* **2021**, *5*, 609097. [CrossRef]
7. Engle, C.R.; van Senten, J. Resilience of Communities and Sustainable Aquaculture: Governance and Regulatory Effects. *Fishes* **2022**, *7*, 268. [CrossRef]

Journal of
Marine Science and Engineering

MDPI

Systematic Review

Ocean Acidification and Aquacultured Seaweeds: Progress and Knowledge Gaps

Tan Hengjie [1], Simon Kumar Das [1], Nur Farah Ain Zainee [2], Raja Yana [3] and Mohammad Rozaimi [1,*]

[1] Department of Earth Sciences and Environment, Faculty of Science and Technology, Universiti Kebangsaan Malaysia, Bangi 43600, Malaysia
[2] Department of Biological Sciences and Biotechnology, Faculty of Science and Technology, Universiti Kebangsaan Malaysia, Bangi 43600, Malaysia
[3] Department of Fisheries Malaysia, No. 30, Persiaran Perdana Precinct 4, Putrajaya 62628, Malaysia
* Correspondence: mdrozaimi@ukm.edu.my

Abstract: This systematic review aimed to synthesise the existing studies regarding the effects of ocean acidification (OA) on seaweed aquaculture. Ocean acidification scenarios may increase the productivity of aquacultured seaweeds, but this depends on species-specific tolerance ranges. Conversely, seaweed productivity may be reduced, with ensuing economic losses. We specifically addressed questions on: how aquacultured seaweeds acclimatise with an increase in oceanic CO_2; the effects of OA on photosynthetic rates and nutrient uptake; and the knowledge gaps in mitigation measures for seaweed farming in OA environments. Articles were searched by using Google Scholar, followed by Scopus and Web of Science databases, limiting the publications from 2001 to 2022. Our review revealed that, among all the OA-related studies on macroalgae, only a relatively small proportion ($n < 85$) have examined the physiological responses of aquacultured seaweeds. However, it is generally agreed that these seaweeds cannot acclimatise when critical biological systems are compromised. The existing knowledge gaps regarding mitigation approaches are unbalanced and have overly focused on monitoring and cultivation methods. Future work should emphasise effective and implementable actions against OA while linking the physiological changes of aquacultured seaweeds with production costs and profits.

Keywords: acclimatisation; macroalgae; physiological response; productivity

Citation: Hengjie, T.; Das, S.K.; Zainee, N.F.A.; Yana, R.; Rozaimi, M. Ocean Acidification and Aquacultured Seaweeds: Progress and Knowledge Gaps. *J. Mar. Sci. Eng.* **2023**, *11*, 78. https://doi.org/10.3390/jmse11010078

Academic Editor: Ana Catarina Matias

Received: 20 October 2022
Revised: 21 November 2022
Accepted: 4 December 2022
Published: 3 January 2023

1. Introduction

The global seaweed aquaculture industry has contributed to the production of various downstream and upstream products such as food, biopolymers, cosmetics, nutraceuticals, bioenergy compounds, and pharmaceuticals [1]. The production of seaweed-based biofuel as an alternative to fossil fuel [2] has managed to reduce up to 1500 tons CO_2 km^{-2} $year^{-1}$ when compared to emissions from fossil fuels [3]. Among its other functions, the open ocean aquaculture of seaweeds provides shoreline protection from storms and waves [3,4]. Seaweed production can also help to reduce ocean eutrophication by absorbing nutrients required for seaweed growth [5]. With a wide distribution of biomass at the global level, Seaweed Aquaculture Beds (SABs) have the potential to at least act as a temporary carbon sink to mitigate the immediate effects of climate change [6]. This is due to the seaweed's capacity for carbon assimilation and accumulation, and CO_2 sequestration in a relatively short period [4,7]. On the other hand, there is evidence indicating that certain naturally growing seaweeds have the capacity for carbon sequestration and accumulation, which can be exported and buried in deep sea regions [8,9]. However, with the elevation in atmospheric CO_2, ocean acidification (OA), as one of the impacts of climate change, will negatively affect entire marine systems. Although this is a globally pressing matter, the discourse on the potential ecological or economic impacts of seaweed production is still limited.

Seaweeds are used in many different industries after they are harvested. As such, seaweed cultivation has grown from constituting 45.9% of global mariculture production in 2004 to 51.3% in 2018. In terms of monetary value, this represents 24.2% of the associated worldwide profits, worth US$ 28.1 billion [10,11]. The combined production output of China, Indonesia, Japan, the Republic of Korea, the Philippines, and Malaysia contributed up to 99.5% of global seaweed mariculture production in 2018 [11–13]. This region led the world's seaweed production, with an estimated at 32.4 million tonnes in 2018, compared to a threefold lower production in 2000 (10.5 million tonnes; [14]). Global seaweed cultivation then increased to 35.1 million tonnes in 2020 for food and non-food uses [15]. The highest output in terms of seaweeds by quantity farmed within temperate and tropical regions are: *Eucheuma* (also known as gusô, 10.2 million tons in 2015), Japanese kelp (8 million tons), *Gracilaria* (sare, 3.9 million tons), *Undaria* (wakame, 2.3 million tons), *Kappaphycus* (elkhorn sea moss, 1.8 million tons), and *Porphyra* (nori, 1.2 million tons) [1,12,16]. From the temperate region, seaweed production in Norway reached a sales value of EUR 74,000 in 2017, while studies have predicted that annual sales will rise further to EUR 4 billion by 2050 [17].

Up to 95.6% of all seaweed used by humans is aquacultured to ensure sustainability in terms of supply and to prevent the overexploitation of the natural population [10]. The remainder is harvested from naturally growing beds. Brown seaweeds, which comprise nearly 64% of the farmed production, are harvested for a variety of uses, including human nutrition and alginate extraction. They are used in various sectors including the medical industry, textile printing, and paper coating [1,6,18,19]. Other seaweeds such as *Chondrus crispus* (Irish moss), *Kappaphycus*, and *Eucheuma* are used as gelling material, emulsifiers, and stabilisers in the pharmaceutical, cosmetic, and food processing industries [1,13,18].

Environmental and biotic stressors can negatively affect the growth and productivity of macroalgae including aquacultured seaweeds [20]. The sources of these stressors vary from marine pollution [21,22], disease outbreaks [23–25], epiphytic infestation, algal parasites [25], periodic storms [26], and global warming [23]. Rising sea-levels due to global warming may cause shifts in shoreline morphology, with the subsequent effect of lowered productivity output through changes in seaweed distributional patterns [20]. Such changes may have more pronounced effects on naturally growing seaweed beds that would be harvested for seaweed industries rather than those in floating cultures [20]. The impact of sea-level changes on floating cultures might not be obvious but changes in pH can still be considered one of the factors which affect their physiological responses [27] (Table 1).

Knowledge on seaweed physiology, especially concerning how environmental stressors affect the productivity of aquacultured seaweed, is clearly essential to ensure the success of seaweed farming [1]. Ocean acidification (OA) is an inherent stress factor for optimal seaweed growth. Ocean acidification, or a reduction in seawater pH, stems from atmospheric CO_2 dissolving in seawater, eventually acidifying the water via the production of carbonic acid (H_2CO_3). This then dissociates into bicarbonate ions (HCO^{3-}) and protons (H^+) [28]. The seawater chemistry is altered when hydrogen ions [H^+] increase and the concentration of carbonate ions ($CaCO_3$) are reduced. This eventually causes a decrease in oceanic pH and leads to OA conditions [29,30]. According to the Representative Concentration Pathway (RCP) 6.0, atmospheric CO_2 emission will rise to 700 ppm by the year 2100. Ocean pH is expected to decline by 0.3–0.5 units towards the end of the century, with an estimated corresponding increase of 100–150% in [H^+] [28,31–33]. Thus, carbonate saturation states are predicted to decline by approximately 45% by 2100 [34–37].

Studies indicate that the metabolic rates and population growths of marine organisms can endure increases in oceanic CO_2 up to their physiological threshold limits because of their ability to acclimatise within an optimum range of pH values [38,39]. However, a failure to acclimatise would put those species at risk of mortality [38,40]. In this regard, although studies regarding the impacts of OA to marine fauna or fisheries aquaculture have been conducted [38,41,42], knowledge gaps remain surrounding the impacts of OA on seaweed aquaculture. While these limits are more discernible for calcifying macroalgae, the tolerance of fleshy macroalgae to OA is still unclear, i.e., whether such environmental changes will positively or negatively affect their productivity [43,44]. The physiological responses of non-calcifying seaweed towards OA are species-specific and inconsistent at different developmental stages [45], mostly due to different carbon-uptake strategies [43,46]. Furthermore, the interactive effects of OA and other environmental variables such as temperature complicates any definitive prediction with regard to the exact impacts of OA effects on fleshy seaweed [45].

In this review, we discuss how the increase in dissolved CO_2 with pH variation will affect the physiological responses of aquacultured seaweeds. In particular, we focused on directions towards answering the following questions: (1) How do aquacultured seaweeds acclimatise to an increase in oceanic CO_2? (2) What are the effects of OA on the photosynthetic rates and nutrient uptakes of aquacultured seaweeds? And: (3) What are the knowledge gaps in the mitigation strategies for future seaweed aquaculture considering an ocean-acidified environment?

2. Materials and Methods

This review follows the guidelines provided by Preferred Reporting Items for Systematic Reviews and Meta-Analyses (PRISMA) and the methodology is presented in a PRISMA flow scheme (Figure 1) [47].

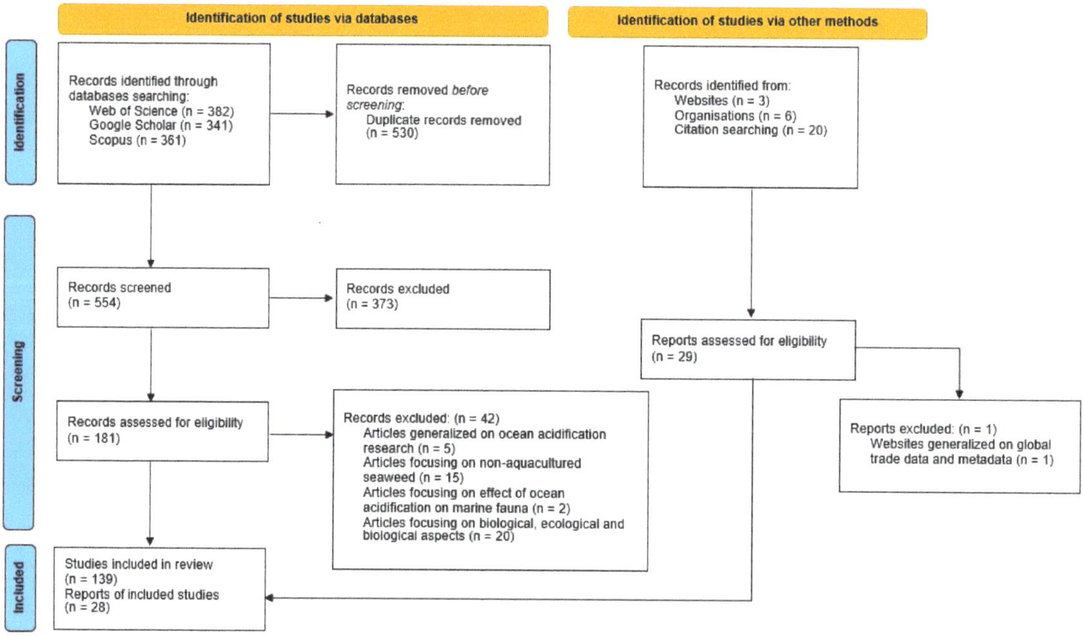

Figure 1. Flow diagram of the systematic review.

2.1. Resource Identification

We broadly approached the review of the literature based on the methodology suggested by Snyder (2019) [48]. We first defined the main purpose and research question using the following scopes: (1) the specific seaweed species aquacultured in (2) scenarios of OA conditions. A stepwise search strategy was started by limiting specific keywords, which were selected based on the two scopes. We trawled Google Scholar, Scopus, and Web of Science as the main literature databases. This step would cover peer-reviewed articles that are accessible through the public domain. The keyword searches in these databases included terms such as "ocean acidification," "elevation of dissolved CO_2," "seaweed aquaculture," "seaweed productivity," "photosynthetic rate", and "physiological response." To achieve a wide coverage in terms of sources for a comprehensive literature search, we added the keyword input by choosing related terms such as "seaweed adaptation", "seaweed acclimatization", "climate change", "global warming", "macroalgae", and "seaweed industry". We applied additional techniques for advanced-phrase searches by combining strings developed and strengthened using Boolean operators (i.e., "AND") in Google Scholar and Scopus. The strings "Ocean acidification" AND "Carbon dioxide" AND "Seaweed" AND "Aquaculture" AND "Physiological responses" returned 702 articles published between 2001 and 2021 (Figure 1). Similarly, when trawling the Web of Science database, the combination of search strings in the field tag to obtain topic hits (TS) were: (TS = (Ocean acidification)) AND (TS = (Seaweed)) AND (TS = (Physiological Response)). This search returned 382 results for papers published between 2012 and 2021. The timeframe is consistent with the methodology for scoping systematic reviews [48], and specifically for this study, and was stipulated to ensure that no species were overlooked. Furthermore, the emphasis on this timeframe for article selection was to minimise the duplication of subject species and to ensure that the published information provided was up to date. To avoid the duplication of articles in the literature search, all articles obtained were imported into a reference manager (Mendeley), as suggested by Li et al. [49].

2.2. Scoping and Screening

Keeping in mind the main scope of the research question (i.e., surrounding aquacultured seaweeds and OA), the search hits were first screened based on the titles and abstracts of the papers to determine their suitability and relevance. During the article trawling, a limited number of non-peer reviewed technical papers and database reports were included in the screening, but only if these articles were highly cited (>100 times), such as those reporting on the economic and management aspects of seaweed aquaculture [22], as well as selected Food and Agriculture Organization (FAO) [12,21,50,51] and Intergovernmental Panel on Climate Change reports [36,37,50,52]. Studies of seaweed species in aquaculture, farmed in both tropical and temperate regions, were shortlisted. Their relevance was further refined to consolidate the literature that directly related to the environmental variables of OA. To ensure a sufficiency of sources obtained for this review, we did not limit our literature search based on studies about the physiological effects of aquacultured seaweeds due to environmental stressors such as global warming and climate change. Rather, we also focused on search results related to the potential of aquacultured seaweeds as a natural mitigation tool for OA. This is in addition to other articles that highlighted any economic valuation—positive or negative—related to OA on the seaweed aquaculture industry. As a result, we made certain that our collection included narratives of the relationship between dissolved CO_2 elevation and aquacultured seaweed species, based primarily on their physiological responses (Table 1), followed by any subsequent impacts on industry-level aquaculture production. Any studies of aquacultured seaweed species that dealt with biological, ecological, and biogeochemical aspects that were not directly related to OA were excluded during the screening. Articles in languages other than English were also excluded (Figure 1).

2.3. Article Analysis

To provide a more insightful perspective to address question (3), a further analysis of the identified articles relating to the mitigation of the effects of OA on aquacultured seaweeds was performed. A first screening was carried out through keyword searches for terms that allude to mitigation approaches, such as site buffering, site selection, framework, and monitoring [53,54]. This was carried out to determine the main keywords that would be analysed, which would then lead to ranking the most discussed mitigation-related strategies in the literature. We found 131 papers which addressed or discussed mitigation to varying extents while focusing on mitigation as the main thematic area. The overarching theme relates to effective global measures involving synergy between policies, regulatory frameworks, and strategy implementation [14,55,56]. To ensure the results were unbiased [48], studies were analysed to determine if the search findings met these specific criteria, in that: (i) the approaches mentioned had been conducted; (ii) the current state of the strategy implementation; or (iii) challenges faced or efforts needed to further develop the mitigation measures. The keywords were searched manually throughout the articles based on occurrence frequency in the main text body of the articles. The total number of specific word hits from each article were recorded accordingly while excluding the same keywords if these appeared in the reference list (see Supplementary Data Table S1). We found a limited number of research articles with the same keywords (Supplementary Data Table S1). Certain keywords appeared less frequently in research articles compared to technical reports, although a similar phrasal context was applied by the authors of the respective articles (Table 2). Based on the article analysis and keyword search, 35 articles were identified as being the most relevant. Up to 10 keywords were finally shortlisted, which were either mentioned as the exact word/phrase in the article or discussed in a similar context. The specific term of 'management' was not considered as a keyword but as a root word since the definition could be further clarified by using the other identified keywords. These keywords and the context used was summarised to avoid an overlapping analysis (Table 2). The amount of specific word hits was summed and converted into percentages representing the proportion of each keyword that appeared out of the overall search (Supplementary Data Table S1) before ranking the 10 components (based on the shortlisted keywords) of mitigation approaches. Lastly, the ranked data was visualised graphically through a radar chart (Figure 2). This chart aims to list the top findings based on the queried article set. Its use allows for a focus on assessing the overall inherent issues and potential mitigation strategies for OA impacts on seaweed aquaculture. The visual presentation and holistic approach of the radar chart summarises the results and highlights the knowledge gaps that would require further studies.

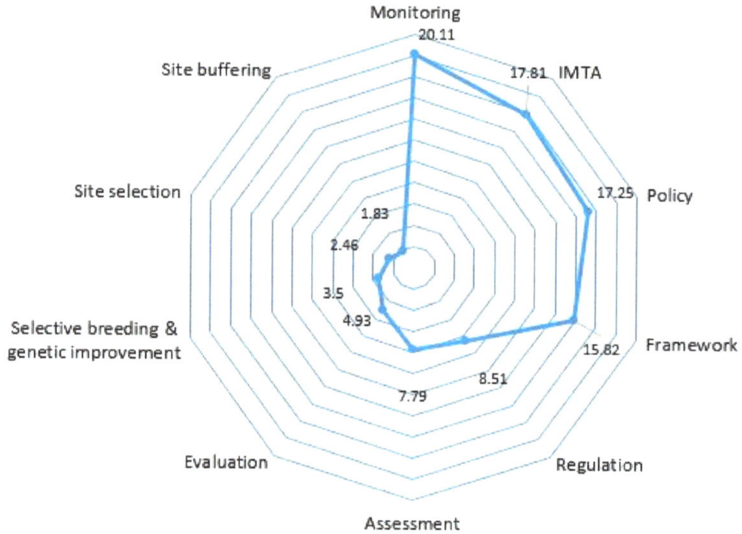

Figure 2. Radar chart that illustrates the existing knowledge gaps in relation to the main thematic areas for mitigating the effects of OA on seaweed aquaculture.

After the compilation was performed, we then identified, analysed, and interpreted any presence of consistent trends in the texts of the shortlisted literature to highlight the most relevant subtopics for further discussion [48]. We discussed these subtopics based on the headings of the following sections (see below). These sections are consistent with the themes of seaweed physiology, environmental stressors, seaweed cultivation, and aquaculture management, based on expert opinions and interpretations in the studies in question, in the context of evaluating the prospects of seaweed production in OA scenarios. The synthesis of studies published in the past 32 years that linked aquacultured seaweeds with OA conditions was then conceptualised (Figure 3) and tabulated (Table 1).

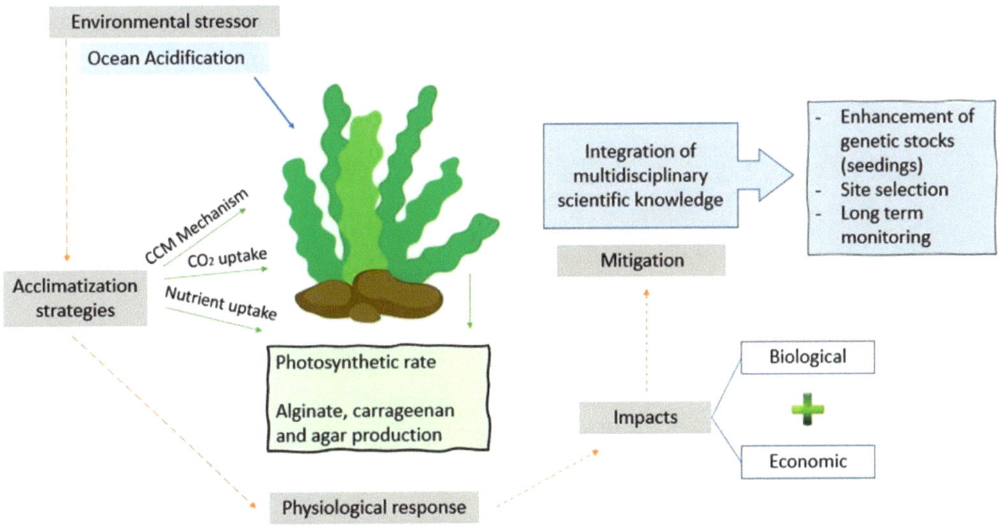

Figure 3. A conceptual diagram, which consists of various aquaculture aspects aimed at sustaining the productivity of aquacultured seaweeds under OA scenarios.

Table 1. The impact of OA on seaweed species in temperate and tropical regions. The CCMs of types 1 to 3 are described in Section 3.

Region	Species	CCM Type	Variable	Main Findings	References
Temperate	*Alaria esculenta*	3	Photosynthetic activity, biochemical composition (lipid content), enzymatic activities (eCA)	Increase in growth, lipid content, and photosynthetic efficiency (F_v/F_m) under elevated CO_2, with lower photon requirements; enzymes are not sensitive to changes in CO_2.	[57–60]
	Saccharina japonica	1, 2, 3	Iodine accumulation, photosynthetic efficiency, photosynthetic oxygen, germination	Tissue growth enhanced under lower pH with a simultaneous increase in iodine accumulation; inhibition of photosynthetic rate is relatively higher under lower pH, and the photosynthetic efficiency (F_v/F_m) is not much affected; reduction in meiospore germination and reproduction rate.	[10,51,61–66]
	Saccharina latissima	2 and 3	Photosynthetic acclimation, pigment composition	Photosynthesis and growth rates are negatively affected; CCMs are deactivated; optimal temperature for growth is 5–15 °C; no effects on biochemical composition.	[17,34,57,67–70]
	Undaria pinnatifida	2 and 3	Gametophyte development	No significant impact on meiospore germination but increase in germling growth rates and gametophyte sizes when seawater pH is reduced from 8.40 to 7.20; rates of net photosynthesis (NPS) of gametophytes and juvenile sporophytes start to decrease when pH drops from 7.20 to 5.5.	[1,10,12,65,71–73]
	Pyropia sp.	1 and 2	Photosynthetic rate, growth rate	Increase in growth and nutrient uptake; growth of thalli enhanced by 30% at pH of 6 and 7; photosynthetic rate increases when pH drops from 7 to 6, but photosynthesis and respiration rate decrease at pH of 4 and 5; tissue death when low pH conditions are prolonged.	[1,4,74–80]

Table 1. *Cont.*

Region	Species	CCM Type	Variable	Main Findings	References
	Gracilaria sp.	2 and 3	Growth rate, photosynthesis, photosynthetic inorganic carbon uptake, iodine accumulation	Growth rate increases through carbon uptake; enhanced carbon/nitrogen ration; photosynthetic pigments remain unchanged; increase in photosynthetic acclimation; increase in iodine accumulation under elevated dissolved CO_2.	[1,4,12,16,61,81–85]
	Chondrus crispus	2 and 3	Photosynthetic rate	Able to acclimatise when oceanic pH decreases, and photosynthetic rate is maintained; carbon fixation rate is highest at pH of 7.	[1,18,29,86–88]
	Sargassum fusiforme	1, 2, 3	Growth rate, nitrogen assimilation, photorespiration	Photosynthetic rate is maintained under increase in CO_2 since the species is able to tolerate pH declines with enhanced relative growths; biomass increase was associated with nitrogen assimilation within tissues.	[12,27,39,89,90]
	Macrocystis pyrifera	1, 2, 3	Germination rate, gametophyte development, iodine accumulation, growth rate, photosynthetic rate	Meiospore germination, gametophyte development, and spore production and recruitment negatively affected in acidified conditions; iodine accumulation slightly increases, and tissue growth exhibited under elevated pCO_2; no changes in growth and photosynthetic rate but increased uptake of CO_2.	[18,61,90–94]
	Sargassum vulgare	1, 2, 3	Alginate content, polysaccharides content, bioactivity (antibacterial activity, antifungal activity), carbohydrate availability, antioxidant capacity, enzymatic activities, photosynthetic rates	This species is able to acclimatise to low pH conditions of 6–6.7; secondary metabolites are lower; bioactive properties grow naturally in acidified conditions; alginate content higher in acidified conditions; increase in dissolved CO_2 results in increased bioactivity, antioxidant capacity, enzyme activity, photosynthetic rate, and polysaccharide content.	[95–99]
	Porphyra sp.	2 and 3	Growth rate	Increase in growth.	[16,94]
	Ulva rigida	2 and 3	Growth rate and assimilation (carbon and nitrogen), HCO3-utilisation, photosynthetic rate, dark respiration rate, soluble protein content, inactivation of CCMs, nitrogen metabolism	Increase in growth rate and nutrient assimilation especially for carbon and nitrogen under acidified conditions; photosynthetic rate, dark respiration, and soluble protein reduced with increased dissolved CO_2; photosynthesis process is negatively affected due to the inactivation of CCMs.	[99–101]
	Fucus vesiculosus	2 and 3	Growth, nutritional quality, carbon and nitrogen content, fertility	Reduction in growth and C:N ratio; no changes in other elemental compositions; increase in pCO_2 alter temporal development of fertility, according to the changes in temperature seasonally.	[45,101–104]
Tropical	*Gracilaria* sp.	2 and 3	Growth rate, photosynthesis, gephotosynthetic inorganic carbon uptake, iodine accumulation	Growth rate increases through carbon uptake; no significant effect on maximum relative electron transport rates (rETRmax); increase in iodine accumulation under elevated dissolved CO_2.	[1,4,12,82,105–107]
	Undaria pinnatifida	2 and 3	Photosynthesis, gametophyte development, germling growth rate	No significant impact on meiospore germination; increase in germling growth rate, and gametophyte size when seawater pH is reduced from 8.40 to 7.20; rates of net photosynthesis of gametophytes and juvenile sporophytes decrease when pH drops from 7.20 to 5.5.	[12,72,73,108]
	Kappaphycus alvarezii	1 and 3	Daily growth rate (DGR), photosynthesis	DGR decreases at pH 6 due to the low availability of photosynthetic carbon sources in low pH conditions; decrease in efficiency of CO_2 accumulation.	[1,12,13,18,93,109]
	Pyropia sp.	1 and 2	Net photosynthesis, growth rate, respiration	Increased growth and nutrient uptake; growth of thalli enhanced by 30% at pH of 6 and 7; photosynthetic rate increases when pH drops from 7 to 6; photosynthesis and respiration rate decrease at pH of 4 and 5; thalli death in prolonged low pH conditions.	[4,74–76,79,80,110–113]
	Eucheuma sp.	2 and 3	Photosynthetic rate	Increase in photosynthetic rate when oceanic pH decreases below 8.	[1,18,94]

Table 1. *Cont.*

Region	Species	CCM Type	Variable	Main Findings	References
	Caulerpa lentillifera	3	Carbon absorption rate	Increase in growth through carbon uptake.	[111,114,115]
	Hypnea spp.	3	Growth rate, maximum quantum yield, chlorophyll a content, antioxidant activity	Decrease in growth rate, maximum quantum yield (fv/fm), and chlorophyll a content; increase in antioxidant activity.	[116–122]
	Gelidium spp.	3	Growth rate, carbohydrate content	Decrease in growth rate; no significant changes in carbohydrate content; reduction in species richness.	[87,123–127]

Table 2. Definition of the terms used in the radar chart, with the selection of words based on the applied contextual meaning to avoid duplication in occurrence counts and their use being overlooked during the keyword search.

Keyword	Definition in Context	References
Policy	The related aims to: mitigate CO_2 emission to the atmosphere;	[8,128]
	inform decision-making at local, regional, and national levels in order to integrate into global goals;	[7,14,16,56,129]
	manage areas that are used for seaweed aquaculture.	
Framework	The provisions for: policy and integrated planning that requires more experimental and innovative practices at different authoritative levels (local, state, or federal jurisdictions);	[16,56]
	legal frameworks, which refer to guidelines in the setting up and management of seaweed aquaculture;	[8,13,54]
	management of fisheries resources and aquaculture governance; conservation and the sustainable use of aquatic living resources.	[14,56,130]
Regulation	The provisions to: adequately manage the resources of coastal aquaculture, including seaweed cultivation;	[13,128]
	coupled with an appropriate monitoring and law enforcement system while banning unsustainable practices.	[14,16]
Monitoring	Efforts to: measure the local environmental and spatial variability in carbonate chemistry within coastal areas or aquaculture farms;	[50,53,54,131]
	track long-term environmental changes through a combination of efforts by various stakeholders globally;	[17,34,128]
	incorporating a Fisheries and Resources Monitoring System (FIRMS) in seaweed aquaculture; improve transparency in fisheries and aquaculture stock and production records.	[14,56,132]
Evaluation	The appraisal of: the application and performance of aquaculture systems that involve several authorities to ensure ecosystem sustainability;	[4,14,54,131,133]
	the interaction with existing resources that are characteristic of coastal areas and suitable for different types of farming.	[13,17,128]
Assessment	The inclusion of: managerial tools for quantifying the risks and benefits associated with seaweed aquaculture;	[8,54]
	diagnosing the current status of stocks in fisheries and aquaculture;	[14,34,133]
	Environmental Impact Assessment;	[25,128]
Site selection	Related efforts to: choose optimal sites for aquaculture activities within each environment;	[4,53,133–135]
	match seaweed species with specific cultivation techniques;	[13]
	include licensing approvals from authorities.	[50]
IMTA	A method of aquaculture that: consist of species components from different trophic levels and serving different ecosystem functions;	[13,53,130,133]
	increases the biomass production and sustainability of aquaculture;	[17,54,130,132]
	mitigates environmental problems caused by specific forms of fed aquaculture	[131]
Site buffering	Related efforts to: buffer seawater pH and carbonate chemistry;	[53]
	quantify the ability of seaweeds to buffer the impacts of climate change including OA.	[3,8,135]
Selective breeding and genetic improvement	A method of aquaculture that is used: to cultivate seaweeds for specific traits to enhance production and resilience under conditions of projected climate change;	[4,13,14,53,54]
	to obtain a culture stock that has increased tolerance against the impacts of OA through strain development.	[34,136]

3. Ocean Acidification and Seaweeds' Photosynthetic Rates and Nutrient Uptakes

When seaweeds with different photosynthetic capacities are exposed to increasing oceanic CO_2 concentrations, biochemical and molecular changes cause varying carbon uptake capacities [39]. This has various implications in terms of alterations in the physiological responses of aquacultured seaweeds (Table 1). Seaweeds that solely obtain CO_2 through diffusive uptake are usually known as non-carbon concentrating mechanism (non-CCM) species [137]. These species will exhibit increases in growth and photosynthetic rates under elevated CO_2, with an eventual increase in species abundance instead of diversity [138]. Despite the availability of more dissolved CO_2 as a result of OA, and the supply of CO_2 as the primary C source is no longer limited, it is worth noting that CO_2 diffuses into water at a slower rate than it does in air [27,139]. Most of the fleshy macroalgae in aquaculture are carbon concentrating mechanism (CCM) species. These seaweeds require an adjustment in terms of their kinetic mechanisms to utilise inorganic carbon such as bicarbonate ions to facilitate CO_2 delivery to the ribulose-1,5-bisphosphate carboxylase/oxygenase (RubisCO) enzyme for carbon fixation [4]. Nonetheless, different CCM species will respond distinctively towards OA in order to acclimatise and survive because not all have the same affinity towards DIC [46,140].

According to Giordano et al. (2005) [141], there are three types of CCMs: (1) CCMs based on C_4 mechanisms and crassulacean acid metabolism (CAM); (2) CCMs based on the active transport of DIC; and (3) CCMs based on changes in CO_2 concentration in compartments adjacent to RubisCO, where HCO^{3-} is mostly found in the chloroplast stroma and is converted to CO_2 when it crosses to the thylakoid. *Macrocystis pyrifera*, for example, is a type of fleshy macroalgae that uses both CO_2 and HCO^{3-} for photosynthesis and growth [91]. Those species with a high affinity for DIC will either exhibit increases in their metabolic processes or will not be obviously affected [140]. In this regard, HCO^{3-} is dehydrated extracellularly by external carbonic anhydrase (CAext), then converted to CO_2 by internal carbonic anhydrase (CAint) to be used by CCM seaweeds, or, alternatively, it is accumulated within an internal inorganic carbon pool [39,91]. This mechanism has been described in *Chondrus crispus* and *Sargassum fusiforme* [39,86,96] as well as in other aquacultured seaweeds that use HCO^{3-}, including *Alaria esculenta*, *S. fusiforme*, *M. pyrifera*, *C. crispus*, and *Saccharina latissima* [91,142]. In contrast, CCM seaweeds with a low affinity for DIC will benefit under elevated CO_2 [137,140].

Aquacultured seaweeds would therefore acclimatise to elevated CO_2 levels through various carbon uptake strategies [3,27], which can be determined by the $\delta^{13}C$ value of seaweed tissues. CCM seaweeds typically have $\delta^{13}C$ values ranging from −30‰ to −10‰. This range, however, varies with changes in pCO_2 in the marine environment [137], indicating that species attempt to acclimatise despite the fluctuating pCO_2 [34,96]. Species such as *S. fusiforme* can tolerate pH drops while maintaining appreciable photosynthetic rates. Subjected to increasing CO_2 levels, this species responds positively based on enhanced relative growth rates and increased biomass associated with significant nitrogen assimilation within its tissues [39,89]. Other studies have shown that *Kappaphycus alvarezii* and *Sargassum vulgare* were able to acclimatise in low-pH conditions (pH 6 and pH 6.7, respectively) [74], while *M. pyrifera* exhibited no changes in growth and photosynthetic rates under elevated oceanic CO_2 because it has the ability to utilise both CO_2 and HCO^{3-} [91]. Other economically important seaweeds such as *Hypnea pseudomusciformis*, *Porphyra yezoensis*, *Gracilaria chilensis*, *G. changii*, and *G. lemaneiformis* exhibited increased growth rates with an increase in dissolved CO_2 concentrations [4,46,81,106,117,143,144]. Hence, the growth of seaweeds partly depends on their HCO^{3-} uptake and utilisation capacity [96], and the impact on their photosynthetic rate is generally species-specific [27]. Certain seaweeds have the potential to mitigate OA through carbon sequestration [3,135,145] due to their respective physiological changes under pH variation, thus leading to high biomass production [7], as was clearly observed for fast-growing fleshy macroalgae (Table 1).

In contrast, other aquacultured seaweeds, notably red seaweeds such as *Gracilaria tenuistipitata*, *Porphyra leucosticta*, and *P. linearis*, experience decreased growth rates in high pCO_2 concentrations due to a reduction in photosynthetic capacity, invariably leading to a decline in photosynthetic efficiency [46,82,146–148]. Decreased growth rates are also possible because CCM seaweeds are sensitive to increased H^+ concentrations during OA, which can disrupt cellular homeostasis [137]. Such effects have been studied in the case of the changes in *M. pyrifera* meiospore germination (Table 1) under extreme OA at a mean pH of 7.60 [92]. However, questions arise as to the precise timeframe during which the seaweed's physiology remains unchanged or in a positively balanced state despite the increase in pCO_2. In a simulated situation of acidified conditions, the photosynthetic rate of *P. linearis* increased but only lasted for approximately 16 h [147]. The low-pH level in an experimental environment resulted in elevated respiration rates and the consumption of accumulated photosynthate, eventually causing a decline in growth rates, especially at a pH of 6.0 [147]. A similar situation was reported for *Porphyra haitanensis*, whereby its thallus growth increased significantly by 30% when the pH decreased by 1–2 units. Lowering pH levels further results in a significant drop in growth rates until the eventual death of its thalli [74]. Additionally, low-pH growth conditions can severely impact alginate and carrageenan production [149,150]. These observations imply that elevations in dissolved CO_2 levels can potentially reduce seaweed productivity.

Not all seaweeds undergo the same consequence in terms of changes in growth rate. This is because the rate of nutrient uptake influences the growth rate of seaweeds under elevated CO_2 conditions. An increase in pCO_2 increased *Ulva rigida*'s nitrogen assimilation rate, which in turn increased its growth rate [82,100,151]. *Sargassum fusiforme* has been shown to have higher NO^{3-} uptake rates when exposed to elevated CO_2 levels [39,152]. In similar experimental conditions, there were no changes in the uptake of inorganic nitrogen by *M. pyrifera* despite an increase in CO_2 [152]. Other factors that influence the growth rate of seaweeds under elevated CO_2 include limitations in nutrient availability in the water column due to the frequent increase in species biomass locally or the stratification that is induced by ocean warming [153]. This occurs through the alteration of ocean carbonate chemistry, which is mainly attributed to the increase in ocean temperature with the increase in CO_2 concentrations [4,154,155]. Stratification hinders nutrient supply to primary producers from deep to surface waters because of the formation of layers that prevent normal water mixing such as during upwelling processes. Such effects have been reported in the Southern Baltic Sea and the Basque coast of Spain, where *Fucus vesiculosus* and *Gelidium corneum* experienced lowered growth rates due to nutrient limitations caused directly by ocean stratification [4,156–158], while *G. corneum* in the Cantabrian Sea became less productive due to a reduction in nitrate availability [157].

4. Potential Mitigation Strategies for Ocean Acidification Impacts on Seaweed Aquaculture

The high production volume of aquacultured seaweed demonstrates industry demands and a significant contribution in terms of farming revenue [159]. The difference between the cost (capital, maintenance, material inputs, and labour) and the income generated, on the other hand, determines the economic sustainability and cost-effectiveness of seaweed cultivation [3,22]. Indeed, if OA has an impact on seaweed production, the market's value chain in terms of supply and demand will be disrupted, resulting in price volatility and declining profit margins in various processing industries and global marketplaces [22,160]. Therefore, a clear motivation to mitigate the effect of OA on seaweed aquaculture revolves around maintaining profitable production output in the circumstances of an ever-changing global climate. Mitigation strategies must be expanded at the regional level [53] and should serve as fundamental frameworks for strengthening policy implementation and to ensure the respective standards for proper seaweed cultivation practice and effective management are delivered [54]. Unfortunately, an overarching management plan that wholly addresses OA, specifically in seaweed aquaculture, is still lacking in many countries [54].

The governance management for seaweed aquaculture—in the flow of policy, framework, regulation, monitoring, followed by assessment and evaluation [14]—ideally combines multiple aspects of mitigation strategies as the essential elements to counteract OA. These key elements would further incorporate the managerial tools for long term implementation, which include site selection, advanced aquaculture techniques, such as integrated multi-trophic aquaculture (IMTA—see below), site buffering, selective breeding, and genetic improvement (Figure 2). These same factors have been emphasised in multiple articles, which provide indirect leads to a more sustainable seaweed aquaculture. Based on the broad framework established, regulations should be developed in accordance with mutual goals such as the Sustainable Development Goals (SDGs), which may include technical guidelines to ensure the long-term growth of the seaweed aquaculture industry [14,55]. It is also recommended to identify and, if necessary, establish authoritative bodies with a functional role at the national level. Their responsibilities should include ensuring the sustainability of aquaculture practice while adhering to existing regulatory policies [129,161]. Based on this approach, carbon tax implementation can be expanded globally to manage CO_2 release in order to reduce greenhouse gas emissions despite a rapidly growing and developing economy [162–164].

5. Multi-Disciplinary Approach to Mitigate the Ocean Acidification Impacts on Seaweed Aquaculture

To counteract the consequences caused by OA, the mitigation steps should ideally be approached in concert. The identification of optimal farming techniques may be a viable first step. Different seaweed aquaculture systems, such as a recirculating land-based system, must be further developed, either through open-water systems or tank cultivation [1,160]. The farming method of choice is primarily determined by the farm's scalable capacity and how to maximise productivity based on local environmental factors [20]. In an open-water cultivation, proper site selection and spatial planning are the key elements to ensure the effectiveness of seaweed aquaculture under elevated CO_2 scenarios [3,6,13,136]. In Malaysia, farmers rely on the monoline method as the main farming technique, which has increased production from 60 thousand tons in 2006 to 261 thousand tons since 2015 [12]. Open-water cultivation may be replaced by land-based seaweed farming. Although the former is not new [165], more research is required to improve cultivation techniques and ensure nutrient uptake efficiency in vitro. Certain seaweeds grew well in the presence of elevated CO_2, but as density increased, nutrient availability became limited due to competition for uptake [132]. Therefore, certain land-based seaweed farming approaches have been implemented through the concept of integrated aquaculture, notably Integrated Multi-trophic Aquaculture (IMTA), to ensure the supply of nutrients required to support sustainability regarding seaweed growth [16]. In general, IMTA, which consists of complementary ecosystems functions within a single aquaculture system, is known as one of the mitigation strategies for marine aquaculture during OA [53,166]. With the co-cultivation of seaweed as a key component, the whole system would lessen the impact of OA through CO_2 buffering while maintaining the possibility of further management applications, such as seaweed-based bioremediation [16,53,159,167]. This would ameliorate the effects of increased CO_2 while also creating a favourable environment for seaweed cultivation.

Sustainable seaweed aquaculture stresses proper monitoring regardless of the production technology. This can be improved with a long-term monitoring programme using networks such as the Global Ocean Acidification Observing Network (GOA-ON) [168] or the more regional Southeast Asian Global Ocean Observing System (SEAGOOS) [169]. The variation of water parameters within each culture should be monitored and controlled to provide optimal conditions for seaweed cultivation [13,134,170,171] using devices that are regularly calibrated to ensure data accuracy [52]. Choosing cultivars with specific traits that exhibit higher growth capacity, thermal tolerance, and disease resistance brings us one step closer to having a comprehensive mitigation strategy [134,136,172,173]. This strategy can be further leveraged through genetic improvement using hybridisation technology, which is a means of modifying and developing existing cultivar strains for higher biomass yields without limiting the choice of aquacultured seaweed species [134]. Concerning the seaweeds selected for a cultivation system, species with high DIC affinity could be prioritised to optimise CO_2 or HCO^{3-} uptake capacity [106,152]. *Gracilaria* may be one of the choice species to extract organic and inorganic components since it is efficient at assimilating ammonia, phosphate, and DIC [106,151].

Risk assessments based on an adaptive framework should be prioritised as part of the steps to address potential problems, such as threats from climate change [8,174]. The

Ecosystem Risk Assessment (ERA) is one viable approach for identifying hazards from adverse events and their consequences for a specific organism. The quantification of pre-impact levels and qualitative studies should be considered based on the projected effects of OA, followed by the design and evaluation of precautionary methods [175]. Indeed, more efforts to conduct risk assessments for marine plants classified by the International Union for the Conservation of Nature (IUCN), particularly macroalgae, are required in all coastal regions [56]. This is necessary to reduce the selection of farmed seaweed species that may exhibit negative growth rates in stressful and disturbed marine environments [56].

Finally, the findings from a broad-based approach will provide information on how to consolidate existing policies while ensuring the sustainability of seaweed aquaculture [175]. While anticipating the ripple effects of declining seaweed production, which may result in economic losses, it is critical to develop a systematic and practical plan to overcome future challenges in seaweed aquaculture. A conceptual framework was outlined (Figure 3), which includes various aspects of seaweed aquaculture geared toward sustaining productivity despite the constraints in OA scenarios.

6. Conclusions

Seaweed aquaculture has the potential to reduce CO_2 emissions while also supporting ecosystem services through CO_2 sequestration; however, elevated CO_2 and OA are likely to have an impact on seaweed production. To mitigate the negative effects of severe OA on aquacultured seaweed, a comprehensive mitigation plan with adequate monitoring is required. Seaweeds, as the largest group of aquacultured species with high productivity by volume, may be affected by regional and global changes in biomass yield. However, the responses of aquacultured seaweeds to OA vary by species, as evidenced by changes in physiological mechanisms such as the photosynthetic rate and nutrient uptake, which affect seaweed productivity. At the same time, the acclimatisation of aquacultured seaweeds to elevated oceanic CO_2 depends on their carbon uptake strategies, while kinetic mechanism adjustments would further determine changes in the photosynthetic rate in each species. If OA has a negative effect on seaweed cultivation, the extent of its impact on seaweed production must be quantified, and because industry profitability is determined by seaweed price and operating expenses, monetary loss can be calculated precisely. More studies are needed to quantify the effects caused by OA on the economy. This includes establishing a link between physiological changes in seaweed and industrial productivity in terms of production costs and potential revenues. As a result, multiple mitigation strategies approached from various angles should be implemented to mitigate the effects of OA on aquacultured seaweed. The emphasis should be on addressing existing knowledge gaps in mitigation approaches, which are still imbalanced and overly skewed toward monitoring- and IMTA-centric efforts. This entails multidisciplinary approaches developed through synergy among various stakeholders—from researchers to aquaculture farmers and policymakers—for a more holistic seaweed aquaculture system that incorporates key mitigation tools. In short, the combined effects of OA on biological and economic factors necessitate the implementation of a more collaborative mitigation strategy that incorporates the various multidisciplinary aspects of OA and seaweed production.

Supplementary Materials: The following supporting information can be downloaded at: https://www.mdpi.com/article/10.3390/jmse11010078/s1, Table S1: Summary of the number of occurrences of the main keywords during the article screening that relate to mitigation strategies of seaweed aquaculture in ocean acidification scenarios.

Author Contributions: Conceptualisation, T.H. and M.R.; methodology, M.R., N.F.A.Z. and T.H.; validation, M.R., R.Y., S.K.D. and N.F.A.Z.; formal analysis, T.H., M.R. and R.Y.; investigation, T.H. and N.F.A.Z.; resources, S.K.D., R.Y. and M.R.; data curation, T.H. and M.R.; writing—original draft preparation, T.H. and M.R.; writing—review and editing, all authors; visualisation, T.H., N.F.A.Z. and M.R.; supervision, M.R., N.F.A.Z. and S.K.D.; project administration, M.R.; funding acquisition, M.R. All authors have read and agreed to the published version of the manuscript.

Funding: This research was funded by Universiti Kebangsaan Malaysia, grant number DIP-2021-021.

Institutional Review Board Statement: Not applicable.

Informed Consent Statement: Not applicable.

Data Availability Statement: Not applicable.

Conflicts of Interest: The authors declare no conflict of interest. The funders had no role in the design of the study; in the collection, analyses, or interpretation of data; in the writing of the manuscript; or in the decision to publish the results.

References

1. Hurd, C.L.; Harrison, P.J.; Bischof, K.; Lobban, C.S. *Seaweed Ecology and Physiology*, 2nd ed.; Cambridge University Press: Cambridge, UK, 2014. [CrossRef]
2. Najiha Badar, S.; Mohammad, M.; Emdadi, Z.; Yaakob, Z. Algae and their growth requirements for bioenergy: A review. *Biofuels* **2021**, *12*, 307–325. [CrossRef]
3. Duarte, C.M.; Wu, J.; Xiao, X.; Bruhn, A.; Krause-Jensen, D. Can seaweed farming play a role in climate change mitigation and adaptation? *Front. Mar. Sci.* **2017**, *4*, 100. [CrossRef]
4. Chung, I.K.; Sondak, C.F.A.; Beardall, J. The future of seaweed aquaculture in a rapidly changing world. *Eur. J. Phycol.* **2017**, *52*, 495–505. [CrossRef]
5. van den Burg, S.W.K.; van Duijn, A.P.; Bartelings, H.; van Krimpen, M.M.; Poelman, M. The economic feasibility of seaweed production in the North Sea. *Aquac. Econ. Manag.* **2016**, *20*, 235–252. [CrossRef]
6. Jagtap, A.S.; Meena, S.N. Chapter 23—Seaweed Farming: A Perspective of Sustainable Agriculture and Socio-Economic Development. In *Natural Resources Conservation and Advances for Sustainability*; Jhariya, M.K., Meena, R.S., Banerjee, A., Meena, S.N., Eds.; Elsevier: Amsterdam, The Netherlands, 2021; pp. 493–501. [CrossRef]
7. Sondak, C.F.A.; Ang, P.O.; Beardall, J.; Bellgrove, A.; Boo, S.M.; Gerung, G.S.; Hepburn, C.D.; Hong, D.D.; Hu, Z.; Kawai, H.; et al. Carbon dioxide mitigation potential of seaweed aquaculture beds (SABs). *J. Appl. Phycol.* **2017**, *29*, 2363–2373. [CrossRef]
8. Froehlich, H.E.; Afflerbach, J.C.; Frazier, M.; Halpern, B.S. Blue growth potential to mitigate climate change through seaweed offsetting. *Curr. Biol.* **2019**, *29*, 3087–3093. [CrossRef]
9. Krause-Jensen, D.; Duarte, C.M. Substantial role of macroalgae in marine carbon sequestration. *Nat. Geosci.* **2016**, *9*, 737–742. [CrossRef]
10. Chopin, T.; Sawhney, M. Seaweeds and Their Mariculture. In *Encyclopedia of Ocean Sciences*, 2nd ed.; Steele, J.H., Thorpe, S.S., Turekian, K.K., Eds.; Elsevier: Oxford, UK, 2009; pp. 4477–4487. [CrossRef]
11. Chopin, T.; Tacon, A. Importance of seaweeds and extractive species in global aquaculture production. *Rev. Fish. Sci. Aquac.* **2021**, *29*, 139–148. [CrossRef]
12. Ferdouse, F.; Holdt, S.L.; Smith, R.; Murúa, P.; Yang, Z. *The Global Status of Seaweed Production, Trade and Utilization*; FAO: Rome, Italy, 2018; pp. 1–114.
13. Radulovich, R.; Neori, A.; Valderrama, D.; Reddy, C.R.K.; Cronin, H.; Forster, J. Chapter 3—Farming of Seaweeds. In *Seaweed Sustainability*; Tiwari, B.K., Troy, D.J., Eds.; Academic Press: San Diego, CA, USA, 2015; pp. 27–59.
14. FAO. *The State of World Fisheries and Aquaculture. Sustainability in Action*; FAO: Rome, Italy, 2020; p. 244. [CrossRef]
15. FAO. *The State of World Fisheries and Aquaculture*; Towards Blue Transformation; FAO: Rome, Italy, 2022; p. 54. [CrossRef]
16. Chopin, T.; Buschmann, A.H.; Halling, C.; Troell, M.; Kautsky, N.; Neori, A.; Kraemer, G.P.; Zertuche-González, J.A.; Yarish, C.; Neefus, C. Integrating seaweeds into marine aquaculture systems: A key toward sustainability. *J. Phycol.* **2001**, *37*, 975–986. [CrossRef]
17. Broch, O.J.; Alver, M.O.; Bekkby, T.; Gundersen, H.; Forbord, S.; Handå, A.; Skjermo, J.; Hancke, K. The kelp cultivation potential in coastal and offshore regions of Norway. *Front. Mar. Sci.* **2019**, *5*, 529. [CrossRef]
18. Bixler, H.J.; Porse, H. A decade of change in the seaweed hydrocolloids industry. *J. Appl. Phycol.* **2011**, *23*, 321–335. [CrossRef]
19. FAO. *World Review of Fisheries Aquaculture*; FAO Fisheries and Aquaculture Department: Rome, Italy, 2009; pp. 1–84.
20. Harley, C.D.G.; Anderson, K.M.; Demes, K.W.; Jorve, J.P.; Kordas, R.L.; Coyle, T.A.; Graham, M.H. Effects of climate change on global seaweed communities. *J. Phycol.* **2012**, *48*, 1064–1078. [CrossRef] [PubMed]
21. Dennis, J.M. *A Guide to the Seaweed Industry*; FAO Fisheries and Aquaculture Technical Paper; FAO: Rome, Italy, 2003; pp. 100–108.
22. FAO. *Social and Economic Dimensions of Carrageenan Seaweed Farming*; FAO Fisheries and Aquaculture Technical Paper; Valderrama, D., Cai, J., Hishamunda, N., Ridler, N., Eds.; FAO: Rome, Italy, 2013; pp. 1–204.
23. Hayashi, L.; Hurtado, A.Q.; Msuya, F.E.; Bleicher-Lhonneur, G.; Critchley, A.T. *A Review of Kappaphycus Farming: Prospects and Constraints BT—Seaweeds and Their Role in Globally Changing Environments*; Seckbach, J., Einav, R., Israel, A., Eds.; Springer: Dordrecht, The Netherlands, 2010; pp. 251–283.
24. Neish, I.C. *Good Agronomy Practices for Kappaphycus and Eucheuma: Including an Overview of Basic Biology*; SEAPlant.net Monograph: Makassar, Indonesia, 2008; pp. 1–72.
25. De San, M. *The Farming of Seaweeds*; Smartfish Programme, Indian Ocean Commission: Ebène, Mauritius, 2012; pp. 1–23.

26. Zainee, N.F.A.; Rozaimi, M. Influence of monsoonal storm disturbance on the diversity of intertidal macroalgae along the eastern coast of Johor (Malaysia). *Reg. Stud. Mar. Sci.* **2020**, *40*, 101481. [CrossRef]
27. Zou, D.; Gao, K. Physiological responses of seaweeds to elevated atmospheric CO_2 concentrations. *Seaweeds Their Role Glob. Chang. Environ.* **2010**, *15*, 115–126. [CrossRef]
28. Orr, J.C.; Fabry, V.J.; Aumont, O.; Bopp, L.; Doney, S.C.; Feely, R.A.; Gnanadesikan, A.; Gruber, N.; Ishida, A.; Joos, F. Anthropogenic ocean acidification over the twenty-first century and its impact on calcifying organisms. *Nature* **2005**, *437*, 681–686. [CrossRef] [PubMed]
29. Hofmann, L.C.; Bischof, K. Ocean acidification effects on calcifying macroalgae. *Aquat. Biol.* **2014**, *22*, 261–279. [CrossRef]
30. Wizemann, A.; Meyer, F.; Hofmann, L.; Wild, C.; Westphal, H. Ocean acidification alters the calcareous microstructure of the green macroalga *Halimeda opuntia*. *Coral Reefs* **2015**, *34*, 941–954. [CrossRef]
31. Jiang, L.Q.; Carter, B.R.; Feely, R.A.; Lauvset, S.K.; Olsen, A. Surface ocean pH and buffer capacity: Past, present and future. *Sci. Rep.* **2019**, *9*, 18624. [CrossRef]
32. Kottmeier, D.M.; Chrachri, A.; Langer, G.; Helliwell, K.E.; Wheeler, G.L.; Brownlee, C. Reduced H^+ channel activity disrupts pH homeostasis and calcification in *coccolithophores* at low ocean pH. *Proc. Natl. Acad. Sci. USA* **2022**, *119*, e2118009119. [CrossRef]
33. Garcia-Soto, C.; Cheng, L.; Caesar, L.; Schmidtko, S.; Jewett, E.B.; Cheripka, A.; Rigor, I.; Caballero, A.; Chiba, S.; Báez, J.C.; et al. An overview of ocean climate change indicators: Sea surface temperature, ocean heat content, ocean pH, dissolved oxygen concentration, arctic sea ice extent, thickness and volume, sea level and strength of the AMOC (Atlantic Meridional Overturning Circula). *Front. Mar. Sci.* **2021**, *8*, 642372. [CrossRef]
34. Callaway, R.; Shinn, A.P.; Grenfell, S.E.; Bron, J.E.; Burnell, G.; Cook, E.J.; Crumlish, M.; Culloty, S.; Davidson, K.; Ellis, R.P.; et al. Review of climate change impacts on marine aquaculture in the UK and Ireland. *Aquat. Conserv. Mar. Freshw. Ecosyst.* **2012**, *22*, 389–421. [CrossRef]
35. Hoegh-Guldberg, O.; Jacob, D.; Taylor, M.; Guillén Bolaños, T.; Bindi, M.; Brown, S.; Camilloni, I.A.; Diedhiou, A.; Djalante, R.; Ebi, K. The human imperative of stabilizing global climate change at 1.5 °C. *Science* **2019**, *365*, eaaw6974. [CrossRef] [PubMed]
36. Pachauri, R.K.; Allen, M.R.; Barros, V.R.; Broome, J.; Cramer, W.; Christ, R.; Church, J.A.; Clarke, L.; Dahe, Q.; Dasgupta, P.; et al. Climate Change 2014: Synthesis Report. In *Contribution of Working Groups I, II and III to the Fifth Assessment Report of the Intergovernmental Panel on Climate Change*; Pachauri, R., Meyer, L., Eds.; IPCC: Geneva, Switzerland, 2014; p. 151.
37. Shukla, P.R.; Skea, J.; Calvo Buendia, E.; Masson-Delmotte, V.; Pörtner, H.O.; Roberts, D.C.; Zhai, P.; Slade, R.; Connors, S.; Van Diemen, R.; et al. (Eds.) *Climate Change and Land: An IPCC Special Report on Climate Change, Desertification, Land Degradation, Sustainable Land Management, Food Security, and Greenhouse Gas Fluxes in Terrestrial Ecosystems*; IPCC: Geneva, Switzerland, 2019; pp. 551–672. [CrossRef]
38. Noor, N.M.; Das, S.K. Effects of elevated carbon dioxide on marine ecosystem and associated fishes. *Thalass. Int. J. Mar. Sci.* **2019**, *35*, 421–429. [CrossRef]
39. Zou, D. Effects of elevated atmospheric CO_2 on growth, photosynthesis and nitrogen metabolism in the economic brown seaweed, *Hizikia fusiforme* (Sargassaceae, Phaeophyta). *Aquaculture* **2005**, *250*, 726–735. [CrossRef]
40. Raven, J.; Caldeira, K.; Elderfield, H.; Hoegh-Guldberg, O.; Liss, P.; Riebesell, U.; Shepherd, J.; Turley, C.; Watson, A. *Ocean Acidification Due to Increasing Atmospheric Carbon Dioxide*; The Royal Society: Cardiff, UK, 2005; pp. 1–68, ISBN 0-85403-617-2.
41. Fabry, V.J.; Seibel, B.A.; Feely, R.A.; Orr, J.C. Impacts of ocean acidification on marine fauna and ecosystem processes. *ICES J. Mar. Sci.* **2008**, *65*, 414–432. [CrossRef]
42. Noor, N.M.; De, M.; Iskandar, A.; Keng, W.L.; Cob, Z.C.; Ghaffar, M.A.; Das, S.K. Effects of elevated carbon dioxide on the growth and welfare of juvenile tiger grouper (*Epinephelus Fuscoguttatus*) × giant grouper (*E. Lanceolatus*) hybrid. *Aquaculture* **2019**, *513*, 734448. [CrossRef]
43. Ho, M.; McBroom, J.; Bergstrom, E.; Diaz-Pulido, G. Physiological responses to temperature and ocean acidification in tropical fleshy macroalgae with varying affinities for inorganic carbon. *ICES J. Mar. Sci.* **2021**, *78*, 89–100. [CrossRef]
44. Johnson, M.D.; Price, N.N.; Smith, J.E. Contrasting effects of ocean acidification on tropical fleshy and calcareous algae. *PeerJ* **2014**, *2*, e411. [CrossRef]
45. Gutow, L.; Rahman, M.M.; Bartl, K.; Saborowski, R.; Bartsch, I.; Wiencke, C. Ocean acidification affects growth but not nutritional quality of the seaweed *Fucus vesiculosus* (phaeophyceae, fucales). *J. Exp. Mar. Biol. Ecol.* **2014**, *453*, 84–90. [CrossRef]
46. Xu, J.; Gao, K. Future CO_2-induced ocean acidification mediates the physiological performance of a green tide alga. *Plant Physiol.* **2012**, *160*, 1762–1769. [CrossRef]
47. Page, M.J.; McKenzie, J.E.; Bossuyt, P.M.; Boutron, I.; Hoffmann, T.C.; Mulrow, C.D.; Shamseer, L.; Tetzlaff, J.M.; Akl, E.A.; Brennan, S.E. The PRISMA 2020 statement: An updated guideline for reporting systematic reviews. *Syst. Rev.* **2021**, *10*, 89. [CrossRef] [PubMed]
48. Snyder, H. Literature review as a research methodology: An overview and guidelines. *J. Bus. Res.* **2019**, *104*, 333–339. [CrossRef]
49. Li, Z.-S.; Hasson, F. Resilience, stress, and psychological well-being in nursing students: A systematic review. *Nurse Educ. Today* **2020**, *90*, 104440. [CrossRef] [PubMed]
50. Bregnballe, J. *A Guide to Recirculation Aquaculture: An Introduction to the New Environmentally Friendly and Highly Productive Closed Fish Farming Systems*; FAO Eurofish International Organisation: Copenhagen, Denmark, 2015; pp. 1–95.
51. Culture Aquaculture Species Information Program: Laminaria japonica Areschoug 1851. Available online: https://www.fao.org/fishery/en/culturedspecies/Laminaria_japonica/en (accessed on 17 August 2020).

52. IPCC. Summary for Policymakers. In *AR4 Climate Change 2007: The Physical Science Basis: Contribution of Working Group I to the Fourth Assessment Report of the Intergovernmental Panel on Climate Change*; Solomon, S., Qin, D., Manning, M., Chen, Z., Marquis, M., Averyt, K.B., Tignor, M., Miller, H.L., Eds.; Cambridge University Press: Cambridge, UK; New York, NY, USA, 2007; pp. 1–18.
53. Clements, J.; Chopin, T. Ocean acidification and marine aquaculture in North America: Potential impacts and mitigation strategies. *Rev. Aquac.* **2016**, *9*, 326–341. [CrossRef]
54. Stévant, P.; Rebours, C.; Chapman, A. Seaweed aquaculture in Norway: Recent industrial developments and future perspectives. *Aquac. Int.* **2017**, *25*, 1373–1390. [CrossRef]
55. Bartley, D.; Menezes, A.; Metzner, R.; Ansah, Y. *The FAO Blue Growth Initiative: Strategy for the Development of Fisheries and Aquaculture in Eastern Africa*; FAO Fisheries and Aquaculture Technical Paper; FAO: Rome, Italy, 2018. [CrossRef]
56. UN Office of Legal Affairs. *The Second World Ocean Assessment*; United Nations: New York, NY, USA, 2021.
57. Gordillo, F.J.L.; Carmona, R.; Viñegla, B.; Wiencke, C.; Jiménez, C. Effects of simultaneous increase in temperature and ocean acidification on biochemical composition and photosynthetic performance of common macroalgae from Kongsfjorden (Svalbard). *Polar Biol.* **2016**, *39*, 1993–2007. [CrossRef]
58. Schiener, P.; Black, K.D.; Stanley, M.S.; Green, D.H. The seasonal variation in the chemical composition of the kelp species *Laminaria gigitata, Laminaria hyperborea, Saccharina latissima* and *Alaria esculenta*. *J. Appl. Phycol.* **2015**, *27*, 363–373. [CrossRef]
59. Iñiguez, C.; Carmona, R.; Lorenzo, M.R.; Niell, F.X.; Wiencke, C.; Gordillo, F.J.L. Increased CO_2 modifies the carbon balance and the photosynthetic yield of two common arctic brown seaweeds: *Desmarestia aculeata* and *Alaria esculenta*. *Polar Biol.* **2016**, *39*, 1979–1991. [CrossRef]
60. Moroney, J.V.; Ynalvez, R.A. Proposed carbon dioxide concentrating mechanism in *Chlamydomonas reinhardtii*. *Eukaryot. Cell* **2007**, *6*, 1251–1259. [CrossRef]
61. Xu, D.; Brennan, G.; Xu, L.; Zhang, X.W.; Fan, X.; Han, W.T.; Mock, T.; McMinn, A.; Hutchins, D.A.; Ye, N. Ocean acidification increases iodine accumulation in kelp-based coastal food webs. *Glob. Chang. Biol.* **2019**, *25*, 629–639. [CrossRef]
62. Kang, J.W.; Chung, I.K. The interactive effects of elevated CO_2 and ammonium enrichment on the physiological performances of *Saccharina japonica* (Laminariales, Phaeophyta). *Ocean Sci. J.* **2018**, *53*, 487–497. [CrossRef]
63. Jiang, Z.; Fang, J.; Mao, Y.; Han, T.; Wang, G. Influence of seaweed aquaculture on marine inorganic carbon dynamics and sea-air CO_2 flux. *J. World Aquac. Soc.* **2013**, *44*, 133–140. [CrossRef]
64. Xu, D.; Wang, D.; Li, B.; Fan, X.; Zhang, X.W.; Ye, N.H.; Wang, Y.; Mou, S.; Zhuang, Z. Effects of CO_2 and seawater acidification on the early stages of *Saccharina japonica* development. *Environ. Sci. Technol.* **2015**, *49*, 3548–3556. [CrossRef] [PubMed]
65. Gao, X.; Kim, J.H.; Park, S.K.; Yu, O.H.; Kim, Y.S.; Choi, H.G. Diverse responses of sporophytic photochemical efficiency and gametophytic growth for two edible kelps, *Saccharina japonica* and *Undaria pinnatifida*, to ocean acidification and warming. *Mar. Pollut. Bull.* **2019**, *142*, 315–320. [CrossRef]
66. Zhang, X.; Xu, D.; Guan, Z.; Wang, S.; Zhang, Y.; Wang, W.; Zhang, X.; Fan, X.; Li, F.; Ye, N. Elevated CO_2 concentrations promote growth and photosynthesis of the brown alga *Saccharina japonica*. *J. Appl. Phycol.* **2020**, *32*, 1949–1959. [CrossRef]
67. Peteiro, C.; Freire, Ó. Offshore cultivation methods affects blade features of the edible seaweed *Saccharina latissima* in a Bay of Galicia, Northwest Spain. *Russ. J. Mar. Biol.* **2011**, *37*, 319. [CrossRef]
68. Pedersen, M. Temperature Effects on the Kelp *Saccharina latissimi*. In Proceedings of the Association for the Science of Limnology and Oceanography (ASLO) Aquatic Sciences Meeting, Grenada, Spain, 22–27 February 2015.
69. Olischläger, M.; Iñiguez, C.; Koch, K.; Wiencke, C.; Gordillo, F.J.L. Increased pCO_2 and temperature reveal ecotypic differences in growth and photosynthetic performance of temperate and arctic populations of *Saccharina latissima*. *Planta* **2017**, *245*, 119–136. [CrossRef]
70. Olischläger, M.; Iñiguez, C.; Gordillo, F.J.L.; Wiencke, C. Biochemical composition of temperate and arctic populations of *Saccharina latissima* after exposure to increased pCO_2 and temperature reveals ecotypic variation. *Planta* **2014**, *240*, 1213–1224. [CrossRef]
71. Shim, J.H.; Kim, J.B.; Hwang, D.-W.W.; Choi, H.-G.G.; Lee, Y. Variations in carbon and nitrogen stable isotopes and in heavy metal contents of mariculture kelp *Undaria pinnatifida* in Gijang, Southeastern Korea. *Algae* **2017**, *32*, 349–357. [CrossRef]
72. Zhang, X.; Hu, H.; Tan, T. Photosynthetic inorganic carbon utilization of gametophytes and sporophytes of *Undaria pinnatifida* (Phaeophyceae). *Phycologia* **2006**, *45*, 642–647. [CrossRef]
73. Leal, P.P.; Hurd, C.L.; Fernández, P.A.; Roleda, M.Y. Meiospore development of the kelps *Macrocystis pyrifera* and *Undaria pinnatifida* under ocean acidification and ocean warming: Independent effects are more important than their interaction. *Mar. Biol.* **2016**, *164*, 7. [CrossRef]
74. Xu, K.; Chen, H.; Wang, W.; Xu, Y.; Ji, D.; Chen, C.; Xie, C. Responses of photosynthesis and CO_2 concentrating mechanisms of marine crop *Pyropia haitanensis* thalli to large pH variations at different time scales. *Algal Res.* **2017**, *28*, 200–210. [CrossRef]
75. Chen, B.; Zou, D.; Ma, J. Interactive effects of elevated CO_2 and nitrogen–phosphorus supply on the physiological properties of *Pyropia haitanensis* (Bangiales, Rhodophyta). *J. Appl. Phycol.* **2016**, *28*, 1235–1243. [CrossRef]
76. Chen, B.; Lin, L.; Ma, Z.; Zhang, T.; Chen, W.; Zou, D. Carbon and nitrogen accumulation and interspecific competition in two algae species, *Pyropia haitanensis* and *Ulva lactuca*, under ocean acidification conditions. *Aquac. Int.* **2019**, *27*, 721–733. [CrossRef]
77. Koh, Y.H.; Kim, M.S. DNA barcoding reveals cryptic diversity of economic red algae, *Pyropia* (Bangiales, Rhodophyta): Description of novel species from Korea. *J. Appl. Phycol.* **2018**, *30*, 3425–3434. [CrossRef]
78. Nelson, W.A. *Pyropia plicata* Sp. Nov. (Bangiales, Rhodophyta): Naming a common intertidal alga from New Zealand. *PhytoKeys* **2013**, *21*, 17–28. [CrossRef]

79. Liu, C.; Zou, D. Effects of elevated CO_2 on the photosynthesis and nitrate reductase activity of *Pyropia haitanensis* (Bangiales, Rhodophyta) grown at different nutrient levels. *Chin. J. Oceanol. Limnol.* **2015**, *33*, 419–429. [CrossRef]
80. Ma, J.; Xu, T.; Bao, M.; Zhou, H.; Zhang, T.; Li, Z.; Gao, G.; Li, X.; Xu, J. Response of the red algae *Pyropia yezoensis* grown at different light intensities to co_2-induced seawater acidification at different life cycle stages. *Algal Res.* **2020**, *49*, 101950. [CrossRef]
81. Kang, J.W.; Chung, I.K. The effects of eutrophication and acidification on the ecophysiology of *Ulva pertusa* Kjellman. *J. Appl. Phycol.* **2017**, *29*, 2675–2683. [CrossRef]
82. Young, C.S.; Gobler, C.J. Ocean acidification accelerates the growth of two bloom-forming macroalgae. *PLoS ONE* **2016**, *11*, e0155152. [CrossRef]
83. Liu, C.; Zou, D.; Yang, Y.; Chen, B.; Jiang, H. Temperature responses of pigment contents, chlorophyll fluorescence characteristics, and antioxidant defenses in *Gracilariopsis lemaneiformis* (Gracilariales, Rhodophyta) under different CO_2 levels. *J. Appl. Phycol.* **2017**, *29*, 983–991. [CrossRef]
84. Wei, Z.; Zhang, Y.; Yang, F.; Liang, J.; Long, L. Increased light availability modulates carbon and nitrogen accumulation in the macroalga *Gracilariopsis lemaneiformis* (Rhodophyta) in response to ocean acidification. *Environ. Exp. Bot.* **2021**, *187*, 104492. [CrossRef]
85. Wang, L.; Jiang, H.; Zou, D.; Ye, C. Effects of increased pH and inorganic carbon on growth and photosynthesis in the macroalga *Gracilaria lemaneiformis* (Gigartinales, Rhodophyta). *Phycologia* **2020**, *59*, 218–226. [CrossRef]
86. Smith, R.G.; Bidwell, R.G.S. Mechanism of photosynthetic carbon dioxide uptake by the red macroalga, *Chondrus crispus*. *Plant Physiol.* **1989**, *89*, 93–99. [CrossRef] [PubMed]
87. Phang, S.-M.; Yeong, H.-Y.; Hussin, H.; Lim, P.-E.; You, H.-C.; Juan, J.-C. Techno-economics of seaweed farming along the coasts of Kelantan, East Coast Peninsular Malaysia. *Malays. J. Sci.* **2017**, *36*, 85–102. [CrossRef]
88. Chopin, T.; Sharp, G.; Belyea, E.; Semple, R.; Jones, D. Open-water aquaculture of the red alga *Chondrus crispus* in Prince Edward Island, Canada. *Hydrobiologia* **1999**, *398–399*, 417–425. [CrossRef]
89. Reidenbach, L.B.; Fernandez, P.A.; Leal, P.P.; Noisette, F.; McGraw, C.M.; Revill, A.T.; Hurd, C.L.; Kübler, J.E. Growth, Ammonium metabolism, and photosynthetic properties of *Ulva australis* (Chlorophyta) under decreasing pH and ammonium enrichment. *PLoS ONE* **2017**, *12*, e0188389. [CrossRef]
90. Bi, Y.; Zhou, Z. Absorption and transport of inorganic carbon in kelps with emphasis on *Saccharina japonica*. In *Applied Photosynthesis*; Najafpour, M.M., Ed.; IntechOpen: Rijeka, Croatia, 2016; Chapter 6. [CrossRef]
91. Fernández, P.A.; Roleda, M.Y.; Hurd, C.L. Effects of ocean acidification on the photosynthetic performance, carbonic anhydrase activity and growth of the giant kelp *Macrocystis pyrifera*. *Photosynth. Res.* **2015**, *124*, 293–304. [CrossRef]
92. Roleda, M.Y.; Morris, J.N.; McGraw, C.M.; Hurd, C.L. Ocean acidification and seaweed reproduction: Increased CO_2 ameliorates the negative effect of lowered pH on meiospore germination in the giant kelp *Macrocystis pyrifera* (Laminariales, Phaeophyceae). *Glob. Chang. Biol.* **2012**, *18*, 854–864. [CrossRef]
93. Tee, M.Z.; Yong, Y.S.; Rodrigues, K.F.; Yong, W.T.L. Growth rate analysis and protein identification of *Kappaphycus alvarezii* (Rhodophyta, Gigartinales) under pH Induced Stress Culture. *Aquac. Rep.* **2015**, *2*, 112–116. [CrossRef]
94. Largo, D.B.; Chung, I.K.; Phang, S.-M.; Gerung, G.S.; Sondak, C.F.A. *Impacts of Climate Change on Eucheuma-Kappaphycus Farming BT—Tropical Seaweed Farming Trends, Problems and Opportunities: Focus on Kappaphycus and Eucheuma of Commerce*; Hurtado, A.Q., Critchley, A.T., Neish, I.C., Eds.; Springer International Publishing: Cham, Switzerland, 2017; pp. 121–129. [CrossRef]
95. Kumar, A.; AbdElgawad, H.; Castellano, I.; Lorenti, M.; Delledonne, M.; Beemster, G.T.S.; Asard, H.; Buia, M.C.; Palumbo, A. Physiological and biochemical analyses shed light on the response of Sargassum vulgare to ocean acidification at different time scales. *Front. Plant Sci.* **2017**, *8*, 570. [CrossRef] [PubMed]
96. Israel, A.; Hophy, M. Growth, photosynthetic properties and Rubisco activities and amounts of marine macroalgae grown under current and elevated seawater CO_2 concentrations. *Glob. Chang. Biol.* **2002**, *8*, 831–840. [CrossRef]
97. Celis-Plá, P.S.M.; Hall-Spencer, J.M.; Horta, P.A.; Milazzo, M.; Korbee, N.; Cornwall, C.E.; Figueroa, F.L. Macroalgal responses to ocean acidification depend on nutrient and light levels. *Front. Mar. Sci.* **2015**, *2*, 26. [CrossRef]
98. Kumar, A.; Buia, M.C.; Palumbo, A.; Mohany, M.; Wadaan, M.A.M.; Hozzein, W.N.; Beemster, G.T.S.; AbdElgawad, H. Ocean acidification affects biological activities of seaweeds: A case study of *Sargassum vulgare* from Ischia Volcanic CO_2 Vents. *Environ. Pollut.* **2020**, *259*, 113765. [CrossRef]
99. Young, C.S.; Lowell, A.; Peterson, B.; Gobler, C.J. Ocean acidification and food limitation combine to suppress herbivory by the gastropod *Lacuna vincta*. *Mar. Ecol. Prog. Ser.* **2019**, *627*, 83–94. [CrossRef]
100. Gordillo, F.J.L.; Niell, F.X.; Figueroa, F.L. Non-photosynthetic enhancement of growth by high CO_2 level in the nitrophilic seaweed *Ulva rigida* C. Agardh (Chlorophyta). *Planta* **2001**, *213*, 64–70. [CrossRef]
101. Meichssner, R.; Krost, P.; Schulz, R. Experimental testing of density-and season-dependent growth in vegetative *Fucus* aquaculture and modelling of growth over one year for different cultivation scenarios. *J. Appl. Phycol.* **2021**, *33*, 3939–3950. [CrossRef]
102. Meichssner, R.; Krost, P.; Schulz, R. Vegetative aquaculture of *Fucus* in the Baltic Sea—Obtaining low-fertility biomass from attached or unattached populations? *J. Appl. Phycol.* **2021**, *33*, 1709–1720. [CrossRef]
103. Graiff, A.; Dankworth, M.; Wahl, M.; Karsten, U.; Bartsch, I. Seasonal variations of *Fucus vesiculosus* fertility under ocean acidification and warming in the western Baltic Sea. *Bot. Mar.* **2017**, *60*, 239–255. [CrossRef]
104. Ober, G.T.; Thornber, C.S. Divergent responses in growth and nutritional quality of coastal macroalgae to the combination of increased pCO_2 and nutrients. *Mar. Environ. Res.* **2017**, *131*, 69–79. [CrossRef]

105. Yang, Y.L.; Li, W.; Chen, W.Z.; Xu, J.T. Photosynthetic responses to solar UV Radiation of *Gracilaria lemaneiformis* cultured under different temperatures and CO_2 concentrations. *Shengtai Xuebao/Acta Ecol. Sin.* **2013**, *33*, 5538–5545. [CrossRef]
106. Lee, W.-K.; Lim, Y.-Y.; Ho, C.-L. pH affects growth, physiology and agar properties of agarophyte *Gracilaria changii* (Rhodophyta) under low light intensity from Morib, Malaysia. *Reg. Stud. Mar. Sci.* **2019**, *30*, 100738. [CrossRef]
107. Kang, J.W.; Kambey, C.; Shen, Z.; Yang, Y.; Chung, I.K. The short-term effects of elevated CO_2 and ammonium concentrations on physiological responses in *Gracilariopsis lemaneiformis* (Rhodophyta). *Fish. Aquat. Sci.* **2017**, *20*, 18. [CrossRef]
108. Gao, G.; Qu, L.; Xu, T.; Burgess, J.G.; Li, X.; Xu, J.; Norkko, J. Future CO_2-induced ocean acidification enhances resilience of a green tide alga to low-salinity stress. *ICES J. Mar. Sci.* **2019**, *76*, 2437–2445. [CrossRef]
109. Thien, V.Y.; Rodrigues, K.F.; Voo, C.L.Y.; Wong, C.M.V.L.; Yong, W.T.L. Comparative transcriptome profiling of *Kappaphycus alvarezii* (Rhodophyta, Solieriaceae) in response to light of different wavelengths and carbon dioxide enrichment. *Plants* **2021**, *10*, 1236. [CrossRef]
110. Chen, B.; Xia, J.; Zou, D.; Zhang, X. Responses to ocean acidification and diurnal temperature variation in a commercially farmed seaweed, *Pyropia haitanensis* (Rhodophyta). *Eur. J. Phycol.* **2019**, *54*, 184–192. [CrossRef]
111. Mashoreng, S.; La Nafie, Y.A.; Isyrini, R. Cultivated seaweed carbon sequestration capacity. In *IOP Conference Series: Earth and Environmental Science*; IOP Publishing: Moscow, Russia, 2019. [CrossRef]
112. Zhang, B.; Liu, X.; Huan, L.; Shao, Z.; Zheng, Z.; Wang, G. Carbonic anhydrase isoforms of *Neopyropia yezoensis*: Intracellular localization and expression profiles in response to inorganic carbon concentration and life stage. *J. Phycol.* **2022**, *58*, 657–668. [CrossRef] [PubMed]
113. Zhang, B.; Xie, X.; Liu, X.; He, L.; Sun, Y.; Wang, G. The carbonate concentration mechanism of *Pyropia yezoensis* (Rhodophyta): Evidence from transcriptomics and biochemical data. *BMC Plant Biol.* **2020**, *20*, 424. [CrossRef]
114. Kevekordes, K.; Holland, D.; Häubner, N.; Jenkins, S.; Koss, R.; Roberts, S.; Raven, J.A.; Scrimgeour, C.M.; Shelly, K.; Stojkovic, S.; et al. Inorganic carbon acquisition by eight species of *Caulerpa* (Caulerpaceae, Chlorophyta). *Phycologia* **2006**, *45*, 442–449. [CrossRef]
115. Zubia, M.; Draisma, S.G.A.; Morrissey, K.L.; Varela-Álvarez, E.; De Clerck, O. Concise review of the genus *Caulerpa* J.V. Lamouroux. *J. Appl. Phycol.* **2020**, *32*, 23–39. [CrossRef]
116. Ganesan, M.; Thiruppathi, S.; Jha, B. Mariculture of *Hypnea musciformis* (Wulfen) Lamouroux in South East Coast of India. *Aquaculture* **2006**, *256*, 201–211. [CrossRef]
117. Pereira, S.A.; Kimpara, J.M.; Valenti, W.C. A bioeconomic analysis of the potential of seaweed *Hypnea pseudomusciformis* farming to different targeted markets. *Aquac. Econ. Manag.* **2020**, *24*, 507–525. [CrossRef]
118. Nauer, F.; Borburema, H.D.S.; Yokoya, N.S.; Fujii, M.T. Effects of ocean acidification on growth, pigment contents and antioxidant potential of the subtropical Atlantic Red Alga *Hypnea pseudomusciformis* Nauer, Cassano & M.C. Oliveira (Gigartinales) in laboratory. *Braz. J. Bot.* **2021**, *44*, 69–77. [CrossRef]
119. Nauer, F.; Yokoya, N.S.; Fujii, M.T. Short-term effects of temperature on morphology and physiology of turf-forming *Hypnea* species (Rhodophyta) from Southeastern Brazil. *Appl. Phycol.* **2021**, *3*, 247–259. [CrossRef]
120. Pereira, S.A.; Kimpara, J.M.; Valenti, W.C. Sustainability of the seaweed *Hypnea pseudomusciformis* farming in the tropical southwestern Atlantic. *Ecol. Indic.* **2021**, *121*, 107101. [CrossRef]
121. Suárez-Álvarez, S.; Gómez-Pinchetti, J.L.; García-Reina, G. Effects of increased CO_2 levels on growth, photosynthesis, ammonium uptake and cell composition in the macroalga *Hypnea spinella* (Gigartinales, Rhodophyta). *J. Appl. Phycol.* **2012**, *24*, 815–823. [CrossRef]
122. Velázquez-Ochoa, R.; Ochoa-Izaguirre, M.J.; Soto-Jiménez, M.F. An analysis of the variability in $\delta^{13}C$ in macroalgae from the Gulf of California: Indicative of carbon concentration mechanisms and isotope discrimination during carbon assimilation. *Biogeosciences* **2022**, *19*, 1–27. [CrossRef]
123. Friedlander, M. Advances in cultivation of Gelidiales. *J. Appl. Phycol.* **2007**, *20*, 1–6. [CrossRef]
124. Mercado, J.M.; Xavier Niell, F.; Candelaria Gil-Rodríguez, M. Photosynthesis might be limited by light, not inorganic carbon availability, in three intertidal Gelidiales species. *New Phytol.* **2001**, *149*, 431–439. [CrossRef]
125. Kawaroe, M.; Sunuddin, A.; Augustine, D.; Lestari, D.F. The effect of CO_2 injection on macroalgae *Gelidium latifolium* biomass growth rate and carbohydrate content. *Indones. J. Mar. Sci. Kelaut.* **2016**, *21*, 85–92. [CrossRef]
126. Melo, R.A.; Harger, B.W.W.; Neushul, M. Gelidium Cultivation in the sea. In *Proceedings of the International Workshop on Gelidium, Santander, Spain, 3–8 September 1991*; Juanes, J.A., Santelices, B., McLachlan, J.L., Eds.; Springer: Dordrecht, The Netherlands, 1991.
127. Porzio, L.; Buia, M.C.; Hall-Spencer, J.M. Effects of ocean acidification on macroalgal communities. *J. Exp. Mar. Bio. Ecol.* **2011**, *400*, 278–287. [CrossRef]
128. Buck, B.H.; Krause, G.; Rosenthal, H. Extensive open ocean aquaculture development within wind farms in Germany: The prospect of offshore co-management and legal constraints. *Ocean Coast. Manag.* **2004**, *47*, 95–122. [CrossRef]
129. Anyanwu, C.N.; Amadi-Eke, A.S.; Nwaka, D.E.; Ezeafulukwe, C.F.; Adaka, G.S. Climate change, effects and mitigation strategies on aquaculture: A review. *Agric. For. Fish.* **2015**, *4*, 70–72. [CrossRef]
130. Ahmed, N.; Thompson, S.; Glaser, M. Global aquaculture productivity, environmental sustainability, and climate change adaptability. *Environ. Manag.* **2019**, *63*, 159–172. [CrossRef]
131. Abreu, M.H.; Pereira, R.; Yarish, C.; Buschmann, A.H.; Sousa-Pinto, I. IMTA with *Gracilaria vermiculophylla*: Productivity and nutrient removal performance of the seaweed in a land-based pilot scale system. *Aquaculture* **2011**, *312*, 77–87. [CrossRef]

132. Troell, M.; Joyce, A.; Chopin, T.; Neori, A.; Buschmann, A.H.; Fang, J.-G. Ecological engineering in aquaculture—Potential for Integrated Multi-Trophic Aquaculture (IMTA) in marine offshore systems. *Aquaculture* **2009**, *297*, 1–9. [CrossRef]
133. Buchholz, C.M.; Krause, G.; Buck, B.H. Seaweed and man. In *Seaweed Biology*; Ecological Studies; Springer: Berlin/Heidelberg, Germany, 2012; pp. 471–493. [CrossRef]
134. Hafting, J.T.; Craigie, J.S.; Stengel, D.B.; Loureiro, R.R.; Buschmann, A.H.; Yarish, C.; Edwards, M.D.; Critchley, A.T. Prospects and challenges for industrial production of seaweed bioactives. *J. Phycol.* **2015**, *51*, 821–837. [CrossRef]
135. Xiao, X.; Agustí, S.; Yu, Y.; Huang, Y.; Chen, W.; Hu, J.; Li, C.; Li, K.; Wei, F.; Lu, Y.; et al. Seaweed Farms Provide Refugia from Ocean Acidification. *Sci. Total Environ.* **2021**, *776*, 145192. [CrossRef]
136. Kim, J.K.; Yarish, C.; Hwang, E.K.; Park, M.; Kim, Y. Seaweed aquaculture: Cultivation technologies, challenges and its ecosystem services. *ALGAE* **2017**, *32*, 1–13. [CrossRef]
137. van der Loos, L.M.; Schmid, M.; Leal, P.P.; McGraw, C.M.; Britton, D.; Revill, A.T.; Virtue, P.; Nichols, P.D.; Hurd, C.L. Responses of macroalgae to CO_2 enrichment cannot be inferred solely from their inorganic carbon uptake strategy. *Ecol. Evol.* **2019**, *9*, 125–140. [CrossRef]
138. Cornwall, C.; Comeau, S.; McCulloch, M. Coralline algae elevate pH at the site of calcification under ocean acidification. *Glob. Chang. Biol.* **2017**, *23*, 4245–4256. [CrossRef] [PubMed]
139. Zweng, R.C.; Koch, M.S.; Bowes, G. The role of irradiance and C-Use strategies in tropical macroalgae photosynthetic response to ocean acidification. *Sci. Rep.* **2018**, *8*, 9479. [CrossRef] [PubMed]
140. Cornwall, C.E.; Revill, A.T.; Hall-Spencer, J.M.; Milazzo, M.; Raven, J.A.; Hurd, C.L. Inorganic carbon physiology underpins macroalgal responses to elevated CO_2. *Sci. Rep.* **2017**, *7*, 46297. [CrossRef]
141. Giordano, M.; Beardall, J.; Raven, J.A. CO_2 Concentrating mechanisms in algae: Mechanisms, environmental modulation, and evolution. *Annu. Rev. Plant Biol.* **2005**, *56*, 99–131. [CrossRef]
142. Koch, M.; Bowes, G.; Ross, C.X. Climate change and ocean acidification effects on seagrasses and marine macroalgae. *Glob. Chang. Biol.* **2013**, *19*, 103–132. [CrossRef]
143. Gao, K.; Aruga, Y.; Asada, K.; Kiyohara, M. Influence of enhanced CO_2 on growth and photosynthesis of the red algae *Gracilaria* sp. and *G. chilensis*. *J. Appl. Phycol.* **1993**, *5*, 563–571. [CrossRef]
144. Gao, K.; Aruga, Y.; Asada, K.; Ishihara, T.; Akano, T.; Kiyohara, M. Enhanced growth of the red alga *Porphyra yezoensis* Ueda in high CO_2 concentrations. *J. Appl. Phycol.* **1991**, *3*, 355–362. [CrossRef]
145. Krause-Jensen, D.; Lavery, P.; Serrano, O.; Marba, N.; Masque, P.; Duarte, C.M. Sequestration of macroalgal carbon: The elephant in the blue carbon room. *Biol. Lett.* **2018**, *14*, 20180236. [CrossRef]
146. García-Sánchez, M.J.; Fernández, J.A.; Niell, X. Effect of inorganic carbon supply on the photosynthetic physiology of *Gracilaria tenuistipitata*. *Planta* **1994**, *194*, 55–61. [CrossRef]
147. Israel, A.; Katz, S.; Dubinsky, Z.; Merrill, J.E.; Friedlander, M. Photosynthetic inorganic carbon utilization and growth of *Porphyra linearis* (Rhodophyta). *J. Appl. Phycol.* **1999**, *11*, 447. [CrossRef]
148. Mercado, J.M.; Javier, F.; Gordillo, L.; Xavier Niell, F.; Figueroa, F.L. Effects of different levels of CO_2 on photosynthesis and cell components of the red alga *Porphyra leucosticta*. *J. Appl. Phycol.* **1999**, *11*, 455–461. [CrossRef]
149. Gnanavel, V.; Roopan, S.M.; Rajeshkumar, S. Aquaculture: An overview of chemical ecology of seaweeds (food species) in natural products. *Aquaculture* **2019**, *507*, 1–6. [CrossRef]
150. Nalamothu, N.; Potluri, A.; Muppalla, M.B. Review on marine alginates and its applications. *J. Pharmacol. Res.* **2014**, *4*, 4006–4015.
151. Razak, A.A.R.; Cob, Z.; Jamari, Z.; Majid, A.M.; Toda, T.; Ross, O. Effect of seaweeds as biofilters for mass culture of rotifer *Brachionus plicatilis*. *Sains Malays.* **2015**, *44*, 891–898. [CrossRef]
152. Fernández, P.A.; Roleda, M.Y.; Leal, P.P.; Hurd, C.L. Seawater pH, and not inorganic nitrogen source, affects pH at the blade surface of *Macrocystis pyrifera*: Implications for responses of the giant kelp to future oceanic conditions. *Physiol. Plant.* **2017**, *159*, 107–119. [CrossRef]
153. Gao, K.; Beardall, J.; Häder, D.-P.; Hall-Spencer, J.M.; Gao, G.; Hutchins, D.A. Effects of ocean acidification on marine photosynthetic organisms under the concurrent influences of warming, UV radiation, and deoxygenation. *Front. Mar. Sci.* **2019**, *6*, 322. [CrossRef]
154. Cao, L.; Caldeira, K.; Jain, A.K. Effects of carbon dioxide and climate change on ocean acidification and carbonate mineral saturation. *Geophys. Res. Lett.* **2007**, *34*, 1–5. [CrossRef]
155. Straub, S.C.; Wernberg, T.; Thomsen, M.S.; Moore, P.J.; Burrows, M.T.; Harvey, B.P.; Smale, D.A. Resistance, extinction, and everything in between—The diverse responses of seaweeds to marine heatwaves. *Front. Mar. Sci.* **2019**, *6*, 763. [CrossRef]
156. Al-Janabi, B.; Kruse, I.; Graiff, A.; Winde, V.; Lenz, M.; Wahl, M. Buffering and amplifying interactions among OAW (Ocean Acidification & Warming) and nutrient enrichment on early life-stage *Fucus vesiculosus* L. (Phaeophyceae) and their carry over effects to hypoxia impact. *PLoS ONE* **2016**, *11*, e0152948. [CrossRef]
157. Díez, I.; Muguerza, N.; Santolaria, A.; Ganzedo, U.; Gorostiaga, J.M. Seaweed assemblage changes in the Eastern Cantabrian Sea and their potential relationship to climate change. *Estuar. Coast. Shelf Sci.* **2012**, *99*, 108–120. [CrossRef]
158. Wahl, M.; Werner, F.J.; Buchholz, B.; Raddatz, S.; Graiff, A.; Matthiessen, B.; Karsten, U.; Hiebenthal, C.; Hamer, J.; Ito, M.; et al. Season affects strength and direction of the interactive impacts of ocean warming and biotic stress in a coastal seaweed ecosystem. *Limnol. Oceanogr.* **2020**, *65*, 807–827. [CrossRef]

159. Ahmad, A.; Abdullah, S.R.S.; Hasan, H.A.; Othman, A.R.; Ismail, N. Aquaculture industry: Supply and demand, best practices, effluent and its current issues and treatment technology. *J. Environ. Manag.* **2021**, *287*, 112271. [CrossRef] [PubMed]
160. Valderrama, D.; Cai, J.; Hishamunda, N.; Ridler, N.; Neish, I.C.; Hurtado, A.Q.; Msuya, F.E.; Krishnan, M.; Narayanakumar, R.; Kronen, M.; et al. The economics of *Kappaphycus* seaweed cultivation in developing countries: A comparative analysis of farming systems. *Aquac. Econ. Manag.* **2015**, *19*, 251–277. [CrossRef]
161. Radzi, S.N.; Osman, K.; Mohd Said, M.N. Progressing towards global citizenship and a sustainable nation: Pillars of climate change education and actions. *Sustainability* **2022**, *14*, 5163. [CrossRef]
162. Geroe, S. Addressing climate change through a low-cost, high-impact carbon tax. *J. Environ. Dev.* **2019**, *28*, 3–27. [CrossRef]
163. Hänsel, M.C.; van den Bergh, J.C.J.M. Taxing interacting externalities of ocean acidification, global warming, and eutrophication. *Nat. Resour. Model.* **2021**, *34*, e12317. [CrossRef]
164. Li, W.; Jia, Z. Carbon tax, emission trading, or the mixed policy: Which is the most effective strategy for climate change mitigation in China? *Mitig. Adapt. Strateg. Glob. Chang.* **2017**, *22*, 973–992. [CrossRef]
165. Craigie, J.S.; Shacklock, P.F. Culture of Irish Moss. In *Cold-Water Aquaculture in Atlantic Canada*, 2nd ed.; Canadian Institute for Research on Regional Development: Moncton, NB, Canada, 1995; pp. 241–270.
166. Chopin, T.; Cooper, J.A.; Reid, G.; Cross, S.; Moore, C. Open-water integrated multi-trophic aquaculture: Environmental biomitigation and economic diversification of fed aquaculture by extractive aquaculture. *Rev. Aquac.* **2012**, *4*, 209–220. [CrossRef]
167. Kang, Y.H.; Shin, J.A.; Kim, M.S.; Chung, I.K. A preliminary study of the bioremediation potential of *Codium fragile* applied to seaweed Integrated Multi-Trophic Aquaculture (IMTA) during the summer. *J. Appl. Phycol.* **2008**, *20*, 183–190. [CrossRef]
168. Global Ocean Acidification Observing Network (GOA-ON). Available online: http://www.goa-on.org/home.php (accessed on 12 October 2022).
169. IOC WESTPAC. Southeast Asian Global Ocean Observing System (SEAGOOS). Available online: https://goosocean.org/ (accessed on 12 October 2022).
170. Hardman-Mountford, N.J.; Moore, G.; Bakker, D.C.E.; Watson, A.J.; Schuster, U.; Barciela, R.; Hines, A.; Moncoiffé, G.; Brown, J.; Dye, S. An operational monitoring system to provide indicators of CO_2-related variables in the ocean. *ICES J. Mar. Sci.* **2008**, *65*, 1498–1503. [CrossRef]
171. Lin, J.-Y.; Tsai, H.-L.; Lyu, W.-H. An integrated wireless multi-sensor system for monitoring the water quality of aquaculture. *MDPI Sens.* **2021**, *21*, 8179. [CrossRef] [PubMed]
172. Kübler, J.E.; Dudgeon, S.R.; Bush, D. Climate change challenges and opportunities for seaweed aquaculture in California, the United States. *J. World Aquac. Soc.* **2021**, *52*, 1069–1080. [CrossRef]
173. Yan, X.-H.; Lv, F.; Liu, C.-J.; Zheng, Y.-F. Selection and characterization of a high-temperature tolerant strain of *Porphyra haitanensis* (Bangiales, Rhodophyta). *J. Appl. Phycol.* **2010**, *22*, 511–516. [CrossRef]
174. Andersen, L.B.; Grefsrud, E.S.; Svåsand, T.; Sandlund, N. Risk understanding and risk acknowledgement: A new approach to environmental risk assessment in marine aquaculture. *ICES J. Mar. Sci.* **2022**, *79*, 987–996. [CrossRef]
175. Holsman, K.; Samhouri, J.; Cook, G.; Hazen, E.; Olsen, E.; Dillard, M.; Kasperski, S.; Gaichas, S.; Kelble, C.R.; Fogarty, M.; et al. An ecosystem-based approach to marine risk assessment. *Ecosyst. Health Sustain.* **2017**, *3*, e01256. [CrossRef]

Journal of
Marine Science and Engineering

MDPI

Review

Understanding Carbon Footprint in Sustainable Land-Based Marine Aquaculture: Exploring Production Techniques

Marta Castilla-Gavilán [1,*], José Manuel Guerra-García [1], Ismael Hachero-Cruzado [2,3] and Marcelino Herrera [3,4]

1 Laboratorio de Biología Marina, Departamento de Zoología, Facultad de Biología, Universidad de Sevilla, Avda. Reina Mercedes 6, 41004 Sevilla, Spain; jmguerra@us.es
2 IFAPA Centro El Toruño, Camino Tiro Pichón s/n, El Puerto de Santa María, 11500 Cádiz, Spain; ismael.hachero@juntadeandalucia.es
3 "Crecimiento Azul", Centro IFAPA el Toruño, Unidad Asociada al CSIC, El Puerto de Santa María, 11500 Cádiz, Spain; marcelino.herrera@juntadeandalucia.es
4 "Crecimiento Azul", IFAPA Centro Agua del Pino, Unidad Asociada al CSIC, Ctra. El Rompido—Punta Umbria, 21459 Cartaya, Spain
* Correspondence: mcasgavilan@gmail.com or mcastilla1@us.es

Abstract: In aquaculture, it is crucial to understand and mitigate the carbon footprint for sustainable production. As demand for seafood increases, various production techniques compete for an eco-friendly status. This review examines the carbon footprint of various land-based marine aquaculture systems, highlighting their environmental impact. Through exploring innovations and best practices, it navigates the complexities of reducing emissions and promoting carbon sequestration. Some proposals for this purpose are based on diversification through low-trophic-level species, the preservation of high-carbon sequestration sites, polyculture, organic aquaculture and improvements in nutrition, feeding, waste and energy management. In this sense, some land-based aquaculture systems are progressively adapting and updating their zootechnical procedures. Recirculating Aquaculture Systems (RASs) offer interesting advantages such as water conservation, pollution reduction and biosecurity. Integrated Multi-Trophic Aquaculture systems (IMTAs) aim to address two major issues in aquaculture: efficient water usage and the environmental impact of effluents, which are rich in organic particles and dissolved nutrients from undigested food and feces; hence, these systems involve cultivating multiple species (polyculture). Biofloc Technology (BFT) is based on the formation of bioflocs in a culture medium. These systems can enhance feeding efficiency and waste management, thus optimizing nutrient utilization and minimizing environmental impact, achieved through reduced water and fertilizer usage. Traditional (extensive) aquaculture systems operate with minimal input of feed and chemicals, relying heavily on the natural productivity of the ecosystems; thus, the need for manufactured feed, the environmental impact associated with feed production and the transportation and overall costs are significantly reduced. Overall, while RASs, BFT and extensive systems in general offer significant sustainability benefits, IMTA's holistic approach to ecosystem management and nutrient recycling makes it, in our estimation, the most effective method in terms of ecological footprint in aquaculture. However, its quantitative evaluation is extremely complex, and there is currently a lack of references about its global carbon footprint. Therefore, further research and development are required, as well as collaboration and knowledge-sharing among stakeholders.

Keywords: best aquaculture practices; greenhouse gasses; carbon footprint; carbon sequestration; RAS; IMTA; BFT; extensive aquaculture; organic aquaculture

Citation: Castilla-Gavilán, M.; Guerra-García, J.M.; Hachero-Cruzado, I.; Herrera, M. Understanding Carbon Footprint in Sustainable Land-Based Marine Aquaculture: Exploring Production Techniques. *J. Mar. Sci. Eng.* **2024**, *12*, 1192. https://doi.org/10.3390/jmse12071192

Academic Editors: Ana Catarina Matias and Carlos Andrade

Received: 11 June 2024
Revised: 9 July 2024
Accepted: 11 July 2024
Published: 16 July 2024

1. Introduction

Over the past few decades, the aquaculture industry has experienced exponential growth, emerging as a vital player in global food production [1]. This surge is attributed to the escalating demand for seafood, driven by population growth, changing dietary habits, and a decline in wild fish stocks [2]. While aquaculture offers a solution to meet the burgeoning seafood demand, its rapid expansion comes with environmental consequences, for instance, those concerning greenhouse gas (GHG) emissions. This has led to several innovative production techniques and policies being improved in Europe during recent years: the EU missions "Restore our Ocean and Waters" or "Green Deal", the Atlantic Action Plan or the Blue Growth strategy [3–6]. More recently, the United Nations Sustainable Development Goals aim to "take urgent actions to combat climate change and its impacts" through Goal 13. These actions imply transformative measures in energy, industrial, transport, food, agricultural and forestry systems to move towards climate-resilient development and achieve net-zero emissions [7]. To understand how the aquaculture industry can contribute to this transformation, it is necessary to understand its contribution to GHG emissions and how they can be mitigated.

According to MacLeod et al. [8], the total emissions in 2017 for all aquaculture were 245 $MtCO_2eq$ (147 $MtCO_2eq$ for marine aquaculture), which would only represent 0.49% (0.27%) of total anthropogenic emissions (53.5 Gt [9]). The low emission intensity of aquaculture, compared to terrestrial agriculture and livestock [8], highlights the importance of maintaining the consumption of fish and seafood over meat. Despite this, aquaculture emissions are lower due to the still greater amount of terrestrial livestock production, but it has been estimated that aquaculture GHG emissions are increasing and that by 2030 they will reach 383 $MtCO_2eq$ [10].

Major global GHG (CO_2, CH_4 and N_2O) emissions come from the electricity and heat production sectors (25%) and the agriculture and livestock sectors (24%) [10]. In the case of the aquaculture industry, emissions are mainly related to the production of aquafeed, energy use and biogeochemical processes occurring in the culture units.

Most aquaculture species are carnivorous or omnivorous, requiring feed with a high protein content, often derived from fishmeal and fish oil [11]. However, sustainability concerns around using wild-caught fish for feeding aquaculture fish have led to an increasing use of cereals (mainly corn, wheat, rice) and oilseeds (mainly soybean) as aquafeed ingredients [12]. The production of these raw ingredients, their transformation and their distribution involve emissions arising from different sources [8,12]. In summary, 57% of aquaculture emissions are related to the utilization of fishmeal. Promoting the culture of noncarnivorous species and replacing fishmeal and fish oil with more sustainable alternatives, such as other plant-based proteins and oils [13] and insect-based proteins [14], aim to mitigate these emissions.

Energy consumption in aquaculture operations also contributes to GHG emissions, and the type of energy source used determines the environmental impact. Fossil fuels are used for the working vessel (when needed) and the distribution of the final product. Electric energy is used throughout the whole industry in production operations and processing [15]. Depending on the farm type and the species cultivated (off-shore, land-based, fish, shellfish, etc.), facilities rely on energy for a range of activities, including harvesting, transport, the collection of juveniles, maintaining water quality, running aeration systems, providing heating or cooling, pumping, lighting and powering vehicles [16,17].

Waste management in aquaculture systems is another aspect influencing GHG emissions due to biogeochemical processes occurring in the culture units. Respiration by organisms and organic matter in the effluents undergoes mineralization, producing CO_2 [18]. In earthen ponds (>40% of global aquaculture production [19]), the sediment is a major site for methanogenic bacteria producing CH_4 from dissolved organic carbon [20]. Moreover, nitrifying and denitrifying bacteria metabolize ammonia and nitrate from effluents and release N_2O through aerobic nitrification and anaerobic denitrification, respectively [21]. The type of aquaculture system also affects GHG emissions. For instance, extensive and semi-intensive systems, where ponds are fertilized to enhance natural productivity, can result in higher emissions due to the microbial breakdown of organic matter [22]. Thus, implementing efficient waste treatment technologies can minimize the release of GHGs into the atmosphere [23]. Water parameters also have to be taken into account; the emission of CO_2 depends on many factors, like air and water temperature and pH [10]. An increase in temperature also stimulates methanogenesis and CH_4 emissions, while high concentrations of dissolved oxygen inhibit these activities [24]. Aquaculture pond sediment pH plays a major role in the production of N_2O. Nitrous oxide production by nitrification increases with pH [25], while denitrification N_2O production increases with a decrease in pH [26]. Additionally, the choice of species influences the environmental impact of aquaculture. Some species are more feed-efficient and have lower waste production [27], contributing to reduced GHG emissions. Concerning algae/phytoplankton cultivation, their productivity contributes to CO_2 flux changes and N_2O synthesis during nitrate assimilation [28].

This introduction outlines some well-known strategies for addressing the challenge of GHG emissions in marine aquaculture. However, several gaps remain in understanding the impact of these strategies and comparing them in terms of carbon footprint. Firstly, there is limited quantitative data on their global carbon footprint, making it challenging to evaluate their overall effectiveness. Moreover, comprehensive life cycle assessments (LCAs) comparing the more sustainable production systems with traditional methods are also sparse, particularly in the marine context, due to the relatively recent application of LCAs in aquaculture [29,30]. Secondly, the long-term impacts of transitioning towards so-called eco-friendly methods have not yet been thoroughly explored. Finally, there is a need for standardized methodologies to measure the carbon footprint of the different aquaculture systems to facilitate more accurate and consistent assessments and comparisons [31]. In this context, and in order to provide a structured analysis, this review seeks to answer the following research questions: (1) What are the primary sources of GHG emissions in land-based marine aquaculture systems? (2) What are the best practices and innovations currently being implemented to mitigate GHG emissions in aquaculture? (3) Which aquaculture system shows the most promise for sustainable production with minimal environmental impact? (4) How can the aquaculture industry further reduce its carbon footprint while maintaining or increasing production?

The following sections present in detail the main mitigation measures and best practices for reducing GHG emissions. It is important to note that many of these strategies are closely interconnected and cannot exist independently. The suitability of current production technologies is then reviewed based on their application of these measures and their carbon footprint according to the literature.

2. Mitigation Strategies

2.1. Diversification through Low-Trophic-Level Species

It has been estimated that in the next decade, fishmeal and fish oil production will not be allowed to meet the demand of the growing aquaculture industry [32]. Indeed, reducing the use of wild-caught fish for this purpose is a well-known strategy for the sustainable development of the sector [33,34]. Other than research on new protein-rich alternative ingredients that can sometimes be nutrient-deficient for carnivorous species [35], diversification through low-trophic-level species has been highlighted to mitigate GHG emissions [34,36,37].

Farming extractive (nonfed) low-trophic-level marine species, such as bivalves, seaweeds or amphipods, represents a sustainable and environmentally friendly approach to aquaculture [38–40]. Gephart et al. [41] and Jones et al. [42] found that among all so-called "blue food", bivalves and seaweeds produce the lowest levels of GHG emissions, land use, N and P emissions, and freshwater use. In general, extractive species can be cultivated with low energy requirements and without inputs since they are autotrophic and/or they can extract dissolved nutrients and particulate matter from the water column, improving water quality and reducing excess nutrients [37]. Bivalves are highly efficient in converting organic matter into protein without the need for external feeds [43]. This process not only mitigates nutrient pollution but also contributes to the overall health of an aquatic ecosystem. Both bivalves and seaweeds play a crucial role in sequestering CO_2, contributing to climate change mitigation [44–46]. Additionally, seaweeds can provide a wide variety of ecosystem services [47] such as improvements in water quality by assimilating excess nutrients and providing habitats for diverse marine organisms. Amphipods constitute a low-trophic-level group and are currently being explored for their potential in aquaculture. As they are opportunistic feeders able to feed on detritus, they enhance water quality while also providing a source of protein and high-quality live food [48,49] without any additional inputs. Furthermore, it has recently been demonstrated that marine amphipods, despite being fed low-omega-3 diets (i.e., detritus), are able to preserve high levels of these dietary lipids, among others [50]. Thus, according to Krause et al. [37], the aquaculture industry provides a more sustainable alternative to red meat consumption by improving the production and consumption of these low-trophic-level marine species.

Penaeids (mostly white shrimp, *Penaeus vannamei*, and giant tiger shrimp, *Penaeus monodon*) are among the largest GHG producers in marine farming, alongside salmon production. However, farmed shrimp produce lower emissions than captured ones, and most of the emissions are associated with their land use and feed production [41]. Therefore, to make penaeid aquaculture a more sustainable blue food source, it is recommended to avoid converting mangrove areas to shrimp farms [51,52] and transition towards herbivore feeds [53].

Concerning fish, it is important to note that only about 5% of all species are herbivorous, and only 30% of these are marine, most of them living in coral reefs [54]. Despite this limitation, it has been demonstrated that they present advantages in energy transformation and resource utilization since they usually have low nutrient requirements, which they can obtain from cultivable seaweeds or microalgae [55]. Moreover, the FAO has recommended the development of aquaculture for these species since they also present high-quality meat and low production costs within a good market demand context, mostly in Asian countries [54,55]. A number of studies have been conducted globally with the objective of developing their culture for commercial or repopulation purposes. Lozano-Muñoz et al. [56] have recently demonstrated that the herbivorous fish *Medialuna ancietae*, a native species from Chile, has great potential for use in a sustainable aquaculture context. They showed that this species has low and fish-free requirements for protein. However, its meat presents a high protein content, demonstrating that *M. ancietae* is an efficient converter of feed into protein. In China, rabbitfish species (*Siganus* spp.) are now widely cultivated with high economic profitability [55] and have been demonstrated to be highly efficient in biosynthesizing long-chain polyunsaturated fatty acids (PUFAs). The milkfish (*Chanos chanos*) is an economically important herbivorous species in southeast Asia [57]. A great variety of protein sources have recently been studied for milkfish meal, such as, for instance, soybeans, insects, polychaetes or seaweeds [58–61]. This trend towards herbivorous fish aquaculture is less widespread in Europe. However, the potential of some species of the family Mugilidae (*Mugil cephalus*, *Liza* spp., *Chelon labrosus*) has been highlighted due to their omnivorous profile, rapid growth and resistance to environmental variations [62]. Martínez et al. [62] also underline the importance of the grey mullet (*M. cephalus*) in current strategies in European aquaculture. These species' culture takes place in extensive or semi-intensive ponds [63], and research efforts to improve it have increased in recent years [64,65].

2.2. *Preservation of High-Carbon Sequestration Sites*

Natural wetlands contribute 20–30% of global terrestrial carbon sequestration [66,67]. However, mangroves, seagrasses, estuaries and salt marshes have been disturbed in many countries by practicing commercial shrimp and fish aquaculture [51,68]. The habitat transformation of these wetlands can increase the emissions of GHGs from organic matter already stored in the system. Many studies have demonstrated greater emissions of greenhouse gasses from sites converted to aquaculture ponds [19,69–71]. For instance, mangrove-converted aquaculture in the Mahakam delta led to a loss of 1925 Mg CO_2eq ha^{-1} [68]. Similarly, conversion from paddy fields to extensive crab ponds increased the 100 yr global warming potential (GWP) from 8.15 to 28 Mg CO_2eq ha^{-1}, mainly due to increased CH_4 emissions with a contribution of 96.3% [19].

In line with the previous section, primary producers play a crucial role in ecosystem restoration and resilience [45,72], which is integral to mitigating the impacts of climate change on aquaculture. Mangroves, seagrasses, estuaries and salt marshes not only sequester carbon [73] but also provide vital habitats for aquatic species, protect coastlines from erosion and support nutrient cycling [74,75]. By restoring and conserving these coastal ecosystems, aquaculture operations can enhance their resilience to environmental stressors and contribute to broader climate change mitigation efforts [76–78].

2.3. Polyculture

The simultaneous cultivation of multiple species in a single aquaculture system emerges as a strategic measure to mitigate GHG emissions in aquaculture [10,79]. Thomas et al. [80] outlined that polyculture requires species compatibility (species sharing the same space without detrimental interactions) and complementarity (compatible species using all available resources). According to these authors, complementarity can be enhanced based on trophic interactions or based on commensalism or mutualism [80]. Multi-trophic interactions provide more advantages for sustainability in a polyculture context: the recycling of water, the preservation of water quality, wastewater valorization through improvements in the nutrient cycle (resources become products, products become waste and waste becomes resources [81]) or the diversification and improvement of production (with economic interest) [80].

Concerning carbon footprint, polyculture systems capitalize on the complementary interactions among different species, improving the nutrient cycle and acting as natural bioremediators [42]. This not only improves water quality within the aquaculture system but also reduces the environmental impact on surrounding ecosystems. This nutrient cycling reduces the need for external inputs, minimizes waste and the need for waste management, and enhances overall system efficiency [82,83]. Additionally, by relying on the ecological interactions among species, the need for supplemental feeds is minimized. This results in cost savings and reduces the overall environmental footprint associated with aquafeed production.

Therefore, selecting suitable species is a key issue in polyculture systems for ensuring sustainability. For instance, the oyster *Crassostrea angulata* seems to be a CO_2 generator through calcification and respiration (153 and 349 g C m^{-2}, respectively), though oyster harvesting sequesters ca. 258 g C $m^{-2}y^{-1}$ due to shell formation [79]. This negative balance can be attenuated by culturing the seaweed *Gracilaria lemaneiformis*, which would act as an efficient sink for CO_2. Zhang et al. [84] reported that a polyculture with the crab *Portunus trituberculatus* and the shrimp *Marsupenaeus japonicus* was much more sustainable than those species and the clam *Ruditapes philippinarum* (−79 versus 194 g CO_2 m^{-2}).

2.4. Nutrition and Feeding Efficiency

Feeding practices in aquaculture are crucial in determining its environmental impact. Inefficient feeding not only increases operational costs for farmers but also contributes significantly to the sector's GHG emissions [85]. Research efforts have been made to reevaluate feeding strategies, with a focus on minimizing the environmental impact of aquafeed production through using new ingredients, optimizing nutrient utilization, and valorizing waste.

As previously mentioned, the production of aquafeed is one of the primary contributors to the carbon footprint in marine aquaculture [8]. Advancements in aquaculture nutrition science have highlighted alternative and sustainable feed ingredients. These ingredients, such as plants and insects, mitigate the impact on wild fish populations and ecosystems and often require less energy in processing [38]. Plant-based proteins have been widely used in aquaculture during the last two decades [86], mostly from soybean, wheat, corn or rice. These plant-based aquafeeds have been demonstrated to present a favorable essential amino acid (EAA) profile, a high content of protein and great digestibility for fish [35,87–89]. Nevertheless, plants also present some challenges and are far from being a net-zero emission ingredient. Some important plant anti-nutrients have been identified [90,91]. Moreover, crops' GHG emission sources are diverse: the use of fertilizers (N_2O), the increasing use of flood-cultivated rice, whose anaerobic conditions are prime sites for CH_4 production [13], or the expansion of soybean cultivation (CO_2 derived from land use change [14,15]). The competition for plant protein resources for human consumption and terrestrial animal feed has led to a rise in prices and the necessity to develop new ingredients [86]. Insect flour has been demonstrated to be highly nutritious and rich in protein, lipids and EAAs [92,93]. From a sustainability point of view [94,95], the main advantages of growing insects for aquafeed purposes are that they can use organic byproducts (wastes), promoting a circular economy, and the environmental benefit associated with land use. However, the mass production of insects is still in a developing stage [35] and needs to be improved, especially in terms of energy use [95]. In addition to the acquisition of any raw ingredient, energy consumption during fishmeal production and distribution is a significant contributor to the carbon footprint of aquafeeds. [8].

The use of additives can improve feeding efficiency [96,97]. Living additives and additives extracted from organisms are preferred over hormones or antibiotics due to their eco-friendly nature [96]. Single-cell protein and oil based on bacterial, microalgal or yeast biomass can also be grown from byproducts from other industries and are widely used as a protein/omega-3 source for aquafeeds [98,99]. These additives have also been demonstrated to have probiotic and inmunostimulant effects on the cultured species [99]. This is also the case with some molecules extracted from seaweeds, such as the polysaccharide ulvan [100]. Enzyme supplementation in feed fish also improves digestibility and feeding efficiency [101]. There have been relatively few studies on the application of enzymes in the diets of marine species [102]. However, the potential of animal byproducts like fish viscera has been highlighted as an enzyme source for marine fish diets [103].

Feeding efficiency also extends beyond the choice of ingredients to the feeding practices employed on aquaculture farms [104]. Precision feeding technologies, such as automated feeding systems and real-time monitoring, allow farmers to tailor feed delivery to the specific nutritional needs of their fish. This not only reduces excess feed and nutrient wastage but also ensures that the fish receive optimal nutrition at the right time, promoting healthier and faster growth. Furthermore, as highlighted above (cf. 2.3), the implementation of polyculture is an example of an approach to improving feeding efficiency.

2.5. Implementing Waste Management Technologies

Aquaculture operations generate significant amounts of inorganic waste, mainly ammonium, nitrate and phosphate, from degraded uneaten feed and fecal matter. When left unmanaged, these wastes contribute to environmental degradation, such as eutrophication and algal blooms [105]. On the other hand, the degradation of organic solids in wastewater can also lead to a high biological oxygen demand and the production of ammonium due to mineralization [106].

Inorganic waste management technologies offer effective solutions to mitigate these potential impacts. One of these technologies is the use of biofilters that allow for the conversion of ammonia wastes [107,108] into less harmful compounds through nitrification [109]. This reduces the need for water exchange and minimizes nutrient discharge into the environment. Additionally, advancements in waste treatment technologies such as UV and ozone treatment [110,111] allow for the more efficient removal of pollutants from aquaculture wastewater. As highlighted above in the *low-trophic-level* and *polyculture* sections, the incorporation of nutrient recovery technologies allows aquaculture facilities to extract valuable nutrients [112], such as phosphorus and nitrate, and recycle them for use as fertilizers in agriculture [113] or in aquaponics systems [81] to produce plant or algal biomass.

Organic waste includes feces and uneaten food, dead organisms, algae waste, etc. [114]. Effluent treatment primarily involves mechanical filtration through sand filters. Sludges are then stored in decantation tanks and recovered for reuse as fertilizers or for transformation into biofuels [115,116]. The valorization of algae, crustacean and fish waste through the recovery of biomolecules such as proteins, polysaccharides and biosurfactants has also been highlighted [114]. Additionally, recent research has focused on the development of eco-friendly enzyme-assisted methods for the extraction of these molecules [117–119]. In conclusion, the management of waste through its valorization reduces the carbon footprint of aquaculture activities while also contributing to the circular economy.

2.6. Energy Efficiency

Some of the previous strategies are linked with energy efficiency. The use of filtration, biofiltration and water treatment technologies minimizes water consumption and reduces the energy needed for water pumping and heating. In addition, implementing precision feeding techniques allows for precise control over feed delivery, minimizing waste and waste management operations.

Transitioning to alternative fuels for vessels (electricity, natural gas, biodiesel, methanol, etc. [120]), fully electric and hybrid trucks and renewable energy sources using solar panels, photovoltaics or wind turbines in aquaculture facilities can significantly reduce the carbon footprint of these operations [15,121]. Aquaculture activities often require heating to maintain optimal water temperatures, especially in cold climates or during the winter months. Heat recovery systems can capture waste heat from the aquaculture process itself (water recirculation or effluent treatment [122]) or from industry [123]. This waste heat can then be reused to heat the culture units. Indeed, the utilization of waste heat from industrial or agricultural sources has been studied, with the resulting energy savings being due to minimal or even no use of carbon-based heat [124]. Furthermore, the controlled production costs resulting from the use of waste heat have been evidenced. Finally, a reduced environmental impact has been observed, with a decrease in CO_2 emissions by more than 26% and the valorization of liquid and solid effluents within a "zero waste" objective [124]. Another promising approach is the utilization of waste-to-energy technologies, such as anaerobic digesters and gasification systems. The production of biogas in methanogenic reactors from sludges through anaerobic digestion to obtain energy and heat [115], which could even be returned to the system to maintain the temperature, promises to make land-based aquaculture even more circular and sustainable [125]. Finally, upgrading to energy-efficient equipment and infrastructure and monitoring energy consumption can result in significant energy savings [126].

2.7. Organic Aquaculture Standards

Organic aquaculture production must be taken into consideration in this section since it has recently attracted significant interest from consumers and investors [127,128]. According to the International Federation of Organic Agriculture Movements (IFOAM), "organic agriculture is a production system that sustains the health of soils, ecosystems, and people. It relies on ecological processes, biodiversity and cycles adapted to local conditions, rather than the use of inputs with adverse effects. Organic Agriculture combines tradition, innovation, and science to benefit the shared environment and promote fair relationships and good quality of life for all involved" [129]. This definition may be applied to aquaculture through the principles of health, ecology, fairness and care [130]. The European Union (EU) Organic Aquaculture Standards, established under Regulation (EC) No. 710/2009 [131], outline specific criteria and obligations for aquaculture operations to be considered organic within the EU. Certification processes and regulatory frameworks may differ between regions. However, common standards and requirements of organic aquaculture include the following [130]:

- System design and location are related to contamination from outside sources, the introduction of exotic species, escapes, contamination with effluent discharges, the use and reuse of water and maintaining water quality.
- Sources of stock, breeds and breeding concern the preference for local species and the prohibition of polyploidy, the use of hormones and the handling of the daylight period.

- The feeding and nutrition of aquaculture animals concern the efficient use of food to minimize loss and the use of organic feed ingredients sourced from certified organic sources whenever possible.
- Health and welfare are related to measures designed to provide adequate space, shelter and environmental enrichment, with the aim of promoting natural behaviors and minimizing stress. Practices such as overcrowding, confinement and the use of stressful handling techniques are prohibited (including during harvest and transportation).
- Processing and labeling operations must maintain detailed records of all inputs, practices and activities related to production. This includes the documentation of feed ingredients, water quality monitoring results, stocking densities and health management practices.

3. Land-Based Aquaculture Farming Systems: In the Race for an Eco-Friendly Status

Traditional intensive flow-through aquaculture systems utilize natural water sources to provide a continuous flow of water to aquaculture ponds or tanks [130]. In the marine context, examples of flow-through systems include raceways and tanks where species such as turbot (*Scophthalmus maximus*), sea bass (*Dicentrarchus labrax*) or sole (*Solea* sp.) are grown [132–134]. These systems operate by diverting water from a natural source, passing it through the culture units and then discharging it back into the environment. The constant flow of water serves to sustain optimal oxygen levels, remove waste and provide a stable environment for the fish [132]. However, this method has a significant impact on the wild environment due to its high densities, which result in high inputs and enriched discharges. Consequently, it has been refined over centuries to enhance production while maintaining water quality, animal welfare and the minimization of ecological disturbances [130]. Here, we discuss some of these improved systems and technologies.

3.1. Recirculating Aquaculture Systems (RASs)

The development of RASs has been characterized by innovation since the 1950s [135], driven by the need to address environmental concerns, optimize resource use and meet the growing demand for seafood production. The main feature and advantage of RASs is water recycling and, consequently, its low water consumption [130]. Figure 1 illustrates an RAS and its main water treatment equipment.

The seawater reservoir tank (1) is equipped with a pH sensing line and a pump to supply a buffer solution that is added to keep the pH stable. Water is first pumped (2) from this tank to a mechanical filter (3), i.e., a sand filter. This filter is regularly backwashed, and the water with sludge is discharged. In other RASs, sludge can be stored in decantation tanks, recovered and reused as fertilizer or for biogas production [115,116]. The water is then pumped into the biological filter (4), where plastic bio-balls are colonized by nitrifying aerobic bacteria [136] that convert ammonia (NH_4^+) to nitrate (NO_3^-). From here, the effluent is also discharged, and the filtered water is (i) returned to the reservoir tank (1) or (ii) pumped into a protein skimmer (5), which removes dissolved fine organic solids [137]. In this filter, small air bubbles are injected at the bottom to attach to the organic compounds and lift them to the top. The skimmer foam is then removed with the effluent, and the skimmed water is returned to the cycle. A final filtration of microparticles and microbial biomass [138] is then carried out by UV irradiation (6). Finally, the water is heated (7) if necessary before it reaches the culture tanks (8).

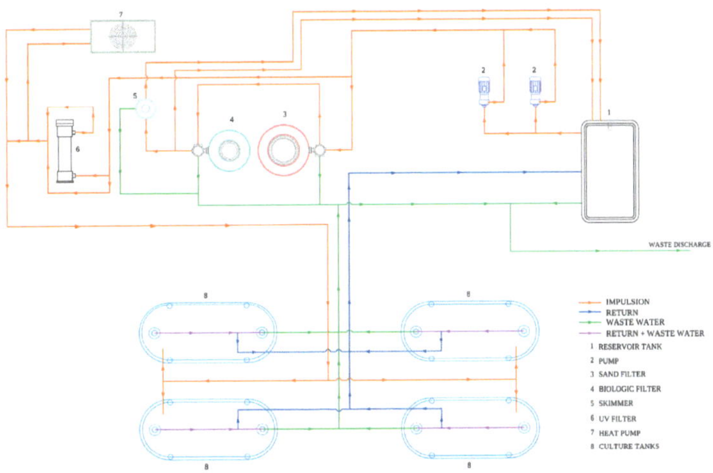

Figure 1. Diagram of a Recirculating Aquaculture System (RAS). Numbers are detailed (in brackets) in the text. Source: SCA Blennius, Puerto Real, Spain.

Recirculating Aquaculture Systems are considered a sustainable production method [108,139]. The advanced water treatment technologies employed in RASs enable the cultivation of high-value species at high stocking densities and high economic profitability while simultaneously ensuring animal welfare and maintaining water quality [130]. Therefore, RASs represent one of the most productive aquaculture methods. As an example, they are the most efficient in terms of land and feed use across major species and production systems in US aquaculture [140]. By treating and recirculating water within the system, an RAS minimizes the discharge of pollutants, such as excess nutrients and fecal matter, into the environment, reducing the risk of water pollution, eutrophication and ecosystem degradation [141]. Its land-based and indoor nature prevents effects on biodiversity due to escapes, diseases and parasite transmission [141]. Concerning the GHG mitigation strategies cited in Section 2, Table 1 summarizes the application of these strategies in RASs. In a marine context, RASs are mainly monospecific and used for carnivorous fish species [142–145]. In terms of their use for rearing low-trophic-level species, these systems have been used with some invertebrates such as shrimps [146–148] or sea urchins [149–151], but this remains a minor use. Regarding feeding strategies, Godoy-Olmos et al. [143] found that feeding by auto-demand feeders led to higher NH_4^+ removal rates, preserving water quality and reducing water consumption. In an RAS context, precision feeding technologies are widely employed [152], as they allow the amount of food and the feeding frequency to be carefully controlled. This is an important advantage of RASs, as it has been shown that overfeeding can lead to an increase in uneaten food, particulate organic matter, ammonia, nitrite and nitrate in the culture units [153,154], which can pollute the water and cause diseases in aquatic organisms [155,156]. Finally, as previously mentioned, the choice of protein sources in the diet for each cultivated species is crucial. A low digestibility of nutrients can again result in an increase in uneaten food and a decrease in water quality [157,158]. The technical complexity of RASs requires high energy consumption, resulting in high costs that may compromise the economic viability of operations and contribute to higher GHG emissions. These are the two most significant constraints of RASs [159]. Therefore, enhancing energy efficiency is critical to ensuring their sustainability. Energy-saving measures, such as energy audits and the use of software with energy performance data, have been identified by some authors [160,161]. These measures can provide valuable information for decision-making. Kucuk et al. [162] recommended upgrading to energy-efficient equipment, particularly pumps, in line with Badiola et al. [163], who identified heat pumps as a major energy consumer in RASs. On the contrary, Bergman et al. [125] show a case study in Sweden, where they cultivate warm-water tilapia and where renewable energy sources are employed. They demonstrated that cultivation in RASs was possible without requiring significant energy compensation. This was accomplished through the valorization of byproducts to produce biogas. In summary, these studies suggest that while further research and improvements are necessary to improve the energy efficiency of RASs, these systems are suitable for implementing advanced technologies that could enhance their sustainability in the future.

Table 1. The synthesis of measures applied in each farming system and estimations of their carbon emissions: ★: few studies address this strategy; ★★: at least half of the studies reviewed address this strategy; ★★★: most of the studies reviewed address this strategy; ★★★★: all studies reviewed address this strategy.

Farming System	Low-Trophic-Level Species	Polyculture	Feeding Efficiency Strategies	Waste Management	Energy Efficiency	Carbon Sequestration	Carbon Emissions
RAS	★☆☆☆	★☆☆☆	★★★☆	★★★★	★★☆☆	☆☆☆☆	6109 kg CO_2e t^{-1} WW [42]
IMTA	★★★★	★★★★	★★★★	★★★★	★★★★	★★★☆	no data
BFT	★★★★	★★★★	★★★★	★★★★	★★☆☆	★★★☆	5945 kg CO_2e t^{-1} of shrimp [164]
Extensive systems	★★★★	★★☆☆	★★★★	★★★☆	★★★★	★★★★	392 kg CO_2e t^{-1} of bivalves WW [42] ~65 t CO_2e ha^{-1} y^{-1} [165]

Carbon sequestration is only achievable through finfish production in off-shore cages, not in RASs, although this capacity has been poorly studied and quantified [164]. Additionally, few studies have calculated the carbon footprint of RAS. According to a review by Jones et al. [42], fed marine finfish cultivation in RASs emits 6109 kg CO_2e per ton of finfish wet weight on average. However, these values can vary between 1382 and 44,400 kg CO_2e per ton of finfish wet weight. These variations are due to different species, foods, locations or energy sources.

In summary, RASs inherently offer advantages such as water conservation, pollution reduction and biosecurity. However, RASs rely on high stocking densities, are disconnected from the natural marine environment and require a high input of external energy. These facts mean that RASs are not in line with the UE organic aquaculture standards [130], and additional measures are needed to ensure compliance with organic principles and regulations. This may include sourcing organic feed ingredients, implementing disease management strategies, maintaining detailed records of inputs and practices, etc.

3.2. Integrated Multi-Trophic Aquaculture Systems (IMTAs)

The origins of Integrated Multi-Trophic Aquaculture (IMTA) systems can be traced back to farmers in ancient civilizations such as China and Egypt, who wanted to raise fish and at the same time use waste to produce complementary products and income [165]. Modern IMTA development gained importance in the 1970s [166], and by the early 2000s, IMTA practices were increasingly adopted globally, promoting ecosystem-based management and resource efficiency [167–169]. Today, IMTAs are a cultivation technology that is currently being driven by the EU due to its efficiency. These systems aim to address two major issues in aquaculture: efficient water usage and the environmental impact of effluents, which are rich in organic particles and dissolved nutrients from undigested food and feces. IMTAs involve cultivating multiple species of different trophic levels (with or without terrestrial organisms) in the same or separate compartments connected by nutrient flows [170]. Thus, the waste (organic and inorganic matter) from the main fed species at a higher trophic level can be utilized by the extractive species at a lower trophic level/levels

(Figure 2). In these "designed ecosystems", nutrient loss and water usage are minimized, and waste is valorized in a circular economy model: resources become products, products become waste and waste becomes resources, in line with UN Sustainable Development Goal 12 [171]. In the example (Figure 2), discarded water from an indoor fish RAS is pumped into an IMTA array placed in a greenhouse. The water with organic solids is first transferred to a solids retention compartment with deposit feeders (e.g., polychaetes) and then to a distribution tank with aeration. Finally, the water reaches the primary producer (e.g., halophytes) compartments for dissolved inorganic nitrogen and phosphorous uptake.

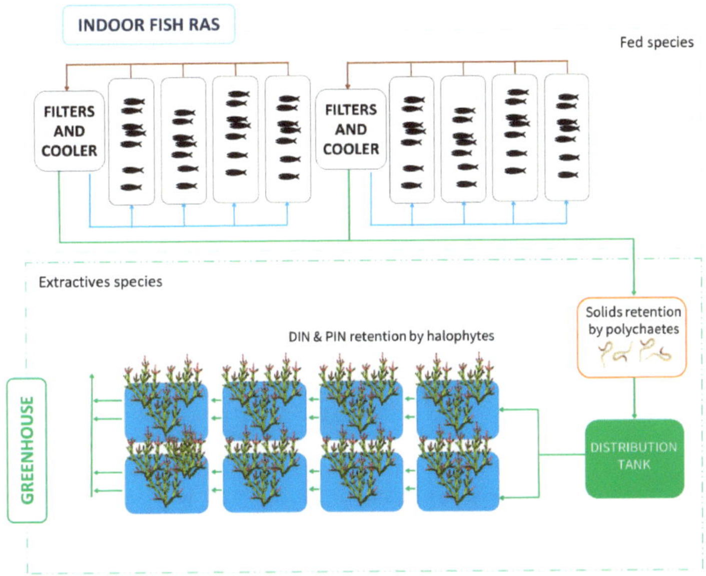

Figure 2. Diagram of a land-based IMTA. Modified from Castilla-Gavilán et al. [81].

There is a wide variety of IMTAs, both offshore (based on net cages, longlines, rafts...) [167,172] and land-based (based on RASs, flow-through systems, extensive earthen ponds...) [81,173–176]: combinations between finfishes, bivalves, gastropods, decapods, amphipods, seaweeds, phytoplankton, echinoderms, sponges, plants, etc., are possible (see Guerra-García et al. [177]). IMTAs can combine high-value species, such as salmon, oyster and sea urchins, with others of lower value, including seaweeds and mussels. From an economic perspective, it is noteworthy that IMTA products are generally well received by consumers, with many willing to pay a premium for products bearing an IMTA label [178]. Numerous studies have demonstrated the high efficiency, productivity and sustainability of IMTAS. It has been suggested that among the different types of IMTAS, recirculation land-based IMTAs have the highest potential for nutrient retention [179]. These authors suggest that a four-species marine IMTA consisting of fish as fed species and algae, bivalves and detritivores as extractive species would achieve the highest theoretical nutrient retention efficiency. This system would absorb between 79% and 94% of the nitrogen, phosphorus and carbon supplied with the aquafeed. This approach has been demonstrated on a large scale in Pacific fish farms producing milkfish (*Chanos chanos*), shrimp (*Penaeus monodon*), clams (*Meretrix lusoria*) and seaweed (*Gracillaria* sp.) [180]. Thus, IMTAs, by definition, use low-trophic-level species and polyculture, which makes these systems highly advantageous in terms of sustainability and GHG mitigation strategies. Feeding efficiency and waste management strategies are enhanced by the use of a deposit feeder link [49,181], which reduces carbon from organic wastes (uneaten aquafeed and feces) and primary producers that maintain and improve water quality, even more so when combined with

RASs [182]. It is important to highlight that, in IMTAs, the feed input is usually similar to that of monoculture systems but with an additional output due to polyculture [183]. Furthermore, in a land-based context, the energy efficiency of IMTAs is mainly related to the energy efficiency of the system used to produce the fed species (i.e., RAS, flow-through system, extensive ponds...). Checa et al. [183] determined that water recirculation for the production of extractive species allows water exchange to be reduced and, consequently, the energy required for pumping and the energy costs. Cunha et al. [184] also showed energy savings for pond aeration in IMTAs with filter feeders. In any case, any lower energy efficiency (or higher energy costs) in IMTAs is offset by the fact that the water treatment produces multiple species, making farms more environmentally and economically resilient [80,185,186].

Carbon sequestration is possible in IMTAs, especially through seaweed and shellfish cultivation [42,184,187,188]. Macroalgae cultivation has been highlighted for its potential to contribute to global blue carbon sequestration due to its capacity for photosynthetic CO_2 assimilation [189,190] and has been recommended as a climate change mitigation strategy [191]. Regarding shellfish, Cunha et al. [184] observed that carbon sequestration was higher in land-based IMTAs (fish + phytoplankton + seaweeds) when oysters were also present (0.50–0.53 mg L^{-1} 8 h^{-1}). Liu et al. [192] also observed that large-scale shellfish–macroalgae IMTAs act as CO_2 sinks and play an important role in the local carbon cycle. However, it is important to note that although these farms can store large amounts of carbon, when the algae and shellfish are harvested, carbon is released through the food chain [44]. In this sense, the benefits of these systems are still being investigated, but their low ecological footprint compared to monospecific aquaculture techniques is unanimous [186,193]. Although IMTAs are currently employed in numerous countries [51] and have been the subject of numerous studies, due to the considerable diversity of these systems (open-water or land-based) and the inherent difficulty in measuring the ecological footprint in circular systems [183], there is currently no global evaluation of their GHG emissions. For instance, Nobre et al. [194] observed a reduction in GHG emissions of between 290 and 350 t CO_2e year^{-1} when abalone were cultivated in IMTAs. In contrast, the work of Chary et al. [195] illustrates the challenges of assessing the impact of IMTAs on climate change. In their study of an open-water IMTA involving red drum and sea cucumbers, they estimated emissions of 2341 kg CO_2eq per t of fresh aquatic product. However, they found few differences between the IMTA and a monospecific red drum farm due to the unbalanced design of the finfish and sea cucumber compartments.

The potential for IMTAs to facilitate the ecointensification of aquaculture has been demonstrated. These systems offer the dual benefits of producing safe products for human consumption and meeting the standards for organic aquaculture [178,196,197].

3.3. Biofloc Technology (BFT) Systems

Biofloc Technology (BFT) originated in the 1970s at Ifremer-COP (French Polynesia) on penaeid species, with the aim of identifying sustainable methods to improve water quality in aquaculture [198]. The initial phase of its development was focused on reducing nitrogen levels through microbial processes [199]. The concept progressed in the 1990s and 2000s, with significant scientific contributions [200,201], promoting its application in shrimp and fish farming in areas where water limitation, land costs and environmental issues were major concerns. Today, BFT is widely adopted for its environmental and economic benefits, particularly in intensive shrimp aquaculture systems [202]. Although shrimp is a species with a high market value, it should be noted that the BFT system can result in the production of off-flavors, which may be attributed to high turbidity, filamentous cyanobacteria and actinomycetes. These factors can negatively impact the quality and market value of BFT-produced shrimps [203]. However, several strategies have been proposed to alleviate this problem, including the introduction of certain microorganisms, such as those from the Bacillaceae family, into BFT system designs as bioreactors [204,205].

BFT systems are based on the formation of bioflocs in a culture medium (Figure 3). These are microscopic ecosystems where uneaten food, excess inorganic nutrients, and feces aggregate along with microorganisms (bacteria, microalgae, diatoms, protozoa…). This aggregation is enabled by a matrix of extracellular polymeric substances secreted by the microorganisms [206]. By maintaining a high carbon/nitrogen ratio [206] through the addition of a carbon source (molasses, glycerol, flours [207]) to the culture medium, the colonies of chemoautotrophic bacteria present in the bioflocs assimilate the ammonium excreted by the cultured species during the nitrification process, transforming it into nitrate. Furthermore, heterotrophic bacteria can also directly assimilate the ammonium into bacterial biomass. Additionally, microalgae contribute to nitrogen absorption during photosynthesis [208]. In this manner, BFT serves to maintain water quality by reducing the ammonium concentration, which is toxic to the cultured species, and transforming it into nitrate, which can accumulate but does not pose a health risk to the cultures. This reduces water consumption, its purification and the associated costs. Moreover, nitrogenous organic wastes are employed in the synthesis of microbial protein, making biofloc a rich source of quality protein that can be used as supplementary feed for the cultured species [209]. Thus, BFT also promotes low-protein diets, reducing the cost and environmental impact associated with the formulation of protein-rich feeds [200]. The addition of organic carbon sources can result in fluctuations in dissolved oxygen levels due to the metabolic processes of aerobic microorganisms [210]. In order to prevent stress on the cultured organisms, it is possible to separate the biofloc reactors from the culture tank [211]. The effluent is then transferred to these reactors, where a carbon source is added in order to stimulate the growth of bioflocs. Subsequently, the water and bioflocs from the reactors are transferred to the culture tanks, where they are consumed [210,212].

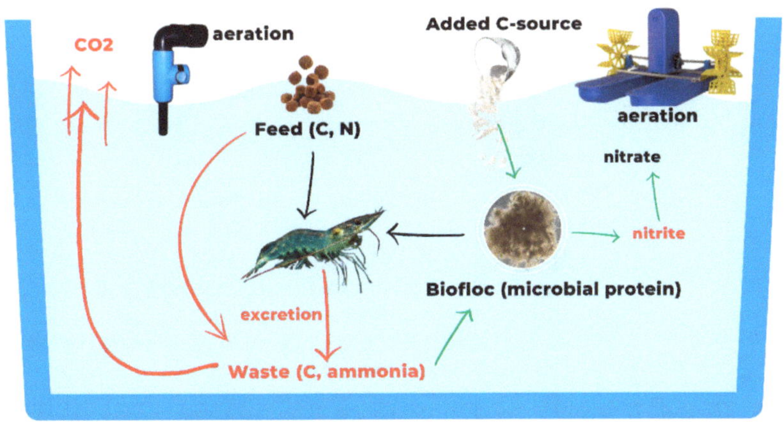

Figure 3. Diagram of a BFT system with a biofloc within the culture unit. Source: authors.

BFT systems have been identified as a promising technology for sustainable food production, contributing to the development of a circular economy. This technology is currently primarily employed in the cultivation of penaeids, as they are omnivorous species that can feed on biofloc [213]. These species are also tolerant to changes in the concentrations of dissolved oxygen in the water and nitrogenous compounds and can withstand high culture densities and high concentrations of suspended solids [214,215]. Like IMTAs, BFT promotes low-trophic-level species cultivation and polyculture. Indeed, some authors have proposed this technology as a form of IMTA [183,197,216] because of the presence of microorganisms in the fed species. A wide variety of integrated BFT systems have been studied, showing higher efficiency than monospecific BFT: shrimp in BFT with Nile tilapia [217], with mullet (*Mugil liza*) and seaweeds (*Ulva fasciata*) [218], with *Ulva lactuca* [219], or with tilapia and the halophyte *Sarcocornia ambigua* [220]. Feeding efficiency and waste management are also enhanced by definition in these systems, as they have the capacity to enhance nutrient utilization and minimize environmental impact, achieved through reduced water and fertilizer usage [178,202]. In BFT systems, food requirements are reduced, and an increase in the survival and growth rates of cultivated species has been observed. Bioflocs present a favorable nutritional profile for numerous aquaculture species, including filter feeders, detritivores and even some herbivorous or omnivorous finfish, which are able to feed directly on particulate organic matter [212]. Other studies have demonstrated that the consumption of bioflocs can enhance growth and fortify the immune system of cultivated species by increasing the activity of their digestive enzymes, resulting in an increase in the feed conversion rate [221]. Concerning energy efficiency, the aeration of BFT systems requires much more energy than conventional tanks and most recirculation systems, which represents a significant cost [206]. However, the need for water exchange is minimized, thereby reducing pumping costs. In general, maintaining the water temperature in BFT systems may not require significant energy input due to the minimal water exchange, which prevents variations in temperature. However, when this technology is employed for the cultivation of tropical species in cold countries, maintaining the temperature can be costly in terms of energy [198,222]. This issue can be resolved by recovering waste heat from the industrial sector or by manufacturing biogas from the farm's own waste [223].

It has been demonstrated that carbon sequestration can occur in BFT ponds, as the phytoplankton present in bioflocs are able to use carbon through photosynthesis, acting as carbon sinks [224,225]. This phenomenon, in conjunction with the reduction in the feed conversion ratio (FCR) observed in BFT, has led to the suggestion that these systems could be employed as a strategy for reducing the GHG emissions associated with shrimp farming, even when a carbon source is added to the ponds [226–228]. Indeed, Huang et al. [226] estimated that GHG emissions in a shrimp BFT system were 5945 kg $CO_2e\ t^{-1}$ of shrimp, which represents one-quarter of the emissions of a super-intensive system, while BFT production represented 60% of the super-intensive yield. Finally, shrimp production is frequently incompatible with organic labels. The application of BFT, the use of organic feed and surveying animal density and welfare could collectively help to achieve organic standards and certification [130].

3.4. Extensive Aquaculture: Earthen Ponds and Intertidal Aquaculture

Extensive aquaculture is the most traditional mode of production and involves raising aquatic organisms in natural or semi-natural settings with minimal human intervention [130]. It includes earthen ponds and intertidal aquaculture of bivalves (Figure 4). Earthen ponds are shallow, man-made ponds where organisms are raised using natural resources such as algae and plankton for food. An example is the use of salt marshes and their seawater reservoirs built by enclosing a piece of salt marsh to guarantee a constant supply of water from the saltworks [229,230]. The macroinvertebrate community naturally inhabiting these ponds includes small molluscs, crustaceans, polychaetes and chironomids, among others, which constitute the main food for nonintensively reared fish [231]. These systems promote the welfare of cultured fish which feed on natural prey. Intertidal bivalve aquaculture is one of the most important aquaculture industries [1] and involves cultivating bivalves in intertidal coastal areas where they filter feed on plankton and organic matter present in the water. These systems operate with minimal input of feed and chemicals, relying heavily on the natural productivity of the ecosystems [83]. Both systems produce high-quality products that can achieve high prices in the market in comparison to products from semi-intensive or intensive systems if they are differentiated through appropriate labeling.

(a)

(b)

Figure 4. (**a**) Oyster trestle tables placed in the intertidal zone; (**b**) marsh ponds in Cadiz Bay.

Concerning our climate change mitigation strategies, extensive aquaculture systems often focus on low-trophic-level species, such as bivalves and herbivorous or omnivorous fish or crustaceans, requiring few resources for growth and reducing the ecological footprint of production [231]. Bivalves, in particular, do not need external feed inputs, relying instead on natural phytoplankton, which enhances their sustainability [232]. Many extensive aquaculture systems employ polyculture or IMTAs [233–235], combining, for example, the extensive culture of fish and crustacean amphipods associated with marsh ponds [48,236]. Fish cultured in marsh ponds are usually produced in polyculture systems (e.g., mugilids, seabream, seabass and sole), especially under extensive conditions, but the monoculture of gilthead seabream (*Sparus aurata*) is being increasingly conducted in marsh ponds. In fact, extensive systems provide a fish product more similar to wild conspecifics, with similarities in the trophic niche and the concentrations of trace metals [231]. Feeding efficiency in extensive aquaculture is inherently high due to the reliance on natural food sources. In these systems, organisms consume naturally occurring plankton and detritus, which are replenished through natural processes like sunlight and nutrient cycling. This reduces the need for manufactured feed, lowering costs and minimizing the environmental impact associated with feed production and transport [75,231]. Extensive earthen ponds can exhibit efficient waste management through natural processes or through their association with constructed wetlands [237]. Waste settles and is broken down by microbial activity, which recycles nutrients back into the pond ecosystem, promoting further algal growth. On the other hand, bivalves filter and clean the water, removing excess nutrients and particulates while promoting biodeposition and enriching the sediment [238]. Regarding energy efficiency, these systems utilize sunlight and natural biological processes for production. Their reliance on natural sunlight for algae growth, which forms the base of the food web, significantly reduces the need for external energy inputs, which are reduced to downstream production stages such as processing, packaging, refrigeration or transport [42].

Extensive aquaculture can also contribute to carbon sequestration, especially in the case of bivalve farming, as previously noted. Bivalves incorporate carbon into their shells, effectively removing it from the atmosphere. However, it is important to be cautious with these statements since some studies affirm that the amount of CO_2 released through respiration is higher than the amount stored in a calcium carbonate shell [79,239]. According to Willer and Aldridge [240], global bivalve production (all systems included) emissions reach 11.1 tons of CO_2e per ton of protein, but excluding transport, emissions range between -5 (when bivalves are a net sink of carbon) and 1874 kg of CO_2e per ton WW (with a median of 392 kg CO_2e). Furthermore, extensive pond systems generally emit fewer GHGs than intensive systems due to their lower energy and feed inputs. However, the precise impact of extensive aquaculture on GHG emissions can vary depending on the specific practices and management of the systems [42]. In China, Zhang et al. [241] have estimated GHG emissions from extensive aquaculture in wetlands and inland ponds to be approximately 65 t CO_2e ha^{-1} y^{-1}. They also showed that coastal wetland systems have the lowest CO_2 emission impact, functioning as a net carbon sink when polyculture is conducted.

These extensive earthen ponds are, above all, providers of ecosystem services, including economic benefits, nutrient absorption (N, C) and habitats for birds, fish and invertebrate species. They also meet the criteria for organic labeling, enabling premium prices and ensuring economic viability [75]. Indeed, in this study, Walton et al. [75] demonstrated that aquaculture can mitigate the ecological degradation of natural wetlands, which represents an ecosystem restoration strategy of the EU [242].

4. Conclusions, Challenges and Implications

In this study, we compared several aquaculture systems in terms of their strategies to reduce GHG emissions. This review demonstrates the proactive approach of the aquaculture industry towards sustainability. The transition from intensive systems to more sustainable and extensive practices and methods, such as RASs, IMTAs and BFT, represents a significant advancement. Nevertheless, the results of our investigation indicate considerable variability between different studies of the same system, which represents an important limitation when evaluating their sustainability and comparing them with other systems.

Despite efforts to quantify emissions accurately, several potential errors and uncertainties must be acknowledged. Methodologically, the reliance on emission factors derived from various sources introduces uncertainties, as these factors can vary based on local conditions, species and management practices [243]. Moreover, these authors highlight that some of these factors are generally not considered (including bubbles released from sediments, aeration-induced emissions, emissions from dry and drying sediments, and emissions from effluent and waste). Data limitations, particularly in certain regions or specific aquaculture practices [164], hindered our ability to provide precise estimates. Future research should focus on improving data quality and homogenizing estimation techniques to enhance the reliability of emission assessments in aquaculture [244,245].

The geographical context also influences emission outcomes in aquaculture [42,241]. Local environmental factors, such as temperature, water quality and ecosystem resilience, play crucial roles in determining emission rates [10,24]. Economic conditions also influence production practices and thus emissions [164,246], with higher-income regions often investing in advanced technologies that may mitigate environmental impacts. Social factors, including regulatory frameworks and community engagement, further shape sustainability outcomes. Techniques like RASs tend to be more prevalent in developed countries [141], whereas extensive systems are more common in developing regions due to economic considerations and resource availability [247,248].

It is widely acknowledged that the systems we have reviewed here are more sustainable than traditional intensive systems [186,193]. In our opinion, while RASs, BFT and extensive systems in general offer significant sustainability benefits, IMTAs represent the most sustainable option due to their holistic approach and the fact that they apply, by definition, all GHG emission mitigation strategies. The FAO supports sustainable aquaculture practices globally and promotes IMTAs as a viable option for improving the environmental and economic performance of aquaculture operations [249]. However, in addition to the methodological limitations encountered in estimating the emissions explained above, there are still limits to the development and expansion of these systems. Polyculture requires technical knowledge about the cycles of two or more species, which can act as a brake in the transition from traditional farms towards these systems. They require complex management practices and tailored system designs, which can be barriers to their adoption [177]. Moreover, the interactions between different species need to be carefully managed to avoid unforeseen ecological impacts [250], and the benefits of IMTAs may vary significantly depending on local environmental conditions [178,251]. In Europe, the regulatory framework for IMTAs has not yet been fully developed or standardized across all member states [177]. However, there are several initiatives and regulations that indirectly support and influence the development of IMTAs (the EU missions "Restore our Ocean and Waters" or "Green Deal", the Atlantic Action Plan, the Blue Growth strategy, the United Nations Sustainable

Development Goal 12 [3–6,171]). Globally, the regulatory frameworks for IMTAs vary significantly, with some regions advancing more rapidly than others. For instance, Canada is one of the leaders in IMTA research and implementation. Their regulatory framework includes provincial and federal oversight, with agencies supporting IMTA through research and development programs [252,253]. In other countries, such as the United States, Chile, China, Japan, South Africa, New Zealand or Australia, the regulatory frameworks are evolving, with increased emphasis on sustainability and environmental protection, which supports the adoption of IMTAs [186,254–257].

In conclusion, while there is growing interest and support for IMTAs worldwide, challenges remain in terms of regulatory harmonization and technical knowledge. Developing specific regulations and guidelines for IMTAs, providing targeted funding and support and fostering international collaboration can help overcome these challenges and promote the wider adoption of IMTAs globally. In addition, improving standardization and transparency in reporting emission data to facilitate more accurate assessments in future studies is crucial for the development of IMTAs and global low-impact aquaculture. Finally, it is important to highlight that sustainable systems (RASs, BFT, IMTAs and other extensive methods) can be combined with one another, contingent on the availability of resources, for enhanced efficiency. Labeling and/or compliance with organic aquaculture standards appear to be crucial factors in achieving profitability with these systems.

Author Contributions: Writing—original draft preparation, M.C.-G.; writing—review and editing, J.M.G.-G., I.H.-C. and M.H.; supervision, M.H. All authors have read and agreed to the published version of the manuscript.

Funding: M.C.-G. was supported by a "Margarita Salas" individual fellowship from the Spanish Ministry of Science, Innovation and Universities, funded by the European Union—NextGeneration EU program.

Institutional Review Board Statement: Not applicable.

Informed Consent Statement: Not applicable.

Data Availability Statement: No new data were created or analyzed in this study. Data sharing is not applicable to this article.

Conflicts of Interest: The authors declare no conflicts of interest.

References

1. FAO. *The State of World Fisheries and Aquaculture 2022: Towards Blue Transformation*; The State of World Fisheries and Aquaculture (SOFIA); FAO: Rome, Italy, 2022; ISBN 978-92-5-136364-5.
2. Blanchard, J.L.; Watson, R.A.; Fulton, E.A.; Cottrell, R.S.; Nash, K.L.; Bryndum-Buchholz, A.; Büchner, M.; Carozza, D.A.; Cheung, W.W.; Elliott, J. Linked Sustainability Challenges and Trade-Offs among Fisheries, Aquaculture and Agriculture. *Nat. Ecol. Evol.* **2017**, *1*, 1240–1249. [CrossRef] [PubMed]
3. European Commission. *EU Mission Restore Our Ocean and Waters Implementation Plan*; European Commission: Brussels, Belgium, 2021.
4. European Commission. *The European Green Deal*; European Commission: Brussels, Belgium, 2019.
5. European Commission. *Action Plan for a Maritime Strategy in the Atlantic Area. Delivering Smart, Sustainable and Inclusive Growth*; European Commission: Brussels, Belgium, 2013.
6. European Commission Blue Growth. Supporting Sustainable Growth of the Marine and Maritime Sectors. Available online: https://s3platform.jrc.ec.europa.eu (accessed on 17 October 2023).
7. United Nations. Goal 13: Climate Action. Available online: https://www.un.org/sustainabledevelopment/climate-change/ (accessed on 14 February 2024).
8. MacLeod, M.J.; Hasan, M.R.; Robb, D.H.F.; Mamun-Ur-Rashid, M. Quantifying Greenhouse Gas Emissions from Global Aquaculture. *Sci. Rep.* **2020**, *10*, 11679. [CrossRef] [PubMed]
9. Christensen, J.M.; Olhoff, A. *Lessons from a Decade of Emissions Gap Assessments*; DTU Library: Delhi, India, 2019.
10. Raul, C.; Pattanaik, S.S.; Prakash, S.; Sreedharan, K.; Bharti, S. Greenhouse Gas Emissions from Aquaculture Systems. *World Aquac.* **2020**, *57*, 57–61.
11. Troell, M.; Naylor, R.L.; Metian, M.; Beveridge, M.; Tyedmers, P.H.; Folke, C.; Arrow, K.J.; Barrett, S.; Crépin, A.-S.; Ehrlich, P.R. Does Aquaculture Add Resilience to the Global Food System? *Proc. Natl. Acad. Sci. USA* **2014**, *111*, 13257–13263. [CrossRef] [PubMed]

12. Ytrestøyl, T.; Aas, T.S.; Åsgård, T. Utilisation of Feed Resources in Production of Atlantic Salmon (*Salmo salar*) in Norway. *Aquaculture* **2015**, *448*, 365–374. [CrossRef]
13. Adamidou, S.; Nengas, I.; Henry, M.; Ioakei Midoy, N.; Rigos, G.; Bell, G.J.; Jauncey, K. Effects of Dietary Inclusion of Peas, Chickpeas and Faba Beans on Growth, Feed Utilization and Health of Gilthead Seabream (*Sparus aurata*). *Aquac. Nutr.* **2011**, *17*, e288–e296. [CrossRef]
14. Rumbos, C.I.; Mente, E.; Karapanagiotidis, I.T.; Vlontzos, G.; Athanassiou, C.G. Insect-Based Feed Ingredients for Aquaculture: A Case Study for Their Acceptance in Greece. *Insects* **2021**, *12*, 586. [CrossRef] [PubMed]
15. Bujas, T.; Koričan, M.; Vukić, M.; Soldo, V.; Vladimir, N.; Fan, A. Review of Energy Consumption by the Fish Farming and Processing Industry in Croatia and the Potential for Zero-Emissions Aquaculture. *Energies* **2022**, *15*, 8197. [CrossRef]
16. Troell, M.; Tyedmers, P.; Kautsky, N.; Rönnbäck, P. Aquaculture and Energy Use. *Encycl. Energy* **2004**, *1*, 97–108.
17. Vo, T.T.E.; Ko, H.; Huh, J.-H.; Park, N. Overview of Solar Energy for Aquaculture: The Potential and Future Trends. *Energies* **2021**, *14*, 6923. [CrossRef]
18. Chen, Y.; Dong, S.; Wang, Z.; Wang, F.; Gao, Q.; Tian, X.; Xiong, Y. Variations in CO_2 Fluxes from Grass Carp *Ctenopharyngodon idella* Aquaculture Polyculture Ponds. *Aquac. Environ. Interact.* **2015**, *8*, 31–40. [CrossRef]
19. Yuan, J.; Xiang, J.; Liu, D.; Kang, H.; He, T.; Kim, S.; Lin, Y.; Freeman, C.; Ding, W. Rapid Growth in Greenhouse Gas Emissions from the Adoption of Industrial-Scale Aquaculture. *Nat. Clim. Chang.* **2019**, *9*, 318–322. [CrossRef]
20. Tan, J.; Lichtfouse, E.; Luo, M.; Liu, Y.; Tan, F.; Zhang, C.; Chen, X.; Huang, J.; Xiao, L. Aquaculture Drastically Increases Methane Production by Favoring Acetoclastic Rather than Hydrogenotrophic Methanogenesis in Shrimp Pond Sediments. *Aquaculture* **2023**, *563*, 738999. [CrossRef]
21. Hu, Z.; Lee, J.W.; Chandran, K.; Kim, S.; Khanal, S.K. Nitrous Oxide (N_2O) Emission from Aquaculture: A Review. *Environ. Sci. Technol.* **2012**, *46*, 6470–6480. [CrossRef] [PubMed]
22. Green, B.M. Fertilizer Use in Aquaculture. In *Feed and Feeding Practices in Aquaculture*; Davis, D.A., Ed.; Woodhead Publishing: Sawston, UK, 2022; pp. 29–63.
23. Tom, A.P.; Jayakumar, J.S.; Biju, M.; Somarajan, J.; Ibrahim, M.A. Aquaculture Wastewater Treatment Technologies and Their Sustainability: A Review. *Energy Nexus* **2021**, *4*, 100022. [CrossRef]
24. Hu, Z.; Wu, S.; Ji, C.; Zou, J.; Zhou, Q.; Liu, S. A Comparison of Methane Emissions Following Rice Paddies Conversion to Crab-Fish Farming Wetlands in Southeast China. *Environ. Sci. Pollut. Res.* **2016**, *23*, 1505–1515. [CrossRef]
25. Hynes, R.K.; Knowles, R. Production of Nitrous Oxide by *Nitrosomonas europaea*: Effects of Acetylene, pH, and Oxygen. *Can. J. Microbiol.* **1984**, *30*, 1397–1404. [CrossRef]
26. Hanaki, K.; Hong, Z.; Matsuo, T. Production of Nitrous Oxide Gas during Denitrification of Wastewater. *Water Sci. Technol.* **1992**, *26*, 1027–1036. [CrossRef]
27. Elvy, J.E.; Symonds, J.E.; Hilton, Z.; Walker, S.P.; Tremblay, L.A.; Casanovas, P.; Herbert, N.A. The Relationship of Feed Intake, Growth, Nutrient Retention, and Oxygen Consumption to Feed Conversion Ratio of Farmed Saltwater Chinook Salmon (*Oncorhynchus tshawytscha*). *Aquaculture* **2022**, *554*, 738184. [CrossRef]
28. Plouviez, M.; Shilton, A.; Packer, M.A.; Guieysse, B. Nitrous Oxide Emissions from Microalgae: Potential Pathways and Significance. *J. Appl. Phycol.* **2019**, *31*, 1–8. [CrossRef]
29. d'Orbcastel, E.R.; Blancheton, J.-P.; Aubin, J. Towards Environmentally Sustainable Aquaculture: Comparison between Two Trout Farming Systems Using Life Cycle Assessment. *Aquac. Eng.* **2009**, *40*, 113–119. [CrossRef]
30. Cao, L.; Diana, J.S.; Keoleian, G.A. Role of Life Cycle Assessment in Sustainable Aquaculture. *Rev. Aquac.* **2013**, *5*, 61–71. [CrossRef]
31. Chang, C.-C.; Chang, K.-C.; Lin, W.-C.; Wu, M.-H. Carbon Footprint Analysis in the Aquaculture Industry: Assessment of an Ecological Shrimp Farm. *J. Clean. Prod.* **2017**, *168*, 1101–1107. [CrossRef]
32. Shah, M.R.; Lutzu, G.A.; Alam, A.; Sarker, P.; Kabir Chowdhury, M.; Parsaeimehr, A.; Liang, Y.; Daroch, M. Microalgae in Aquafeeds for a Sustainable Aquaculture Industry. *J. Appl. Phycol.* **2018**, *30*, 197–213. [CrossRef]
33. Cottrell, R.S.; Metian, M.; Froehlich, H.E.; Blanchard, J.L.; Sand Jacobsen, N.; McIntyre, P.B.; Nash, K.L.; Williams, D.R.; Bouwman, L.; Gephart, J.A. Time to Rethink Trophic Levels in Aquaculture Policy. *Rev. Aquac.* **2021**, *13*, 1583–1593. [CrossRef]
34. Froehlich, H.E.; Jacobsen, N.S.; Essington, T.E.; Clavelle, T.; Halpern, B.S. Avoiding the Ecological Limits of Forage Fish for Fed Aquaculture. *Nat. Sustain.* **2018**, *1*, 298–303. [CrossRef]
35. Bandara, T. Alternative Feed Ingredients in Aquaculture: Opportunities and Challenges. *J. Entomol. Zool. Stud.* **2018**, *6*, 3087–3094.
36. Naylor, R.L.; Hardy, R.W.; Bureau, D.P.; Chiu, A.; Elliott, M.; Farrell, A.P.; Forster, I.; Gatlin, D.M.; Goldburg, R.J.; Hua, K. Feeding Aquaculture in an Era of Finite Resources. *Proc. Natl. Acad. Sci. USA* **2009**, *106*, 15103–15110. [CrossRef] [PubMed]
37. Krause, G.; Le Vay, L.; Buck, B.H.; Costa-Pierce, B.A.; Dewhurst, T.; Heasman, K.G.; Nevejan, N.; Nielsen, P.; Nielsen, K.N.; Park, K.; et al. Prospects of Low Trophic Marine Aquaculture Contributing to Food Security in a Net Zero-Carbon World. *Front. Sustain. Food Syst.* **2022**, *6*, 875509. [CrossRef]
38. Nijdam, D.; Rood, T.; Westhoek, H. The Price of Protein: Review of Land Use and Carbon Footprints from Life Cycle Assessments of Animal Food Products and Their Substitutes. *Food Policy* **2012**, *37*, 760–770. [CrossRef]
39. Hilborn, R.; Banobi, J.; Hall, S.J.; Pucylowski, T.; Walsworth, T.E. The Environmental Cost of Animal Source Foods. *Front. Ecol. Environ.* **2018**, *16*, 329–335. [CrossRef]

40. Guerra-García, J.M.; Hachero-Cruzado, I.; González-Romero, P.; Jiménez-Prada, P.; Cassell, C.; Ros, M. Towards Integrated Multi-Trophic Aquaculture: Lessons from Caprellids (Crustacea: Amphipoda). *PLoS ONE* **2016**, *11*, e0154776. [CrossRef] [PubMed]
41. Gephart, J.A.; Henriksson, P.J.; Parker, R.W.; Shepon, A.; Gorospe, K.D.; Bergman, K.; Eshel, G.; Golden, C.D.; Halpern, B.S.; Hornborg, S. Environmental Performance of Blue Foods. *Nature* **2021**, *597*, 360–365. [CrossRef]
42. Jones, A.R.; Alleway, H.K.; McAfee, D.; Reis-Santos, P.; Theuerkauf, S.J.; Jones, R.C. Climate-Friendly Seafood: The Potential for Emissions Reduction and Carbon Capture in Marine Aquaculture. *BioScience* **2022**, *72*, 123–143. [CrossRef] [PubMed]
43. Wright, A.C.; Fan, Y.; Baker, G.L. Nutritional Value and Food Safety of Bivalve Molluscan Shellfish. *J. Shellfish Res.* **2018**, *37*, 695–708. [CrossRef]
44. Duarte, C.M.; Wu, J.; Xiao, X.; Bruhn, A.; Krause-Jensen, D. Can Seaweed Farming Play a Role in Climate Change Mitigation and Adaptation? *Front. Mar. Sci.* **2017**, *4*, 100. [CrossRef]
45. Duarte, C.M.; Delgado-Huertas, A.; Marti, E.; Gasser, B.; Martin, I.S.; Cousteau, A.; Neumeyer, F.; Reilly-Cayten, M.; Boyce, J.; Kuwae, T.; et al. Carbon Burial in Sediments below Seaweed Farms. *bioRxiv* **2023**. [CrossRef]
46. Feng, J.-C.; Sun, L.; Yan, J. Carbon Sequestration via Shellfish Farming: A Potential Negative Emissions Technology. *Renew. Sustain. Energy Rev.* **2023**, *171*, 113018. [CrossRef]
47. Cotas, J.; Gomes, L.; Pacheco, D.; Pereira, L. Ecosystem Services Provided by Seaweeds. *Hydrobiology* **2023**, *2*, 75–96. [CrossRef]
48. Jiménez-Prada, P.; Hachero-Cruzado, I.; Guerra-García, J. Aquaculture Waste as Food for Amphipods: The Case of *Gammarus insensibilis* in Marsh Ponds from Southern Spain. *Aquac. Int.* **2020**, *29*, 139–153. [CrossRef]
49. Castilla-Gavilán, M.; Guerra-García, J.; Moreno-Oliva, J.; Hachero-Cruzado, I. How Much Waste Can the Amphipod *Gammarus insensibilis* Remove from Aquaculture Effluents? A First Step toward IMTA. *Aquaculture* **2023**, *573*, 739552. [CrossRef]
50. Ribes-Navarro, A.; Navarro, J.C.; Hontoria, F.; Kabeya, N.; Standal, I.B.; Evjemo, J.O.; Monroig, Ó. Biosynthesis of Long-Chain Polyunsaturated Fatty Acids in Marine Gammarids: Molecular Cloning and Functional Characterisation of Three Fatty Acyl Elongases. *Mar. Drugs* **2021**, *19*, 226. [CrossRef]
51. Ahmed, N.; Cheung, W.W.; Thompson, S.; Glaser, M. Solutions to Blue Carbon Emissions: Shrimp Cultivation, Mangrove Deforestation and Climate Change in Coastal Bangladesh. *Mar. Policy* **2017**, *82*, 68–75. [CrossRef]
52. Ahmed, N.; Thompson, S.; Glaser, M. Integrated Mangrove-Shrimp Cultivation: Potential for Blue Carbon Sequestration. *Ambio* **2018**, *47*, 441–452. [CrossRef]
53. Maldonado, C.; Cuzon, G.; Guzmán, E.; Brito, R.; Soto, L.; Arena, L.; Gaxiola, G. Effect of an Herbivorous Diet on Energy Balance of *Litopenaeus vannamei* at Selected Ontogenetic Stages. *Aquaculture* **2009**, *296*, 123–128. [CrossRef]
54. Tolentino-Pablico, G.; Bailly, N.; Froese, R.; Elloran, C. Seaweeds Preferred by Herbivorous Fishes. In *Nineteenth International Seaweed Symposium, Proceedings of the 19th International Seaweed Symposium, Kobe, Japan, 26–31 March 2007*; Borowitzka, M.A., Critchley, A.T., Kraan, S., Peters, A., Sjøtun, K., Notoya, M., Eds.; Developments in Applied Phycology; Springer: Dordrecht, The Netherlands, 2009; pp. 483–488, ISBN 978-1-4020-9619-8.
55. Li, Y.; Zhang, Q.; Liu, Y. Rabbitfish—An Emerging Herbivorous Marine Aquaculture Species. In *Aquaculture in China*; John Wiley & Sons, Ltd.: Hoboken, NJ, USA, 2018; pp. 329–334, ISBN 978-1-119-12075-9.
56. Lozano-Muñoz, I.; Castellaro, G.; Bueno, G.; Wacyk, J. Herbivorous Fish (*Medialuna ancietae*) as a Sustainable Alternative for Nutrition Security in Northern Chile. *Sci. Rep.* **2022**, *12*, 1619. [CrossRef]
57. Jose, D.M.; Divya, P. A Review on Aquaculture Important Fish *Chanos chanos*, Forsskal 1775, the Milkfish. *J. Aquac. Trop.* **2022**, *37*, 1–26. [CrossRef]
58. Putra, D.; Muhsinah, M.; Arisa, I. *The Substitution of Soybean Meal by Fermented Tofu Dregs in the Milkfish (Chanos chanos) Diet*; IOP Publishing: Bristol, UK, 2021; Volume 674, p. 012102.
59. Herawati, V.E.; Pinandoyo, P.; Windarto, S.; Hariyadi, P.; Hutabarat, J.; Darmanto, Y.; Rismaningsih, N.; Prayitno, S.B.; Radjasa, O.K. Maggot Meal (Hermetia illucens) Substitution on Fish Meal to Growth Performance, and Nutrient Content of Milkfish (Chanos chanos). *HAYATI J. Biosci.* **2020**, *27*, 154. [CrossRef]
60. Aslamyah, S.; Umam, M.K.; Lestari, A.D. *The Effect of Protein Levels in Fermentation Feed Supplemented Lumbricus sp. Extract as Feed Additive on Growth Performance and Body Chemical Composition of Milkfish, Chanos chanos Forskal 1775*; IOP Publishing: Bristol, UK, 2020; Volume 575, p. 012036.
61. Nurfadillah, N.; Ningsih, H.; Rahimi, S.; Dewiyanti, I.; Mellisa, S.; Syahril, A. *The Effect of Ethanolic Extracts Ulva lactuca on Growth Performance and Survival Rate of Milk Fish (Chanos chanos)*; IOP Publishing: Bristol, UK, 2021; Volume 674, p. 012049.
62. Martínez, F.P.; Bermúdez, L.; Aznar, M.J.; Moyano, F.J. Evaluation of Enzyme Additives on the Nutritional Use of Feeds with a High Content of Plant Ingredients for *Mugil cephalus*. *Fishes* **2019**, *4*, 56. [CrossRef]
63. Biswas, G.; De, D.; Thirunavukkarasu, A.; Natarajan, M.; Sundaray, J.; Kailasam, M.; Kumar, P.; Ghoshal, T.; Ponniah, A.; Sarkar, A. Effects of Stocking Density, Feeding, Fertilization and Combined Fertilization-Feeding on the Performances of Striped Grey Mullet (*Mugil cephalus* L.) Fingerlings in Brackishwater Pond Rearing Systems. *Aquaculture* **2012**, *338*, 284–292. [CrossRef]
64. Jana, S.N.; Sudesh; Garg, S.K.; Sabhlok, V.P.; Bhatnagar, A. Nutritive Evaluation of Lysine-and Methionine-Supplemented Raw vs Heat-Processed Soybean to Replace Fishmeal as a Dietary Protein Source for Grey Mullet, *Mugil cephalus*, and Milkfish, *Chanos chanos*. *J. Appl. Aquac.* **2012**, *24*, 69–80. [CrossRef]
65. Gisbert, E.; Mozanzadeh, M.T.; Kotzamanis, Y.; Estévez, A. Weaning Wild Flathead Grey Mullet (*Mugil cephalus*) Fry with Diets with Different Levels of Fish Meal Substitution. *Aquaculture* **2016**, *462*, 92–100. [CrossRef]

66. Lal, R. Carbon Sequestration. *Philos. Trans. R. Soc. B Biol. Sci.* **2007**, *363*, 815–830. [CrossRef]
67. Mitsch, W.J.; Bernal, B.; Nahlik, A.M.; Mander, Ü.; Zhang, L.; Anderson, C.J.; Jørgensen, S.E.; Brix, H. Wetlands, Carbon, and Climate Change. *Landsc. Ecol.* **2013**, *28*, 583–597. [CrossRef]
68. Arifanti, V.B.; Kauffman, J.B.; Hadriyanto, D.; Murdiyarso, D.; Diana, R. Carbon Dynamics and Land Use Carbon Footprints in Mangrove-Converted Aquaculture: The Case of the Mahakam Delta, Indonesia. *For. Ecol. Manag.* **2019**, *432*, 17–29. [CrossRef]
69. Yang, P.; Lai, D.Y.; Yang, H.; Lin, Y.; Tong, C.; Hong, Y.; Tian, Y.; Tang, C.; Tang, K.W. Large Increase in CH_4 Emission Following Conversion of Coastal Marsh to Aquaculture Ponds Caused by Changing Gas Transport Pathways. *Water Res.* **2022**, *222*, 118882. [CrossRef]
70. Yang, P.; Bastviken, D.; Lai, D.; Jin, B.; Mou, X.; Tong, C.; Yao, Y. Effects of Coastal Marsh Conversion to Shrimp Aquaculture Ponds on CH_4 and N_2O Emissions. *Estuar. Coast. Shelf Sci.* **2017**, *199*, 125–131. [CrossRef]
71. Ferrón, S.; Ortega, T.; Forja, J.M. Benthic Fluxes in a Tidal Salt Marsh Creek Affected by Fish Farm Activities: Río San Pedro (Bay of Cádiz, SW Spain). *Mar. Chem.* **2009**, *113*, 50–62. [CrossRef]
72. Duarte, C.M.; Losada, I.J.; Hendriks, I.E.; Mazarrasa, I.; Marbà, N. The Role of Coastal Plant Communities for Climate Change Mitigation and Adaptation. *Nat. Clim. Chang.* **2013**, *3*, 961–968. [CrossRef]
73. Morris, E.P.; Flecha, S.; Figuerola, J.; Costas, E.; Navarro, G.; Ruiz, J.; Rodriguez, P.; Huertas, E. Contribution of Doñana Wetlands to Carbon Sequestration. *PLoS ONE* **2013**, *8*, e71456. [CrossRef] [PubMed]
74. Friess, D.A.; Yando, E.S.; Alemu, J.B.; Wong, L.-W.; Soto, S.D.; Bhatia, N. Ecosystem Services and Disservices of Mangrove Forests and Salt Marshes. In *Oceanography and Marine Biology*; Taylor & Francis: Abingdon, UK, 2020.
75. Walton, M.; Vilas, C.; Cañavate, J.P.; González-Ortegón, E.; Prieto, A.; Van Bergeijk, S.; Green, A.J.; Librero, M.; Mazuelos, N.; Le Vay, L. A Model for the Future: Ecosystem Services Provided by the Aquaculture Activities of Veta La Palma, Southern Spain. *Aquaculture* **2015**, *448*, 382–390. [CrossRef]
76. Erwin, K.L. Wetlands and Global Climate Change: The Role of Wetland Restoration in a Changing World. *Wetl. Ecol. Manag.* **2009**, *17*, 71–84. [CrossRef]
77. Čížková, H.; Květ, J.; Comin, F.A.; Laiho, R.; Pokorný, J.; Pithart, D. Actual State of European Wetlands and Their Possible Future in the Context of Global Climate Change. *Aquat. Sci.* **2013**, *75*, 3–26. [CrossRef]
78. Green, A.J.; Alcorlo, P.; Peeters, E.T.; Morris, E.P.; Espinar, J.L.; Bravo-Utrera, M.A.; Bustamante, J.; Díaz-Delgado, R.; Koelmans, A.A.; Mateo, R. Creating a Safe Operating Space for Wetlands in a Changing Climate. *Front. Ecol. Environ.* **2017**, *15*, 99–107. [CrossRef]
79. Han, T.; Shi, R.; Qi, Z.; Huang, H.; Liang, Q.; Liu, H. Interactive Effects of Oyster and Seaweed on Seawater Dissolved Inorganic Carbon Systems: Implications for Integrated Multi-Trophic Aquaculture. *Aquac. Environ. Interact.* **2017**, *9*, 469–478. [CrossRef]
80. Thomas, M.; Pasquet, A.; Aubin, J.; Nahon, S.; Lecocq, T. When More Is More: Taking Advantage of Species Diversity to Move towards Sustainable Aquaculture. *Biol. Rev.* **2021**, *96*, 767–784. [CrossRef] [PubMed]
81. Castilla-Gavilán, M.; Muñoz-Martínez, M.; Zuasti, E.; Canoura-Baldonado, J.; Mondoñedo, R.; Hachero-Cruzado, I. Yield, Nutrients Uptake and Lipid Profile of the Halophyte *Salicornia ramosissima* Cultivated in Two Different Integrated Multi-Trophic Aquaculture Systems (IMTA). *Aquaculture* **2024**, *583*, 740547. [CrossRef]
82. Wang, X.; Broch, O.J.; Forbord, S.; Handå, A.; Skjermo, J.; Reitan, K.I.; Vadstein, O.; Olsen, Y. Assimilation of Inorganic Nutrients from Salmon (*Salmo salar*) Farming by the Macroalgae (*Saccharina latissima*) in an Exposed Coastal Environment: Implications for Integrated Multi-Trophic Aquaculture. *J. Appl. Phycol.* **2014**, *26*, 1869–1878. [CrossRef]
83. Strand, Ø.; Jansen, H.M.; Jiang, Z.; Robinson, S.M. Perspectives on Bivalves Providing Regulating Services in Integrated Multi-Trophic Aquaculture. In *Goods and Services of Marine Bivalves*; Springer: Berlin/Heidelberg, Germany, 2019; pp. 209–230.
84. Zhang, D.; Tian, X.; Dong, S.; Chen, Y.; Feng, J.; He, R.-P.; Zhang, K. Carbon Budgets of Two Typical Polyculture Pond Systems in Coastal China and Their Potential Roles in the Global Carbon Cycle. *Aquac. Environ. Interact.* **2020**, *12*, 105–115. [CrossRef]
85. Davis, D.A. *Feed and Feeding Practices in Aquaculture*; Woodhead Publishing: Sawston, UK, 2022; ISBN 0-12-822992-6.
86. Hardy, R.W. Utilization of Plant Proteins in Fish Diets: Effects of Global Demand and Supplies of Fishmeal. *Aquac. Res.* **2010**, *41*, 770–776. [CrossRef]
87. Martínez-Llorens, S.; Moñino, A.V.; Tomás Vidal, A.; Salvador, V.J.M.; Pla Torres, M.; Jover Cerdá, M. Soybean Meal as a Protein Source in Gilthead Sea Bream (*Sparus aurata* L.) Diets: Effects on Growth and Nutrient Utilization. *Aquac. Res.* **2007**, *38*, 82–90. [CrossRef]
88. Pereira, T.; Oliva-Teles, A. Evaluation of Corn Gluten Meal as a Protein Source in Diets for Gilthead Sea Bream (*Sparus aurata* L.) Juveniles. *Aquac. Res.* **2003**, *34*, 1111–1117. [CrossRef]
89. Gómez-Requeni, P.; Mingarro, M.; Kirchner, S.; Calduch-Giner, J.; Médale, F.; Corraze, G.; Panserat, S.; Martin, S.; Houlihan, D.; Kaushik, S. Effects of Dietary Amino Acid Profile on Growth Performance, Key Metabolic Enzymes and Somatotropic Axis Responsiveness of Gilthead Sea Bream (*Sparus aurata*). *Aquaculture* **2003**, *220*, 749–767. [CrossRef]
90. Alarcón, F.J.; Moyano, F.J.; Díaz, M. Effect of Inhibitors Present in Protein Sources on Digestive Proteases of Juvenile Sea Bream (*Sparus aurata*). *Aquat. Living Resour.* **1999**, *12*, 233–238. [CrossRef]
91. Francis, G.; Makkar, H.P.; Becker, K. Antinutritional Factors Present in Plant-Derived Alternate Fish Feed Ingredients and Their Effects in Fish. *Aquaculture* **2001**, *199*, 197–227. [CrossRef]
92. Gasco, L.; Finke, M.; Van Huis, A. Can Diets Containing Insects Promote Animal Health? *J. Insects Food Feed.* **2018**, *4*, 1–4. [CrossRef]

93. Hachero-Cruzado, I.; Betancor, M.B.; Coronel-Dominguez, A.J.; Manchado, M.; Alarcón-López, F.J. Assessment of Full-Fat *Tenebrio molitor* as Feed Ingredient for *Solea senegalensis*: Effects on Growth Performance and Lipid Profile. *Animals* **2024**, *14*, 595. [CrossRef]
94. Sogari, G.; Bellezza Oddon, S.; Gasco, L.; van Huis, A.; Spranghers, T.; Mancini, S. Review: Recent Advances in Insect-Based Feeds: From Animal Farming to the Acceptance of Consumers and Stakeholders. *Animals* **2023**, *17*, 100904. [CrossRef]
95. Quang Tran, H.; Van Doan, H.; Stejskal, V. Environmental Consequences of Using Insect Meal as an Ingredient in Aquafeeds: A Systematic View. *Rev. Aquac.* **2022**, *14*, 237–251. [CrossRef]
96. Ogunkalu, O. Effects of Feed Additives in Fish Feed for Improvement of Aquaculture. *Eurasian J. Food Sci. Technol.* **2019**, *3*, 49–57.
97. Encarnação, P. 5—Functional Feed Additives in Aquaculture Feeds. In *Aquafeed Formulation*; Nates, S.F., Ed.; Academic Press: San Diego, CA, USA, 2016; pp. 217–237, ISBN 978-0-12-800873-7.
98. Pereira, A.G.; Fraga-Corral, M.; Garcia-Oliveira, P.; Otero, P.; Soria-Lopez, A.; Cassani, L.; Cao, H.; Xiao, J.; Prieto, M.A.; Simal-Gandara, J. Single-Cell Proteins Obtained by Circular Economy Intended as a Feed Ingredient in Aquaculture. *Foods* **2022**, *11*, 2831. [CrossRef] [PubMed]
99. Glencross, B.D.; Huyben, D.; Schrama, J.W. The Application of Single-Cell Ingredients in Aquaculture Feeds—A Review. *Fishes* **2020**, *5*, 22. [CrossRef]
100. Ponce, M.; Anguís, V.; Fernández-Díaz, C. Assessing the Role of Ulvan as Immunonutrient in *Solea senegalensis*. *Fish Shellfish Immunol.* **2024**, *146*, 109399. [CrossRef]
101. Liang, Q.; Yuan, M.; Xu, L.; Lio, E.; Zhang, F.; Mou, H.; Secundo, F. Application of Enzymes as a Feed Additive in Aquaculture. *Mar. Life Sci. Technol.* **2022**, *4*, 208–221. [CrossRef] [PubMed]
102. Kalhoro, H.; Zhou, J.; Hua, Y.; Ng, W.-K.; Ye, L.; Zhang, J.; Shao, Q. Soy Protein Concentrate as a Substitute for Fish Meal in Diets for Juvenile *Acanthopagrus schlegelii*: Effects on Growth, Phosphorus Discharge and Digestive Enzyme Activity. *Aquac. Res.* **2018**, *49*, 1896–1906. [CrossRef]
103. González-Riopedre, M.; Márquez, L.; Sieiro, M.; Vázquez, U.; Maroto, J.; Barcia, R.; Moyano, F. Use of Purified Extracts from Fish Viscera as an Enzyme Additive in Feeds for Juvenile Marine Fish. In *New Additives and Ingredients in the Formulation of Aquafeeds*; Centro Tecnologico del Mar-Fundacion (CETMAR): Vigo, Spain, 2013.
104. Robb, D.H.; Crampton, V.O.; Robb, D.; Crampton, V. On-Farm Feeding and Feed Management: Perspectives from the Fish Feed Industry. *On-Farm Feed. Feed Manag. Aquac.* **2013**, *489*, 518.
105. Farmer, A. Phosphate Pollution: A Global Overview of the Problem. In *Phosphorus: Polluter and Resource of the Future—Removal and Recovery from Wastewater*; Schaum, C., Ed.; IWA Publishing: London, UK, 2018; pp. 35–55.
106. Chen, S.; Timmons, M.B.; Aneshansley, D.J.; Bisogni, J.J., Jr. Suspended Solids Characteristics from Recirculating Aquacultural Systems and Design Implications. *Aquaculture* **1993**, *112*, 143–155. [CrossRef]
107. Badiola, M.; Mendiola, D.; Bostock, J. Recirculating Aquaculture Systems (RAS) Analysis: Main Issues on Management and Future Challenges. *Aquac. Eng.* **2012**, *51*, 26–35. [CrossRef]
108. Martins, C.; Eding, E.H.; Verdegem, M.C.; Heinsbroek, L.T.; Schneider, O.; Blancheton, J.-P.; d'Orbcastel, E.R.; Verreth, J. New Developments in Recirculating Aquaculture Systems in Europe: A Perspective on Environmental Sustainability. *Aquac. Eng.* **2010**, *43*, 83–93. [CrossRef]
109. Camargo, J.A.; Alonso, A.; Salamanca, A. Nitrate Toxicity to Aquatic Animals: A Review with New Data for Freshwater Invertebrates. *Chemosphere* **2005**, *58*, 1255–1267. [CrossRef] [PubMed]
110. Tripathi, S.; Hussain, T. Chapter 7—Water and Wastewater Treatment through Ozone-Based Technologies. In *Development in Wastewater Treatment Research and Processes*; Shah, M., Rodriguez-Couto, S., Biswas, J., Eds.; Elsevier: Amsterdam, The Netherlands, 2022; pp. 139–172, ISBN 978-0-323-85583-9.
111. Kurniawan, S.; Novarini; Yuliwati, E.; Ariyanto, E.; Morsin, M.; Sanudin, R.; Nafisah, S. Greywater Treatment Technologies for Aquaculture Safety: Review. *J. King Saud. Univ.-Eng. Sci.* **2023**, *35*, 327–334. [CrossRef]
112. Kurniawan, S.B.; Abdullah, S.R.S.; Imron, M.F.; Ahmad, A.; Mohd Said, N.S.; Mohd Rahim, N.F.; Mohammad Alnawajha, M.; Abu Hasan, H.; Othman, A.R.; Purwanti, I.F. Potential of Valuable Materials Recovery from Aquaculture Wastewater: An Introduction to Resource Reclamation. *Aquac. Res.* **2021**, *52*, 2954–2962. [CrossRef]
113. Ezziddine, M.; Liltved, H. Nutrients Recovery from Aquaculture Waste for Use as Fertilizer in Soilless Growth Systems. *Int. Symp. Grow. Media Compost. Substrate Anal.* **2019**, *1305*, 399–406. [CrossRef]
114. Alvarado-Ramírez, L.; Santiesteban-Romero, B.; Poss, G.; Sosa-Hernández, J.E.; Iqbal, H.M.; Parra-Saldívar, R.; Bonaccorso, A.D.; Melchor-Martínez, E.M. Sustainable Production of Biofuels and Bioderivatives from Aquaculture and Marine Waste. *Front. Chem. Eng.* **2023**, *4*, 1072761. [CrossRef]
115. Mirzoyan, N.; Tal, Y.; Gross, A. Anaerobic Digestion of Sludge from Intensive Recirculating Aquaculture Systems: Review. *Aquaculture* **2010**, *306*, 1–6. [CrossRef]
116. Del Campo, L.M.; Ibarra, P.; Gutiérrez, X.; Takle, H.R. Utilization of Sludge from Recirculation Aquaculture Systems. *Nofima Rapp.* **2010**, *9*.
117. Dumay, J.; Clément, N.; Morançais, M.; Fleurence, J. Optimization of Hydrolysis Conditions of *Palmaria palmata* to Enhance R-Phycoerythrin Extraction. *Bioresour. Technol.* **2013**, *131*, 21–27. [CrossRef]
118. Le Strat, Y.; Ruiz, N.; Fleurence, J.; Pouchus, Y.-F.; Déléris, P.; Dumay, J. Marine Fungal Abilities to Enzymatically Degrade Algal Polysaccharides, Proteins and Lipids: A Review. *J. Appl. Phycol.* **2022**, *34*, 1131–1162. [CrossRef]

119. Le Guillard, C.; Bergé, J.-P.; Donnay-Moreno, C.; Cornet, J.; Ragon, J.-Y.; Fleurence, J.; Dumay, J. Optimization of R-Phycoerythrin Extraction by Ultrasound-Assisted Enzymatic Hydrolysis: A Comprehensive Study on the Wet Seaweed *Grateloupia turuturu*. *Mar. Drugs* **2023**, *21*, 213. [CrossRef] [PubMed]
120. Perčić, M.; Vladimir, N.; Fan, A. Techno-Economic Assessment of Alternative Marine Fuels for Inland Shipping in Croatia. *Renew. Sustain. Energy Rev.* **2021**, *148*, 111363. [CrossRef]
121. Folke, C.; Kautsky, N. Aquaculture with Its Environment: Prospects for Sustainability. *Ocean Coast. Manag.* **1992**, *17*, 5–24. [CrossRef]
122. Ion, I.V.; Popescu, F.; Coman, G.; Frățița, M. Heat Requirement in an Indoor Recirculating Aquaculture System. *Energy Rep.* **2022**, *8*, 11707–11714. [CrossRef]
123. Kuyumcu, M.E.; Tutumlu, H.; Yumrutaş, R. Performance of a Swimming Pool Heating System by Utilizing Waste Energy Rejected from an Ice Rink with an Energy Storage Tank. *Energy Convers. Manag.* **2016**, *121*, 349–357. [CrossRef]
124. Via-Aqua Conclusion Du Programme AQUAGRINERGIE. Available online: https://www.via-aqua.fr/ (accessed on 3 June 2024).
125. Bergman, K.; Henriksson, P.J.G.; Hornborg, S.; Troell, M.; Borthwick, L.; Jonell, M.; Philis, G.; Ziegler, F. Recirculating Aquaculture Is Possible without Major Energy Tradeoff: Life Cycle Assessment of Warmwater Fish Farming in Sweden. *Environ. Sci. Technol.* **2020**, *54*, 16062–16070. [CrossRef] [PubMed]
126. He, P.; Davy, D.; Sciortino, J.; Beveridge, M.C.; Arnason, R.; Gudmundsson, A. Countering Climate Change: Measures and Tools to Reduce Energy Use and Greenhouse Gas Emission in Fisheries and Aquaculture. In *Impacts of Climate Change on Fisheries and Aquaculture*; FAO: Rome, Italy, 2018; p. 585.
127. Perdikaris, C.; Paschos, I. Organic Aquaculture in Greece: A Brief Review. *Rev. Aquac.* **2010**, *2*, 102–105. [CrossRef]
128. Lagutkina, L.Y.; Ponomarev, S. Organic Aquaculture as Promising Trend of the Fish Industry Development. *Agric. Biol.* **2018**, *53*, 326–336. [CrossRef]
129. IFOAM (International Federation of Organic Agriculture Movements). *IFOAM General Assembly*; IFOAM: Vignola, Italy, 2008.
130. Angel, D.; Jokumsen, A.; Lembo, G. Aquaculture Production Systems and Environmental Interactions. In *Organic Aquaculture: Impacts and Future Developments*; Lembo, G., Mente, E., Eds.; Springer International Publishing: Cham, Switzerland, 2019; pp. 103–118, ISBN 978-3-030-05603-2.
131. Sicuro, B. An Overview of Organic Aquaculture in Italy. *Aquaculture* **2019**, *509*, 134–139. [CrossRef]
132. Labatut, R.A.; Olivares, J.F. Culture of Turbot (*Scophthalmus maximus*) Juveniles Using Shallow Raceways Tanks and Recirculation. *Aquac. Eng.* **2004**, *32*, 113–127. [CrossRef]
133. d'Orbcastel, E.R.; Lemarié, G.; Breuil, G.; Petochi, T.; Marino, G.; Triplet, S.; Dutto, G.; Fivelstad, S.; Coeurdacier, J.-L.; Blancheton, J.-P. Effects of Rearing Density on Sea Bass (*Dicentrarchus labrax*) Biological Performance, Blood Parameters and Disease Resistance in a Flow through System. *Aquat. Living Resour.* **2010**, *23*, 109–117. [CrossRef]
134. Salas-Leiton, E.; Anguis, V.; Manchado, M.; Cañavate, J.P. Growth, Feeding and Oxygen Consumption of Senegalese Sole (*Solea senegalensis*) Juveniles Stocked at Different Densities. *Aquaculture* **2008**, *285*, 84–89. [CrossRef]
135. Murray, F.; Bostock, J.; Fletcher, D. *Review of Recirculation Aquaculture System Technologies and Their Commercial Application*; University of Stirling Aquaculture: Scotland, UK, 2014.
136. Foysal, M.; Fotedar, R.; Gupta, S.; Chaklader, M. Biological Ball Filters Regulate Bacterial Communities in Marron (*Cherax cainii*) Culture System. *Lett. Appl. Microbiol.* **2019**, *68*, 455–463. [CrossRef] [PubMed]
137. Kovács, B.D.; de Jesus Gregersen, K.J.; Rüppel, F.; von Danwitz, A.; Pedersen, L.-F. Evaluating Protein Skimmer Performance in a Commercial Seawater Recirculating Aquaculture System (RAS). *Aquac. Eng.* **2023**, *103*, 102369. [CrossRef]
138. de Jesus Gregersen, K.J.; Pedersen, P.B.; Pedersen, L.-F.; Liu, D.; Dalsgaard, J. UV Irradiation and Micro Filtration Effects on Micro Particle Development and Microbial Water Quality in Recirculation Aquaculture Systems. *Aquaculture* **2020**, *518*, 734785. [CrossRef]
139. Ebeling, J.M.; Timmons, M.B. Recirculating Aquaculture Systems. In *Aquaculture Production Systems*; John Wiley & Sons: Hoboken, NJ, USA, 2012; pp. 245–277.
140. Engle, C.R.; Kumar, G.; van Senten, J. Resource-Use Efficiency in US Aquaculture: Farm-Level Comparisons across Fish Species and Production Systems. *Aquac. Environ. Interact.* **2021**, *13*, 259–275. [CrossRef]
141. Ahmed, N.; Turchini, G.M. Recirculating Aquaculture Systems (RAS): Environmental Solution and Climate Change Adaptation. *J. Clean. Prod.* **2021**, *297*, 126604. [CrossRef]
142. Ranjan, R.; Megarajan, S.; Xavier, B.; Raju, S.S.; Ghosh, S.; Gopalakrishnan, A. Design and Performance of Recirculating Aquaculture System for Marine Finfish Broodstock Development. *Aquac. Eng.* **2019**, *85*, 90–97. [CrossRef]
143. Godoy-Olmos, S.; Jauralde, I.; Monge-Ortiz, R.; Milián-Sorribes, M.C.; Jover-Cerdá, M.; Tomás-Vidal, A.; Martínez-Llorens, S. Influence of Diet and Feeding Strategy on the Performance of Nitrifying Trickling Filter, Oxygen Consumption and Ammonia Excretion of Gilthead Sea Bream (*Sparus aurata*) Raised in Recirculating Aquaculture Systems. *Aquac. Int.* **2022**, *30*, 581–606. [CrossRef]
144. Almeida, D.B.; Magalhães, C.; Sousa, Z.; Borges, M.T.; Silva, E.; Blanquet, I.; Mucha, A.P. Microbial Community Dynamics in a Hatchery Recirculating Aquaculture System (RAS) of Sole (*Solea senegalensis*). *Aquaculture* **2021**, *539*, 736592. [CrossRef]
145. Meriç, İ. Mineral Element and Nutrient Composition of Two Newly-Introduced Fish Species (*Dentex dentex* and *Seriola dumerili*) in Recirculating Aquaculture System (RAS). *GIDA/J. Food* **2017**, *42*, 163–168. [CrossRef]

146. Yang, J.; Ni, Q.; Zhang, Y.; Xu, B. Construction Technology on RAS for Shrimp Culture. *Trans. Chin. Soc. Agric. Eng.* **2010**, *26*, 136–140.
147. Suantika, G.; Situmorang, M.L.; Kurniawan, J.B.; Pratiwi, S.A.; Aditiawati, P.; Astuti, D.I.; Azizah, F.F.N.; Djohan, Y.A.; Zuhri, U.; Simatupang, T.M. Development of a Zero Water Discharge (ZWD)—Recirculating Aquaculture System (RAS) Hybrid System for Super Intensive White Shrimp (*Litopenaeus vannamei*) Culture under Low Salinity Conditions and Its Industrial Trial in Commercial Shrimp Urban Farming in Gresik, East Java, Indonesia. *Aquac. Eng.* **2018**, *82*, 12–24.
148. Fauzi, R.L.; Pamungkas, A.P.; Purwadi, D. *White Shrimp Litopenaeus vannamei Based Agroindustry through Recirculating Aquaculture System to Increase Competitiveness*; EDP Sciences: Les Ulis, France, 2020; Volume 147, p. 01002.
149. Grosso, L.; Rakaj, A.; Fianchini, A.; Morroni, L.; Cataudella, S.; Scardi, M. Integrated Multi-Trophic Aquaculture (IMTA) System Combining the Sea Urchin *Paracentrotus lividus*, as Primary Species, and the Sea Cucumber *Holothuria tubulosa* as Extractive Species. *Aquaculture* **2021**, *534*, 736268. [CrossRef]
150. Sartori, D.; Scuderi, A.; Sansone, G.; Gaion, A. Echinoculture: The Rearing of *Paracentrotus lividus* in a Recirculating Aquaculture System—Experiments of Artificial Diets for the Maintenance of Sexual Maturation. *Aquac. Int.* **2015**, *23*, 111–125. [CrossRef]
151. Sartori, D.; Gaion, A. Can Sea Urchins Benefit from an Artificial Diet? Physiological and Histological Assessment for Echinoculture Feasibility Evaluation. *Aquac. Nutr.* **2016**, *22*, 1214–1221. [CrossRef]
152. Li, H.; Cui, Z.; Cui, H.; Bai, Y.; Yin, Z.; Qu, K. A Review of Influencing Factors on a Recirculating Aquaculture System: Environmental Conditions, Feeding Strategies, and Disinfection Methods. *J. World Aquac. Soc.* **2023**, *54*, 566–602. [CrossRef]
153. Chang, C.; Fang, W.; Jao, R.-C.; Shyu, C.; Liao, I.-C. Development of an Intelligent Feeding Controller for Indoor Intensive Culturing of Eel. *Aquac. Eng.* **2005**, *32*, 343–353. [CrossRef]
154. Steinberg, K.; Zimmermann, J.; Meyer, S.; Schulz, C. Start-up of Recirculating Aquaculture Systems: How Do Water Exchange Rates Influence Pikeperch (*Sander lucioperca*) and Water Composition? *Aquac. Eng.* **2018**, *83*, 151–159. [CrossRef]
155. Davidson, J.; Good, C.; Welsh, C.; Summerfelt, S.T. Abnormal Swimming Behavior and Increased Deformities in Rainbow Trout *Oncorhynchus mykiss* Cultured in Low Exchange Water Recirculating Aquaculture Systems. *Aquac. Eng.* **2011**, *45*, 109–117. [CrossRef]
156. Pedersen, P.B.; von Ahnen, M.; Fernandes, P.; Naas, C.; Pedersen, L.-F.; Dalsgaard, J. Particle Surface Area and Bacterial Activity in Recirculating Aquaculture Systems. *Aquac. Eng.* **2017**, *78*, 18–23. [CrossRef]
157. Davidson, J.; Barrows, F.T.; Kenney, P.B.; Good, C.; Schroyer, K.; Summerfelt, S.T. Effects of Feeding a Fishmeal-Free versus a Fishmeal-Based Diet on Post-Smolt Atlantic Salmon *Salmo salar* Performance, Water Quality, and Waste Production in Recirculation Aquaculture Systems. *Aquac. Eng.* **2016**, *74*, 38–51. [CrossRef]
158. Davidson, J.; Good, C.; Barrows, F.T.; Welsh, C.; Kenney, P.B.; Summerfelt, S.T. Comparing the Effects of Feeding a Grain-or a Fish Meal-Based Diet on Water Quality, Waste Production, and Rainbow Trout *Oncorhynchus mykiss* Performance within Low Exchange Water Recirculating Aquaculture Systems. *Aquac. Eng.* **2013**, *52*, 45–57. [CrossRef]
159. Goddek, S.; Joyce, A.; Kotzen, B.; Burnell, G.M. (Eds.) *Aquaponics Food Production Systems: Combined Aquaculture and Hydroponic Production Technologies for the Future*; Springer Nature: Berlin/Heidelberg, Germany, 2019.
160. Badiola, M.; Basurko, O.C.; Piedrahita, R.; Hundley, P.; Mendiola, D. Energy Use in Recirculating Aquaculture Systems (RAS): A Review. *Aquac. Eng.* **2018**, *81*, 57–70. [CrossRef]
161. Badiola, M.; Basurko, O.C.; Gabiña, G.; Mendiola, D. Integration of Energy Audits in the Life Cycle Assessment Methodology to Improve the Environmental Performance Assessment of Recirculating Aquaculture Systems. *J. Clean. Prod.* **2017**, *157*, 155–166. [CrossRef]
162. Kucuk, H.; Midilli, A.; Özdemir, A.; Çakmak, E.; Dincer, I. Exergetic Performance Analysis of a Recirculating Aquaculture System. *Energy Convers. Manag.* **2010**, *51*, 1033–1043. [CrossRef]
163. Badiola, M.; Cabezas, O.; Curtin, R.; García, M.; Gartzia, I.; Mendiola, D. *Land Based On-Growing of Atlantic Cod (Gadus morhua) to Marketable Size–a Feasibility Study from the Basque Country (Northern Spain)*; AZTI: San Sebastian, Spain, 2014.
164. MacLeod, M.; Hasan, M.R.; Robb, D.H.; Mamun-Ur-Rashid, M. *Quantifying and Mitigating Greenhouse Gas Emissions from Global Aquaculture*; FAO Fisheries and Aquaculture Technical Paper; Food and Agriculture Organization of the United Nations: Rome, Italy, 2019; ISBN 978-92-5-131992-5.
165. Kodama, M. *Overview and History of IMTA, from Ancient to Modern Times*; Aquaculture Department, Southeast Asian Fisheries Development Center: Bangkok, Thailand, 2022; pp. 2–5.
166. Ryther, J.H.; Goldman, J.C.; Gifford, C.E.; Huguenin, J.E.; Wing, A.S.; Clarner, J.P.; Williams, L.D.; Lapointe, B.E. Physical Models of Integrated Waste Recycling- Marine Polyculture Systems. *Aquaculture* **1975**, *5*, 163–177. [CrossRef]
167. Troell, M.; Joyce, A.; Chopin, T.; Neori, A.; Buschmann, A.H.; Fang, J.-G. Ecological Engineering in Aquaculture—Potential for Integrated Multi-Trophic Aquaculture (IMTA) in Marine Offshore Systems. *Aquaculture* **2009**, *297*, 1–9. [CrossRef]
168. Chopin, T.; Yarish, C.; Wilkes, R.; Belyea, E.; Lu, S.; Mathieson, A. Developing Porphyra/Salmon Integrated Aquaculture for Bioremediation and Diversification of the Aquaculture Industry. *J. Appl. Phycol.* **1999**, *11*, 463–472. [CrossRef]
169. Chopin, T. Integrated Multi-Trophic Aquaculture. *N. Aquac.* **2006**, *12*, 4.
170. Dunbar, M.B.; Malta, E.; Brunner, L.; Hughes, A.; Ratcliff, J.; Johnson, M.; Jacquemin, B.; Michel, R.; Cunha, M.; Oliveira, G.; et al. Defining Integrated Multi-Trophic Aquaculture: A Consensus. *Aquac. Eur.* **2020**, *45*, 22–27.
171. United Nations Goal 12: Ensure Sustainable Consumption and Production Patterns. Available online: https://sdgs.un.org/goals/goal12 (accessed on 17 October 2023).

172. Buck, B.H.; Troell, M.F.; Krause, G.; Angel, D.L.; Grote, B.; Chopin, T. State of the Art and Challenges for Offshore Integrated Multi-Trophic Aquaculture (IMTA). *Front. Mar. Sci.* **2018**, *5*, 165. [CrossRef]
173. Abreu, M.H.; Pereira, R.; Yarish, C.; Buschmann, A.H.; Sousa-Pinto, I. IMTA with *Gracilaria vermiculophylla*: Productivity and Nutrient Removal Performance of the Seaweed in a Land-Based Pilot Scale System. *Aquaculture* **2011**, *312*, 77–87. [CrossRef]
174. Shpigel, M. Mariculture Systems, Integrated Land-Based. In *Encyclopedia of Sustainability Science and Technology*; Meyers, R.A., Ed.; Springer: New York, NY, USA, 2012; pp. 6309–6318, ISBN 978-1-4419-0851-3.
175. Shpigel, M.; Shauli, L.; Odintsov, V.; Ashkenazi, N.; Ben-Ezra, D. *Ulva lactuca* Biofilter from a Land-Based Integrated Multi Trophic Aquaculture (IMTA) System as a Sole Food Source for the Tropical Sea Urchin *Tripneustes gratilla elatensis*. *Aquaculture* **2018**, *496*, 221–231. [CrossRef]
176. Pereira, R.; Yarish, C.; Critchley, A.T. Seaweed Aquaculture for Human Foods in Land Based and IMTA Systems. In *Applications of Seaweeds in Food and Nutrition*; Elsevier: Amsterdam, The Netherlands, 2024; pp. 77–99.
177. Guerra-García, J.M.; Martínez-Pita, I.; Šegvić-Bubić, T.; Manchado, M.; Arechavala-Lopez, P.; Calado, R.; Marchio, E.; Gentry, R.; Tlusty, M.F.; Rhyne, A.; et al. Chapter 5—Aquaculture and Conservation. In *Coastal Habitat Conservation*; Espinosa, F., Ed.; Academic Press: Cambridge, MA, USA, 2023; pp. 111–146, ISBN 978-0-323-85613-3.
178. Knowler, D.; Chopin, T.; Martínez-Espiñeira, R.; Neori, A.; Nobre, A.; Noce, A.; Reid, G. The Economics of Integrated Multi-Trophic Aquaculture: Where Are We Now and Where Do We Need to Go? *Rev. Aquac.* **2020**, *12*, 1579–1594. [CrossRef]
179. Nederlof, M.A.J.; Verdegem, M.C.J.; Smaal, A.C.; Jansen, H.M. Nutrient Retention Efficiencies in Integrated Multi-Trophic Aquaculture. *Rev. Aquac.* **2022**, *14*, 1194–1212. [CrossRef]
180. Chang, B.-V.; Liao, C.-S.; Chang, Y.-T.; Chao, W.-L.; Yeh, S.-L.; Kuo, D.-L.; Yang, C.-W. Investigation of a Farm-Scale Multitrophic Recirculating Aquaculture System with the Addition of *Rhodovulum sulfidophilum* for Milkfish (*Chanos chanos*) Coastal Aquaculture. *Sustainability* **2019**, *11*, 1880. [CrossRef]
181. Jerónimo, D.; Lillebø, A.I.; Santos, A.; Cremades, J.; Calado, R. Performance of Polychaete Assisted Sand Filters under Contrasting Nutrient Loads in an Integrated Multi-Trophic Aquaculture (IMTA) System. *Sci. Rep.* **2020**, *10*, 20871. [CrossRef]
182. Ramli, N.M.; Verreth, J.A.J.; Yusoff, F.M.; Nurulhuda, K.; Nagao, N.; Verdegem, M.C.J. Integration of Algae to Improve Nitrogenous Waste Management in Recirculating Aquaculture Systems: A Review. *Front. Bioeng. Biotechnol.* **2020**, *8*, 1004. [CrossRef]
183. Checa, D.; Macey, B.M.; Bolton, J.J.; Brink-Hull, M.; O'Donohoe, P.; Cardozo, A.; Poersch, L.H.; Sánchez, I. Circularity Assessment in Aquaculture: The Case of Integrated Multi-Trophic Aquaculture (IMTA) Systems. *Fishes* **2024**, *9*, 165. [CrossRef]
184. Cunha, M.E.; Quental-Ferreira, H.; Parejo, A.; Gamito, S.; Ribeiro, L.; Moreira, M.; Monteiro, I.; Soares, F.; Pousão-Ferreira, P. Understanding the Individual Role of Fish, Oyster, Phytoplankton and Macroalgae in the Ecology of Integrated Production in Earthen Ponds. *Aquaculture* **2019**, *512*, 734297. [CrossRef]
185. Aubin, J.; Hussenot, J. *Integrated Aquaculture-Agriculture and Agroecology in Aquaculture: Views from Europe-Advancing Integrated Agriculture-Aquaculture through Agroecology*; FAO Workshop: Montpellier, France, 2018.
186. Hossain, A.; Senff, P.; Glaser, M. Lessons for Coastal Applications of IMTA as a Way towards Sustainable Development: A Review. *Appl. Sci.* **2022**, *12*, 11920. [CrossRef]
187. Ahmed, N.; Bunting, S.W.; Glaser, M.; Flaherty, M.S.; Diana, J.S. Can Greening of Aquaculture Sequester Blue Carbon? *Ambio* **2017**, *46*, 468–477. [CrossRef] [PubMed]
188. Song, J.; Wang, Y.; Huang, L.; Peng, Y.; Tan, K.; Tan, K. The Effects of Bivalve Aquaculture on Carbon Storage in the Water Column and Sediment of Aquaculture Areas. *Sci. Total Environ.* **2024**, *937*, 173538. [CrossRef]
189. Hill, R.; Bellgrove, A.; Macreadie, P.I.; Petrou, K.; Beardall, J.; Steven, A.; Ralph, P.J. Can Macroalgae Contribute to Blue Carbon? An Australian Perspective. *Limnol. Oceanogr.* **2015**, *60*, 1689–1706. [CrossRef]
190. Chung, I.K.; Beardall, J.; Mehta, S.; Sahoo, D.; Stojkovic, S. Using Marine Macroalgae for Carbon Sequestration: A Critical Appraisal. *J. Appl. Phycol.* **2011**, *23*, 877–886. [CrossRef]
191. IPCC (Intergovernmental Panel on Climate Change). *Summary for Policymakers: Special Report on the Ocean and Cryosphere in a Changing Climate*; IPCC: Geneva, Switzerland, 2019.
192. Liu, Y.; Zhang, J.; Wu, W.; Zhong, Y.; Li, H.; Wang, X.; Yang, J.; Zhang, Y. Effects of Shellfish and Macro-Algae IMTA in North China on the Environment, Inorganic Carbon System, Organic Carbon System, and Sea–Air CO_2 Fluxes. *Front. Mar. Sci.* **2022**, *9*, 864306. [CrossRef]
193. Holdt, S.L.; Edwards, M.D. Cost-Effective IMTA: A Comparison of the Production Efficiencies of Mussels and Seaweed. *J. Appl. Phycol.* **2014**, *26*, 933–945. [CrossRef]
194. Nobre, A.M.; Robertson-Andersson, D.; Neori, A.; Sankar, K. Ecological–Economic Assessment of Aquaculture Options: Comparison between Abalone Monoculture and Integrated Multi-Trophic Aquaculture of Abalone and Seaweeds. *Aquaculture* **2010**, *306*, 116–126. [CrossRef]
195. Chary, K.; Aubin, J.; Sadoul, B.; Fiandrino, A.; Covès, D.; Callier, M.D. Integrated Multi-Trophic Aquaculture of Red Drum (*Sciaenops Ocellatus*) and Sea Cucumber (*Holothuria scabra*): Assessing Bioremediation and Life-Cycle Impacts. *Aquaculture* **2020**, *516*, 734621. [CrossRef]
196. Chopin, T.; Lively, A.; Wiper, J.; Totten, L. Canadian IMTA Kelps Get Organic Certification and Are Ready to Hit the Marketplace. *Aquac. Eur.* **2014**, *39*, 14–15.

197. Khanjani, M.H.; Zahedi, S.; Mohammadi, A. Integrated Multitrophic Aquaculture (IMTA) as an Environmentally Friendly System for Sustainable Aquaculture: Functionality, Species, and Application of Biofloc Technology (BFT). *Environ. Sci. Pollut. Res.* **2022**, *29*, 67513–67531. [CrossRef] [PubMed]
198. Emerenciano, M.; Gaxiola, G.; Cuzon, G. Biofloc Technology (BFT): A Review for Aquaculture Application and Animal Food Industry. *Biomass Now-Cultiv. Util.* **2013**, 12, 301–328.
199. Samocha, T.M. *Sustainable Biofloc Systems for Marine Shrimp*; Academic Press: Cambridge, MA, USA, 2019; ISBN 0-12-818239-3.
200. Avnimelech, Y. *Biofloc Technology: A Practical Guide Book*; World Aquaculture Society: Baton Rouge, LA, USA, 2009; ISBN 978-1-888807-16-5.
201. Crab, R. Bioflocs Technology: An Integrated System for the Removal of Nutrients and Simultaneous Production of Feed in Aquaculture. Ph.D. Thesis, Ghent University, Ghent, Belgium, 2010.
202. Zimmermann, S.; Kiessling, A.; Zhang, J. The Future of Intensive Tilapia Production and the Circular Bioeconomy without Effluents: Biofloc Technology, Recirculation Aquaculture Systems, Bio-RAS, Partitioned Aquaculture Systems and Integrated Multitrophic Aquaculture. *Rev. Aquac.* **2023**, *15*, 22–31. [CrossRef]
203. Mugwanya, M.; Dawood, M.A.O.; Kimera, F.; Sewilam, H. Biofloc Systems for Sustainable Production of Economically Important Aquatic Species: A Review. *Sustainability* **2021**, *13*, 7255. [CrossRef]
204. Lauderdale, C.V.; Aldrich, H.C.; Lindner, A.S. Isolation and Characterization of a Bacterium Capable of Removing Taste- and Odor-Causing 2-Methylisoborneol from Water. *Water Res.* **2004**, *38*, 4135–4142. [CrossRef]
205. Guttman, L.; van Rijn, J. Isolation of Bacteria Capable of Growth with 2-Methylisoborneol and Geosmin as the Sole Carbon and Energy Sources. *Appl. Environ. Microbiol.* **2012**, *78*, 363–370. [CrossRef] [PubMed]
206. Hargreaves, J.A. *Biofloc Production Systems for Aquaculture*; Southern Regional Aquaculture Center: Stoneville, MS, USA, 2013; Volume 4503.
207. Zhao, D.; Pan, L.; Huang, F.; Wang, C.; Xu, W. Effects of Different Carbon Sources on Bioactive Compound Production of Biofloc, Immune Response, Antioxidant Level, and Growth Performance of *Litopenaeus vannamei* in Zero-water Exchange Culture Tanks. *J. World Aquac. Soc.* **2016**, *47*, 566–576. [CrossRef]
208. Ebeling, J.M.; Timmons, M.B.; Bisogni, J. Engineering Analysis of the Stoichiometry of Photoautotrophic, Autotrophic, and Heterotrophic Removal of Ammonia–Nitrogen in Aquaculture Systems. *Aquaculture* **2006**, *257*, 346–358. [CrossRef]
209. Burford, M.A.; Thompson, P.J.; McIntosh, R.P.; Bauman, R.H.; Pearson, D.C. The Contribution of Flocculated Material to Shrimp (*Litopenaeus vannamei*) Nutrition in a High-Intensity, Zero-Exchange System. *Aquaculture* **2004**, *232*, 525–537. [CrossRef]
210. De Schryver, P.; Crab, R.; Defoirdt, T.; Boon, N.; Verstraete, W. The Basics of Bio-Flocs Technology: The Added Value for Aquaculture. *Aquaculture* **2008**, *277*, 125–137. [CrossRef]
211. Ulloa Walker, D.A.; Morales Suazo, M.C.; Emerenciano, M.G.C. Biofloc Technology: Principles Focused on Potential Species and the Case Study of Chilean River Shrimp Cryphiops Caementarius. *Rev. Aquac.* **2020**, *12*, 1759–1782. [CrossRef]
212. Crab, R.; Defoirdt, T.; Bossier, P.; Verstraete, W. Biofloc Technology in Aquaculture: Beneficial Effects and Future Challenges. *Aquaculture* **2012**, *356*, 351–356. [CrossRef]
213. Crab, R.; Lambert, A.; Defoirdt, T.; Bossier, P.; Verstraete, W. The Application of Bioflocs Technology to Protect Brine Shrimp (*Artemia franciscana*) from Pathogenic Vibrio Harveyi. *J. Appl. Microbiol.* **2010**, *109*, 1643–1649. [CrossRef]
214. El-Sayed, A.M. Use of Biofloc Technology in Shrimp Aquaculture: A Comprehensive Review, with Emphasis on the Last Decade. *Rev. Aquac.* **2021**, *13*, 676–705. [CrossRef]
215. Emerenciano, M.G.; Miranda-Baeza, A.; Martínez-Porchas, M.; Poli, M.A.; do Vieira, F.N. Biofloc Technology (BFT) in Shrimp Farming: Past and Present Shaping the Future. *Front. Mar. Sci.* **2021**, *8*, 813091. [CrossRef]
216. Bossier, P.; Ekasari, J. Biofloc Technology Application in Aquaculture to Support Sustainable Development Goals. *Microb. Biotechnol.* **2017**, *10*, 1012–1016. [CrossRef] [PubMed]
217. Holanda, M.; Ravagnan, E.; Lara, G.; Santana, G.; Furtado, P.; Cardozo, A.; Wasielesky, W., Jr.; Poersch, L.H. Integrated Multitrophic Culture of Shrimp *Litopenaeus vannamei* and Tilapia *Oreochromis niloticus* in Biofloc System: A Pilot Scale Study. *Front. Mar. Sci.* **2023**, *10*, 1060846. [CrossRef]
218. Legarda, E.C.; da Silva, D.; Miranda, C.S.; Pereira, P.K.M.; Martins, M.A.; Machado, C.; de Lorenzo, M.A.; Hayashi, L.; do Nascimento Vieira, F. Sea Lettuce Integrated with Pacific White Shrimp and Mullet Cultivation in Biofloc Impact System Performance and the Sea Lettuce Nutritional Composition. *Aquaculture* **2021**, *534*, 736265. [CrossRef]
219. Carvalho, A.; Costa, L.C.d.O.; Holanda, M.; Poersch, L.H.; Turan, G. Influence of Total Suspended Solids on the Growth of the Sea Lettuce *Ulva lactuca* Integrated with the Pacific White Shrimp *Litopenaeus vannamei* in a Biofloc System. *Fishes* **2023**, *8*, 163. [CrossRef]
220. Poli, M.A.; Legarda, E.C.; de Lorenzo, M.A.; Pinheiro, I.; Martins, M.A.; Seiffert, W.Q.; do Nascimento Vieira, F. Integrated Multitrophic Aquaculture Applied to Shrimp Rearing in a Biofloc System. *Aquaculture* **2019**, *511*, 734274. [CrossRef]
221. Xu, W.-J.; Pan, L.-Q. Enhancement of Immune Response and Antioxidant Status of *Litopenaeus vannamei* Juvenile in Biofloc-Based Culture Tanks Manipulating High C/N Ratio of Feed Input. *Aquaculture* **2013**, *412*, 117–124. [CrossRef]
222. Crab, R.; Kochva, M.; Verstraete, W.; Avnimelech, Y. Bio-Flocs Technology Application in over-Wintering of Tilapia. *Aquac. Eng.* **2009**, *40*, 105–112. [CrossRef]

223. Choudhury, A.; Lepine, C.; Witarsa, F.; Good, C. Anaerobic Digestion Challenges and Resource Recovery Opportunities from Land-Based Aquaculture Waste and Seafood Processing Byproducts: A Review. *Bioresour. Technol.* **2022**, *354*, 127144. [CrossRef] [PubMed]
224. Ogello, E.O.; Outa, N.O.; Obiero, K.O.; Kyule, D.N.; Munguti, J.M. The Prospects of Biofloc Technology (BFT) for Sustainable Aquaculture Development. *Sci. Afr.* **2021**, *14*, e01053. [CrossRef]
225. Manan, H.; Kasan, N.A.; Ikhwanuddin, M.; Kamaruzzan, A.S.; Jalilah, M.; Fauzan, F.; Suloma, A.; Amin-Safwan, A. Biofloc Technology in Improving Shellfish Aquaculture Production—A Review. *Ann. Anim. Sci.* **2023**. [CrossRef]
226. Huang, M.; Zhou, Y.; Tian, H.; Pan, S.; Yang, X.; Gao, Q.; Dong, S. Rapidly Increased Greenhouse Gas Emissions by Pacific White Shrimp Aquacultural Intensification and Potential Solutions for Mitigation in China. *Aquaculture* **2024**, *587*, 740825. [CrossRef]
227. Hu, Z.; Lee, J.W.; Chandran, K.; Kim, S.; Sharma, K.; Brotto, A.C.; Khanal, S.K. Nitrogen Transformations in Intensive Aquaculture System and Its Implication to Climate Change through Nitrous Oxide Emission. *Bioresour. Technol.* **2013**, *130*, 314–320. [CrossRef]
228. Soares, D.C.; Henry-Silva, G.G. Emission and Absorption of Greenhouse Gases Generated from Marine Shrimp Production (*Litopeneaus vannamei*) in High Salinity. *J. Clean. Prod.* **2019**, *218*, 367–376. [CrossRef]
229. Izquierdo, C.; Usero, J.; Gracia, I. Speciation of Heavy Metals in Sediments from Salt Marshes on the Southern Atlantic Coast of Spain. *Mar. Pollut. Bull.* **1997**, *34*, 123–128. [CrossRef]
230. Yúfera, M.; Arias, A.M. *Traditional Polyculture in "Esteros" in the Bay of Cádiz (Spain). Hopes and Expectancies for the Prevalence of a Unique Activity in Europe*; European Aquaculture Society: Stavanger, Norway, 2010.
231. Guerra-García, J.; Calero-Cano, S.; Donázar-Aramendía, I.; Giráldez, I.; Morales, E.; Arechavala-Lopez, P.; Cervera-Currado, J. Farming *Sparus aurata* (Teleostei: Sparidae) in Marsh Ponds: Trophic Characterization and Trace Metal Accumulation. *Mar. Environ. Res.* **2023**, *188*, 106007. [CrossRef] [PubMed]
232. Forrest, B.M.; Keeley, N.B.; Hopkins, G.A.; Webb, S.C.; Clement, D.M. Bivalve Aquaculture in Estuaries: Review and Synthesis of Oyster Cultivation Effects. *Aquaculture* **2009**, *298*, 1–15. [CrossRef]
233. Costa, L.C.D.O.; Xavier, J.A.A.; Neves, L.F.D.M.; De Azambuja, A.M.V.; Junior, W.W.; Figueiredo, M.R.C. Polyculture of *Litopenaeus vannamei* Shrimp and *Mugil platanus* Mullet in Earthen Ponds. *Rev. Bras. de Zootec.* **2013**, *42*, 605–611. [CrossRef]
234. Sipaúba-Tavares, L.; Donadon, A.; Milan, R. Water Quality and Plankton Populations in an Earthen Polyculture Pond. *Braz. J. Biol.* **2011**, *71*, 845–855. [CrossRef]
235. Cunha, M.E.; Quental-Ferreira, H.; Leão, A.C.; Matias, A.; Matias, D.; Joaquim, S.; Soares, F.; Pousão-Ferreira, P. *Integrated Multi-Trophic Aquaculture in Earthen Ponds*; Proceddings of the AQUA 2012; Global Aquaculture: Securing our Future: Prague, Czech Republic, 2012.
236. Jiménez-Prada, P.; Hachero-Cruzado, I.; Giráldez, I.; Fernández-Diaz, C.; Vilas, C.; Cañavate, J.P.; Guerra-García, J.M. Crustacean Amphipods from Marsh Ponds: A Nutritious Feed Resource with Potential for Application in Integrated Multi-Trophic Aquaculture. *PeerJ* **2018**, *6*, e4194. [CrossRef]
237. Miller, D.; Semmens, K. Waste Management in Aquaculture. *West Va. Univ. Ext. Serv. Publ. No. AQ02-1. USA* **2002**, *8*. Available online: https://freshwater-aquaculture.extension.org/wp-content/uploads/2019/08/WasteManagemetninAquaculture.pdf (accessed on 1 June 2024).
238. Mitchell, I.M. In Situ Biodeposition Rates of Pacific Oysters (*Crassostrea gigas*) on a Marine Farm in Southern Tasmania (Australia). *Aquaculture* **2006**, *257*, 194–203. [CrossRef]
239. Jiang, Z.J.; Fang, J.G.; Han, T.T.; Mao, Y.Z.; Li, J.Q.; Du, M.R. The Role of *Gracilaria lemaneiformis* in Eliminating the Dissolved Inorganic Carbon Released from Calcification and Respiration Process of *Chlamys farreri*. *J. Appl. Phycol.* **2014**, *26*, 545–550. [CrossRef]
240. Willer, D.F.; Aldridge, D.C. Sustainable Bivalve Farming Can Deliver Food Security in the Tropics. *Nat. Food* **2020**, *1*, 384–388. [CrossRef]
241. Zhang, Y.; Tang, K.W.; Yang, P.; Yang, H.; Tong, C.; Song, C.; Tan, L.; Zhao, G.; Zhou, X.; Sun, D. Assessing Carbon Greenhouse Gas Emissions from Aquaculture in China Based on Aquaculture System Types, Species, Environmental Conditions and Management Practices. *Agric. Ecosyst. Environ.* **2022**, *338*, 108110. [CrossRef]
242. Terrados, J.; Otero, M.; Bacci, T.; Didderen, K.; La Porta, B.; Teunis, M.; Bouma, T.J. Blue Carbon Ecosystem Restoration. In *Manual for the Creation of Blue Carbon Projects in Europe and the Mediterranean*; International Union for Conservation of Nature and Natural Resources: Gland, Switzerland, 2021.
243. Kosten, S.; Almeida, R.M.; Barbosa, I.; Mendonça, R.; Santos Muzitano, I.; Sobreira Oliveira-Junior, E.; Vroom, R.J.E.; Wang, H.-J.; Barros, N. Better Assessments of Greenhouse Gas Emissions from Global Fish Ponds Needed to Adequately Evaluate Aquaculture Footprint. *Sci. Total Environ.* **2020**, *748*, 141247. [CrossRef]
244. Vasanth, M.; Muralidhar, M.; Saraswathy, R.; Nagavel, A.; Dayal, J.S.; Jayanthi, M.; Lalitha, N.; Kumararaja, P.; Vijayan, K.K. Methodological Approach for the Collection and Simultaneous Estimation of Greenhouse Gases Emission from Aquaculture Ponds. *Env. Monit. Assess.* **2016**, *188*, 671. [CrossRef]
245. Chen, G.; Bai, J.; Bi, C.; Wang, Y.; Cui, B. Global Greenhouse Gas Emissions from Aquaculture: A Bibliometric Analysis. *Agric. Ecosyst. Environ.* **2023**, *348*, 108405. [CrossRef]
246. Maulu, S.; Hasimuna, O.J.; Haambiya, L.H.; Monde, C.; Musuka, C.G.; Makorwa, T.H.; Munganga, B.P.; Phiri, K.J.; Nsekanabo, J.D. Climate Change Effects on Aquaculture Production: Sustainability Implications, Mitigation, and Adaptations. *Front. Sustain. Food Syst.* **2021**, *5*, 609097. [CrossRef]

247. Edwards, P. *Aquaculture, Poverty Impacts and Livelihoods*; ODI: London, UK, 2000.
248. Bostock, J.; McAndrew, B.; Richards, R.; Jauncey, K.; Telfer, T.; Lorenzen, K.; Little, D.; Ross, L.; Handisyde, N.; Gatward, I. Aquaculture: Global Status and Trends. *Philos. Trans. R. Soc. B Biol. Sci.* **2010**, *365*, 2897–2912. [CrossRef]
249. Barrington, K.; Chopin, T.; Robinson, S. Integrated Multi-Trophic Aquaculture (IMTA) in Marine Temperate Waters. *Integr. Maric. Glob. Rev. FAO Fish. Aquac. Tech. Pap.* **2009**, *529*, 7–46.
250. Rosa, J.; Lemos, M.F.L.; Crespo, D.; Nunes, M.; Freitas, A.; Ramos, F.; Pardal, M.Â.; Leston, S. Integrated Multitrophic Aquaculture Systems—Potential Risks for Food Safety. *Trends Food Sci. Technol.* **2020**, *96*, 79–90. [CrossRef]
251. Gunning, D.; Maguire, J.; Burnell, G. The Development of Sustainable Saltwater-Based Food Production Systems: A Review of Established and Novel Concepts. *Water* **2016**, *8*, 598. [CrossRef]
252. Chopin, T.; MacDonald, B.; Robinson, S.; Cross, S.; Pearce, C.; Knowler, D.; Noce, A.; Reid, G.; Cooper, A.; Speare, D.; et al. The Canadian Integrated Multi-Trophic Aquaculture Network (CIMTAN)—A Network for a New Era of Ecosystem Responsible Aquaculture. *Fisheries* **2013**, *38*, 297–308. [CrossRef]
253. Sidey, M. Aquaculture Regulation in Canada: A Case for Modernization, Standardization and Collaboration. Master's Thesis, Memorial University of Newfoundland, St. John's, NL, Canada, 2018.
254. Bueno, P.B. Widening the Horizon of Asian Mariculture with IMTA. *J. Indian Soc. Coast. Agric. Res.* **2021**, *39*, 239–248. [CrossRef]
255. Buschmann, A.H.; Cabello, F.; Young, K.; Carvajal, J.; Varela, D.A.; Henríquez, L. Salmon Aquaculture and Coastal Ecosystem Health in Chile: Analysis of Regulations, Environmental Impacts and Bioremediation Systems. *Ocean Coast. Manag.* **2009**, *52*, 243–249. [CrossRef]
256. Camelo-Guarin, S.; Molinet, C.; Soto, D. Recommendations for Implementing Integrated Multitrophic Aquaculture in Commercial Farms at the Landscape Scale in Southern Chile. *Aquaculture* **2021**, *544*, 737116. [CrossRef]
257. Stenton-Dozey, J.M.E.; Heath, P.; Ren, J.S.; Zamora, L.N. New Zealand Aquaculture Industry: Research, Opportunities and Constraints for Integrative Multitrophic Farming. *N. Z. J. Mar. Freshw. Res.* **2021**, *55*, 265–285. [CrossRef]

Journal of
Marine Science and Engineering

MDPI

Article

Express Diagnosis and Prediction of Remote Mass Mortality of Scallop *Mizuhopecten yessoensis* in Mariculture Farms Using Biomarkers

Valentina Vladimirovna Slobodskova, Nadezhda Vladimirovna Dovzhenko, Sergey Petrovich Kukla, Victor Pavlovich Chelomin and Andrey Alexandrovich Mazur *

V.I. Il'ichev Pacific Oceanological Institute, Far Eastern Branch, Russian Academy of Sciences, 690041 Vladivostok, Russia
* Correspondence: mazur.aa@poi.dvo.ru

Abstract: The cage method for the cultivation of the seaside scallop *Mizuhopecten yessoensis* is the most developed and popular method at sea farms in Primorsky Krai (Sea of Japan). However, this method of mollusk cultivation requires the careful planning of farming activities. Recently, mariculture farms in different countries have often encountered the mass mortality of cultured hydrobionts. The causes of such diseases are not quite clear, and often their identification requires a large amount of time and financial expenditure. Therefore, the use of predictive mechanisms based on biomarkers can help identify hidden threats in cultured scallop organisms that lead to mass mortality. In this study, we propose a rapid diagnostic method for predicting the distant mass mortality of *M. yessoensis* cultured in cages using biomarkers. The assessment of the pathological state of cultured mollusks at earlier developmental stages using the DNA comet method and oxidative stress markers (malondialdehyde) will allow the diagnosis and prediction of significant losses of marketable individuals in marine farms. In this study, we evaluated different age groups of mollusks cultured in the different water areas of Peter the Great Bay (Sea of Japan). During the study, we found that the death of cultured mollusks increased with increasing DNA damage and the active accumulation of malondialdehyde in tissues. It was observed that in scallops aged 1+ cultured in Severnaya Bay, high levels of DNA molecule damage and malondialdehyde were registered in the digestive glands and gills, which subsequently led to the death of almost all marketable individuals aged 3+. Therefore, the work is of significant value in assisting the aquaculture industry in solving the emerging problems of scallop farming and preserving marketable products. The proposed markers effectively reflect the condition of molluscs under extreme conditions caused by various factors, making them highly suitable for monitoring studies and forecasts on aquaculture farms.

Keywords: aquaculture; DNA damage; malondialdehyde; mass mortality; scallop

Citation: Slobodskova, V.V.; Dovzhenko, N.V.; Kukla, S.P.; Chelomin, V.P.; Mazur, A.A. Express Diagnosis and Prediction of Remote Mass Mortality of Scallop *Mizuhopecten yessoensis* in Mariculture Farms Using Biomarkers. *J. Mar. Sci. Eng.* **2024**, *12*, 1151. https://doi.org/10.3390/jmse12071151

Academic Editors: Ana Catarina Matias and Carlos Andrade

Received: 30 May 2024
Revised: 5 July 2024
Accepted: 6 July 2024
Published: 9 July 2024

1. Introduction

Aquaculture is an important and promising sector of global fisheries, contributing to the increase in and conservation of biological resources. According to the United Nations Food and Agriculture Organization (FAO), marine and coastal aquaculture production approached almost 30 million tons (USD 67.4 billion) in 2016, of which bivalves accounted for 16.9 million tons (58.8 percent of total production) [1]. In 2018, global aquaculture production rose to a record 114.5 million tons, amounting to USD 263.6 billion. Total aquaculture production consisted of 82.1 million tons of aquatic animals (USD 250.1 billion), 32.4 million tons of algae (USD 13.3 billion), and 26,000 tons of ornamental shells and pearls (USD 179,000). Shellfish, including bivalves, accounted for 17.7 million tons (USD 34.6 billion). The total value of shellfish was USD 34.6 billion [2]. Despite the impact of the COVID-19 pandemic, global aquaculture produced a record 122.6 million tons in 2020, including 87.5 million tons of aquatic animals valued at USD 264.8 billion [3]. In 2022, the world produced a record 130.9 million tons of aquaculture products, up 6.6 percent from 2020 [4]. Meanwhile, experts have estimated that total fishery and aquaculture production could rise to 201 million tons by 2030 [2].

Recently, many mariculture farms using the method of cage cultivation have begun to face the mass mortality of cultivated hydrobionts [5], which is accompanied by significant financial losses. Mass mortality has been noted in different species of cultured bivalves (e.g., oysters, scallops, and mussels), and is often associated with exposure to parasites, rickettsia-like organisms [6], viruses, bacteria [7,8], and others. There are also other prerequisites affecting the survival of hydrobionts, such as environmental factors, excessive planting density, and inbreeding [6]. In addition, we cannot rule out the negative impact on the condition of cultivated mollusks from the use of various plastic devices, such as cages, buoys, etc. It is known that a large amount of plastic waste is generated in the running of mariculture farms due to the degradation of plastic devices in the aggressive marine environment [9]. When plastic products degrade, various toxic substances can be released into the water, which in turn can cause various negative changes in the bodies of cultivated aquatic organisms [10]. In this regard, the further sustainable development of mariculture farms requires a thorough study of the effects of the industrial cultivation of hydrobionts on the environment, as well as the continuous assessment of the physiological state of the cultured hydrobionts, allowing the early detection of pathological changes, which in turn can lead to the mass death of cultured mollusks [7].

The coastal strip of Primorsky Krai is one of the most promising regions of Russia for mariculture development due to its good climatic conditions and availability of considerable water areas, especially in the southern part of the region, which is suitable for hydrobiont cultivation. More than 50 operating mariculture farms are currently engaged in the cultivation of the most valuable species of mollusks, echinoderms, and algae [11]. In recent decades, the seaside scallop *Mizuhopecten yessoensis* (Jay, 1857) has been one of the traditional cultivated marine bivalves in Primorsky Krai. However, in recent years, some farms have begun to encounter the mass mortality of mollusks during cultivation.

The aim of this study was to use biomarkers as prognostic signals and compare them with the mortality of *M. yessoensis* from different water areas. To achieve this aim, indicators of the levels of damage to DNA molecules and malondialdehyde in the tissues of the gills and digestive glands of cultured scallops were determined. This approach is a promising, modern, diagnostic and prognostic tool for detecting hidden pathological shifts in any organism [12]. The obtained biomarker indicators were compared with data on the mortality of scallops from the studied aquafarms.

2. Materials and Methods

2.1. Site of Bivalve Collection and Materials

A comparative study was carried out on scallops grown in seafood farms located in different waters of Peter the Great Bay (Sea of Japan) in Severnaya Bay (Slavyansky Bay), and in the waters of Vostochnaya Bay (Rikord Island) (Figure 1). These water areas have different hydrological regimes, for example, Severnaya Bay is relatively shallow and characterized by weak water exchange, while the farm in Vostochnaya Bay on Rikord Island is located in a deeper water area with intense water exchange. The mollusks were selected in the post-spawning period (November) at ages 1+, 2+, and 3+ from the waters of Rikord Island, and 1+ and 2+ from Severnaya Bay, since after the summer period a mass death of mollusks aged 3+ was registered in this water area.

For the study, scallops were removed from their cages and dissected on ice, and the gills and digestive glands were isolated. The determination of the level of DNA molecule damage using the comet assay was carried out immediately after tissue extraction from mollusks in specialized premises located at the coastal base of Severnaya Bay and Vostochnaya Bay. Residues of tissues intended for the determination of malondialdehyde were immediately frozen in liquid nitrogen (Dewar vessel SDS-6M (JSC "NPO Geliymash", Moscow, Russia)) and transported to the laboratory for further processing.

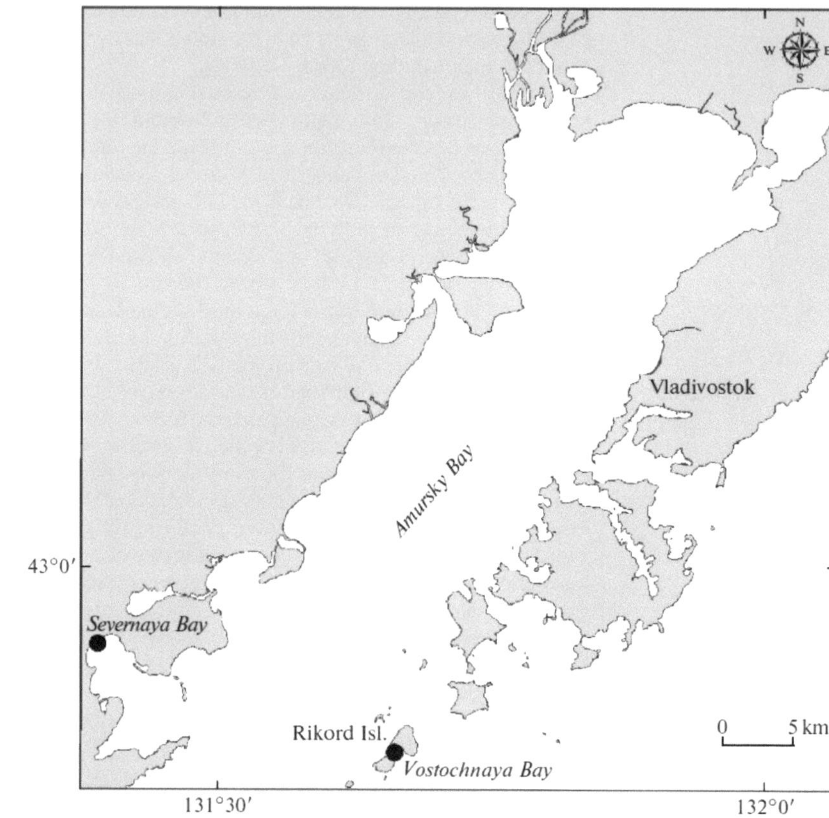

Figure 1. The areas of Peter the Great Bay, Sea of Japan, where the scallops *Mizuhopecten yessoensis* were collected.

2.2. Determination of the Damage to DNA Molecules

To determine the extent of the damage in the DNA molecules, we used an alkaline variant of the comet assay [11,13] successfully adapted to marine organisms [14]. For this purpose, individual cells were isolated using an isotonic solution (500 mM NaCl, 12.5 mM KCl, 5 mM EDTA-Na2, and 20 mM Tris-HCl; pH 7.4). The working concentration was 105 cells/mL. Then, 50 μL of cell suspension was added to 100 μL of 1% low-melting-point agarose (MP Biomedicals, Eschwege, Germany) in 0.04 M phosphate buffer (pH 7.4) at 37 °C, mixed thoroughly, applied to a slide previously coated with 1% agarose solution for better adhesion, and covered with a coverslip. The sample was placed in the refrigerator for 3 min for agarose curing. The coverslip was carefully removed, and the slide was submerged in a lysis solution (2.5 M NaCl; 0.1 M EDTA-Na (1% Triton X-100 and 10% DMSO); 0.02 M Tris; pH 10) for 1 h in the dark at 4 °C. After washing with cold distilled water, the slides were placed in electrophoresis buffer (300 mM NaOH and 1 mM EDTA-Na2) and incubated for 40 min. Electrophoresis was performed at 2 V/cm for 15 min. After neutralization (0.4 M Tris-HCl; pH 7.4), the slides were stained with SYBR Green I. The visualization and registration of DNA comets was performed using a scanning fluorescence microscope (Zeiss, Oberkochen, Germany, AxioImager A1) equipped with an AxioCam MRc digital camera. For digital image processing, we used the computer program CASP software v 1.2.2. (CASPLab, Wroclaw, Poland; https://casplab.com, accessed on 24 April 2022), which allows for the calculation of different parameters of comets, indicating the degree of cellular DNA damage. For each comet, the fraction of DNA in the comet tail

(DNAt) was calculated. In the scallop groups studied, 15 gel slides (1 slide = 1 individual) containing at least 50 comets in each were analyzed.

2.3. Determination of Malondialdehyde (MDA) Content

The content of MDA, a product of the oxidative degradation of fatty acids, was determined in tissues and subcellular fractions by color reaction with 2-thiobarbituric acid (TBA) [15]. The tissues were homogenized in a homogenizer (SILENTCRUSHER S; Heidolph Instruments GmbH & Co., Schwabach, Germany) in 0.1 M phosphate buffer (pH = 7.5). To prevent lipid peroxidation during the determination of MDA, an alcoholic solution of butylhydroxytoluene (Merck KGaA, Darmstadt, Germany. CAS-no 128-37-0) was added to the samples to a final concentration of 5 mM. The content of malondialdehyde was determined by color reaction with 2-thiobarbituric acid (TBA, Merck KGaA, Darmstadt, Germany. CAS-no 504-17-6).

Then, 30% trichloroacetic acid (AppliChem GmbH, Darmstadt, Germany. CAS-no 76-03-9) and 0.75% TBA solution were added sequentially to the tissue homogenate. The mixture was thoroughly mixed and heated for 20 min in a water bath (Memmert WNB 7, Memmert GmbH + Co., Schwabach, Germany) at a temperature of 95 °C. After cooling, sediments were separated from the samples by centrifugation at 3000 rpm for 20 min. The measurements were carried out at a wavelength of 580 nm and 532 nm; then, the difference in the readings of the optical density was found. To calculate the MDA content, the molar extinction coefficient was used: 1.56 × 105/cm/M. The relative content of MDA was expressed in nmol per g fresh wet weight of the tissue. The measurements were carried out on a Shimadzu UV-2550 spectrophotometer (Scinteck Instruments LLC, Manassas, VA, USA).

2.4. Statistics

Statistical processing of the results was performed using the application software package STATISTICA 6.0 and Microsoft Excel 2016. The results were evaluated for each experiment by comparing group mean values ($p < 0.01$ using Mann–Whitney test). Correlation coefficients between MDA and % DNA damage were estimated using Pearson regression.

3. Results

The results of our study showed that one-year-old (1+) and two-year-old (2+) scallops sampled from Severnaya Bay had significantly higher DNA molecule degradation in the gills and digestive glands than those from Vostochnaya Bay (Figure 2).

The level of DNA molecule damage in mollusks cultured in the area of Vostochnaya Bay in both the gills and digestive glands in all investigated age groups was practically at the same level and did not exceed 20%. At the same time, no significant differences between different age groups of mollusks from this water area were revealed. At the same time, in *M. yessoensis* from Severnaya Bay, the levels of gill and digestive gland DNA damage also did not differ significantly among different-aged individuals. Significant differences were observed in one- and two-year-old individuals of generations from different water areas. Thus, in the gills and digestive glands of scallops from Severnaya Bay, the level of DNA molecule damage was almost three times higher than in scallops from Vostochnaya Bay in both age groups (1+, 2+). The content of MDA in scallop tissue (Figure 3) had a similar pattern to the level of DNA damage. In *M. yessoensis* from Severnaya Bay, this index was significantly higher, almost three times higher in the gills and more than five times higher in the digestive glands than in scallops from the waters of Vostochnaya Bay (Figure 3).

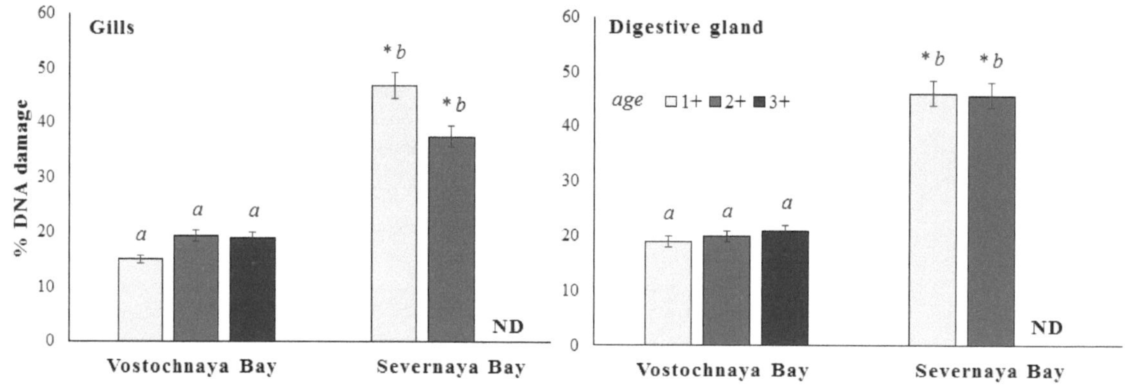

Figure 2. Degree of DNA damage in the gills and digestive glands of *M. yessoensis* cultured in Severnaya Bay and Vostochnaya Bay (* significant differences between scallops from different water areas, $p = 99\%$; *a*, *b*—marked significant differences between different ages of scallops from the same water area, $p < 0.01$. All differences were assessed using the Mann–Whitney test). "ND"—no data.

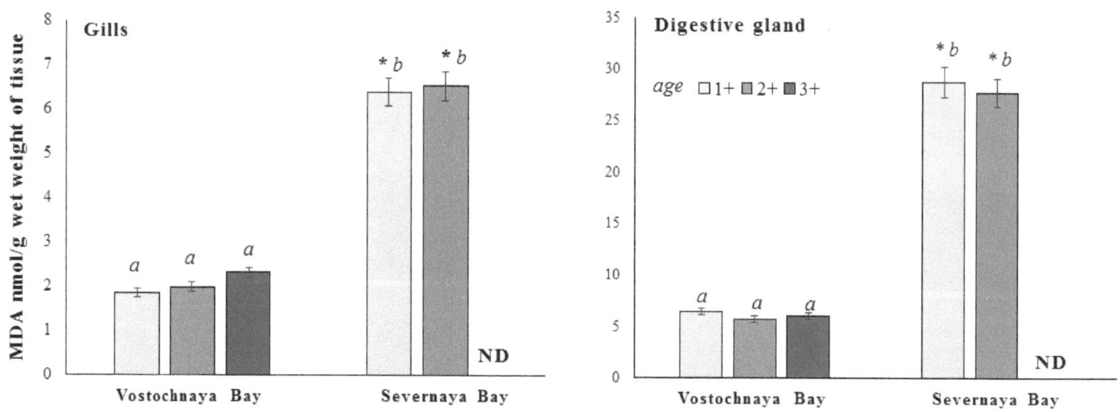

Figure 3. Levels of MDA in the gills and digestive glands of *M. yessoensis* cultured in Severnaya Bay and Vostochnaya Bay (* significant differences between scallops from different water areas, $p = 99\%$; a, b—marked significant differences between different ages of scallops from the same water area, $p < 0.01$. All differences were assessed using the Mann–Whitney test). "ND"—no data.

Thus, in the Severnaya Bay generation, the MDA concentration in the gills was within 2–2.5 nmol/g raw weight, and in the digestive gland was within 5–7 nmol/g. As can be seen from the results obtained, there were no significant differences between age groups within one generation in each water area. At the same time, significant differences were observed between the same-age groups of mollusks from different water areas. Also, there was a significant positive correlation between the level of MDA in tissues and % DNA damage in the tissues studied in two different generations of scallops ($p < 0.01$) (Vostochnaya Bay—DNA/MDA ratio: gills—$R^2 = 0.4943$; digestive gland—$R^2 = 0.2504$. Severnaya Bay—DNA/MDA ratio: gills—$R^2 = 1$; digestive gland—$R^2 = 1$) (Figure S1).

4. Discussion

We would like to emphasize in particular that, in scallops cultured in Severnaya Bay, we recorded a mortality rate of scallops aged 2+ of more than 50%, and a 95% rate at the age of 3+ years. At the same time, in the mollusks in the farm located in the waters of

Vostochnaya Bay, mass mortality was not observed: mortality was less than 20% in all age groups (Table S1). Based on the results obtained, we can conclude that the damage to the DNA molecules of the gills and digestive gland cells of *M. yessoensis* at values over 40% (Figure 2), and high levels of MDA (Figure 3) in tissues, are signals of irreversible effects occurring in the body, resulting in the mass death of hydrobionts. It is known that MDA is a widely used indicator of the development of oxidative stress, and the content of high concentrations of MDA in tissues (Figure 3) can itself lead to DNA molecule damage and mutations [16]. Most likely, during periods of mass mortality, mollusks are in a state of oxidative stress, which is confirmed by a significant increase in MDA content in the tissues of mollusks, with DNA molecule damage reaching critical values of almost 50%. Based on the results (Figures 1 and 2), it can be assumed that the level of damage to DNA molecules is 40% or higher in the gills and digestive glands of scallops aged 2+ from Severnaya Bay; moreover, the concentration of MDA in the gills is higher than 6 nmol/g wet weight, and above 25 nmol/g wet weight in the digestive gland, indicating irreversible processes in the bodies of scallops that lead to mass death, as evidenced by the lack of data for this species at the age of 3+.

Various researchers have noted that the peak mortality in scallop farms corresponds to the late stage of spawning [6,8]. It is believed that the stress associated with spawning is the main factor of shellfish mortality, due to the weakening of the main protective systems of the body's immune, antioxidant, and reparative systems, and it is during this period that the body is prone to various infections and invasions [6]. As another prerequisite for the high mortality of hydrobionts, some authors refer to excessive planting density, which contributes to the development of infectious processes, as well as inbreeding, which weakens the protective physiological potential of cultured mollusks [5–8]. In addition, plastic hydraulic devices (e.g., cages, ropes, and buoys), which are widely used in aquaculture, can pose a serious risk to cultivated aquatic organisms [9,10]. Plastic is a complex mixture in which chemical additives are loosely bound and can be washed into the environment [9,17,18]. Also, in the aquatic environment, especially the marine environment, which is characterized by highly dynamic physical and hydrochemical factors, the initial processes of degradation of plastic products occur, accelerating the leaching of toxic chemicals into the environment [19–21]. In addition to various diseases and chemical pollutants, the physiological parameters (including mortality) of cultured scallops are directly influenced by the environmental variables (temperature, salinity, oxygen, etc.) of a particular water area. Some researchers note that various types of polymers that interact with marine organisms, as a rule, do not have a noticeable effect on survival, but cause various sublethal effects at the molecular biochemical level associated with the generation of reactive oxygen species (ROS) and the development of oxidative stress processes, which in turn, and in combination with other acting factors, can be prerequisites for the mass mortality of aquatic organisms [22–24]. Based on the above examples and our results, we can suggest that, in mariculture farms where there is a high mortality rate, cultivated aquatic organisms are in a state of oxidative stress, which can be caused by the combined effects of negative factors. As is known, oxidative stress results in a weakening of the body's antioxidant, repair, and immune systems, which in turn entails a cascade of biochemical changes that can lead to irreversible pathological consequences.

The most important manifestation of the negative impact of environmental factors on cultivated aquatic organisms at the molecular level is genotoxicity, which manifests as the accumulation of damage to the DNA molecule. According to the generally accepted perspective, accumulating damage in the cell genome can trigger a cascade of biochemical changes leading to cell apoptosis [24–27].

Our results confirm these conclusions, since it is the failure of the DNA molecule repair system that leads to the accumulation of damage in its structure, and the weakening of antioxidant protection contributes to the accumulation of lipid peroxidation products in tissues.

5. Conclusions

The analysis of the experimental data indicated that mollusks cultured in Severnaya Bay experienced pronounced oxidative stress, leading to increased MDA content in tissues and significant DNA damage. We believe that these biomarkers are early and sensitive indicators of the physiological state of cultured mollusks, as well as signals of the development of pathologies leading to the mass mortality of hydrobionts. Thus, when the level of DNA molecule damage in scallops aged 1+ is around 40–50%, we can foresee upcoming economic losses due to the mass mortality of hydrobionts in the next year. Of course, to make such predictions with confidence, more data on genome damage and MDA levels at different and earlier age stages of shellfish development must be collected. The results obtained in this work can be used to undertake new and promising research on predicting sustainability using biomarkers, not only for mariculture farms, but also for natural ecosystems in marine areas that are used for the commercial reproduction of hydrobionts.

Supplementary Materials: The following supporting information can be downloaded at: https://www.mdpi.com/article/10.3390/jmse12071151/s1, Figure S1: Correlation between the level of DNA damage and MDA concentration in scallop tissues between groups of different ages in each water area. A—gills, B — digestive gland; Table S1: Data on mortality of the studied groups of scallops were provided by employees of aquafarms.

Author Contributions: Conceptualization, V.V.S. and V.P.C.; Methodology, N.V.D., S.P.K. and A.A.M.; Software, S.P.K.; Formal analysis, N.V.D. and A.A.M.; Data curation, V.V.S. and S.P.K.; Writing—original draft, V.V.S.; Writing—review & editing, V.P.C.; Visualization, N.V.D. and A.A.M.; Funding acquisition, V.P.C. All authors have read and agreed to the published version of the manuscript.

Funding: The research was supported by the State assignment for research work of V.I. Il'ichev Pacific Oceanological Institute, FEB RAS (no. 124022100077-0).

Institutional Review Board Statement: Not applicable.

Informed Consent Statement: Not applicable.

Data Availability Statement: The original contributions presented in the study are included in the article/Supplementary Materials, further inquiries can be directed to the corresponding author.

Conflicts of Interest: The authors declare that the research was conducted in the absence of any commercial or financial relationships that could be construed as a potential conflict of interest.

References

1. FAO. The State of World Fisheries and Aquaculture 2018. In *Meeting the Sustainable Development Goals*; Licence: CC BY-NC-SA 3.0 IGO; FAO: Rome, Italy, 2018.
2. FAO. The State of World Fisheries and Aquaculture 2020. In *Sustainability in Action*; FAO: Rome, Italy, 2020. [CrossRef]
3. FAO. The State of World Fisheries and Aquaculture 2022. In *Towards Blue Transformation*; FAO: Rome, Italy, 2022. [CrossRef]
4. FAO. The State of World Fisheries and Aquaculture 2024. In *Blue Transformation in Action*; FAO: Rome, Italy, 2024. [CrossRef]
5. Xiao, J.; Ford, S.E.; Yang, H.; Zhang, G.; Zhang, F.; Guo, X. Studies on mass summer mortality of cultured zhikong scallops (*Chlamys farreri* Jones et Preston) in China. *Aquaculture* **2005**, *250*, 602–615. [CrossRef]
6. Lan, Y.; Ye, T.; Xue, Y.; Liu, H.; Zhang, H.; Cheng, D.; Zhao, M.; Zhang, Y.; Li, S.; Ma, H.; et al. Phyiological and immunological responses to mass mortality in noble scallop *Chlamys nobilis* cultured in Nan'ao waters of Shantou, China. *Fish Shellfish Immunol.* **2018**, *82*, 453–459. [CrossRef] [PubMed]
7. Read, P.A.; Fernandes, T.F. Management of environmental impacts of marine aquaculture in Europe. *Aquaculture* **2003**, *226*, 139–163. [CrossRef]
8. Zhang, X.; Zhu, M.; Li, R.; Wang, Z.; Xia, B.; Zhang, L. Density-dependent mortality of the scallop *Chlamys farreri* in grow-out cuture. *Aquac Res.* **2006**, *37*, 842–844. [CrossRef]
9. Hingant, M.; Mallarino, S.; Conforto, E.; Dubillot, E.; Barbier, P.; Bringer, A.; Thomas, H. Artificial weathering of plastics used in oyster farming. *Sci. Total Environ.* **2023**, *868*, 161638. [CrossRef] [PubMed]
10. Mohsen, M.; Lin, C.; Tu, C.; Zhang, C.; Xu, S.; Yang, H. Association of heavy metals with plastics used in aquaculture. *Mar. Poll. Bull.* **2022**, *174*, 113312. [CrossRef] [PubMed]
11. Koval', I.V.; Ovchinnikova, I.A. Sostoyanie i tendencii razvitiya akvakul'tury v Primorskom krae. *Vestnik TGU* **2013**, *1*, 36–47.
12. Collins, A.R. Investigating oxidative DNA damage and its pepair using the comet assay. *Mutat. Res.* **2009**, *681*, 24–32. [CrossRef] [PubMed]

13. Singh, N.P.; McCoy, M.T.; Tice, R.R.; Schneider, E.L. A simple technique for quantitation of low levels of DNA damage in individual cells. *Exp. Cell Res.* **1988**, *175*, 184–191. [CrossRef]
14. Chelomin, V.P.; Mazur, A.A.; Slobodskova, V.V.; Kukla, S.P.; Dovzhenko, N.V. Genotoxic properties of polystyrene (PS) microspheres in the filter-feeder mollusk Mytilus trossulus (Gould, 1850). *J. Mar. Sci. Eng.* **2022**, *10*, 273. [CrossRef]
15. Buege, J.A.; Aust, S.D. Microsomal lipid peroxidation. *Methods Enzymol.* **1978**, *52*, 302–310. [CrossRef] [PubMed]
16. Cooke, M.S.; Evans, M.D.; Dizdaroglu, M.; Lunec, J. Oxidative DNA damage: Mechanisms, mutation, and disease. *FASEB J.* **2003**, *17*, 1195–1214. [CrossRef]
17. Kwon, B.G.; Amamiya, K.; Sato, H.; Chung, S.Y.; Kodera, Y.; Kim, S.K.; Lee, E.J.; Saido, K. Monitoring of styrene oligomers as indicators of polystyrene plastic pollution in the North-West Pacific Ocean. *Chemosphere* **2017**, *180*, 500–505. [CrossRef]
18. Lee, S.; Alam, M.B.; Lee, S.; Jung, M.; Shim, W.J.; Kim, S. Identification and quantification of photodegradation products of disposed expanded polystyrene buoy used in aquaculture. *Mar. Poll. Bull.* **2023**, *192*, 114998. [CrossRef]
19. Rani, M.; Shim, W.J.; Jang, M.; Han, G.M.; Hong, S.H. Releasing of hexabromocyclododecanes from expanded polystyrenes in seawater -field and laboratory experiments. *Chemosphere* **2017**, *185*, 798–805. [CrossRef]
20. Song, Y.K.; Hong, S.H.; Eo, S.; Han, G.M.; Joon, S.W. Rapid Production of Micro- and Nanoplastics by Fragmentation of Expanded Polystyrene Exposed to Sunlight. *Environ. Sci. Technol.* **2020**, *54*, 11191–11200. [CrossRef]
21. Mattsson, K.; Björkroth, F.; Karlsson, T.; Hassellöv, M. Nanofragmentation of Expanded Polystyrene under Simulated Environmental Weathering (Thermooxidative Degradation and Hydrodynamic Turbulence). *Front. Mar. Sci.* **2021**, *7*, 578178. [CrossRef]
22. Dvoretsky, A.; Dvoretsky, V. Biological Aspects, Fisheries, and Aquaculture of Yesso Scallops in Russian Waters of the Sea of Japan. *Diversity* **2022**, *14*, 399. [CrossRef]
23. Gambardella, C.; Morgana, S.; Bramini, M.; Rotini, A.; Manfra, L.; Migliore, L.; Piazza, V.; Garaventa, F.; Faimali, M. Ecotoxicological effects of polystyrene microbeads in a battery of marine organisms belonging to different trophic levels. *Mar. Environ. Res.* **2018**, *141*, 313–321. [CrossRef] [PubMed]
24. Prokic, M.D.; Radovanovic, T.B.; Gavric, J.P.; Faggio, C. Ecotoxicological effects of microplastics: Examination of biomarkers, current state and future perspectives. *Anal. Chem.* **2019**, *111*, 37–46. [CrossRef]
25. Dovzhenko, N.V.; Chelomin, V.P.; Mazur, A.A.; Kukla, S.P.; Slobodskova, V.V.; Istomina, A.A.; Zhukovskaya, A.F. Oxidative Stress in Far Eastern Mussel *Mytilus trossulus* (Gould, 1850) Exposed to Combined Polystyrene Microspheres (μPSs) and CuO-Nanoparticles (CuO-NPs). *J. Mar. Sci. Eng.* **2022**, *10*, 707. [CrossRef]
26. Zacchi, F.L.; Flores-Nunes, F.; Mattos, J.J.; Lima, D.; Lüchmann, K.H.; Sasaki, S.T.; Bícego, M.C.; Taniguchi, S.; Montone, R.C.; Almeida, E.A.; et al. Biochemical and molecular responses in oysters *Crassostrea brasiliana* collected from estuarine aquaculture areas in Southern Brazil. *Mar. Poll. Bull.* **2018**, *135*, 110–118. [CrossRef] [PubMed]
27. Jeong, C.B.; Won, E.J.; Kang, H.M.; Lee, M.C.; Hwang, D.S.; Hwang, U.K.; Zhou Souissi, S.; Lee, S.J.; Lee, J.S. Microplastic size-dependent toxicity, oxidative stress induction, and p-JNK and p-p38 activation in the Monogonont rotifer (*Brachionus koreanus*). *Environ. Sci. Technol.* **2016**, *50*, 8849–8857. [CrossRef]

Journal of *Marine Science and Engineering*

MDPI

Brief Report

Offshore Wind Farms in South Korea: A Potential Site for Scallop Culture

Dae-Won Lee [1,†], Sung-Yong Oh [1,†], Jordan Jun Chul Park [2], Yun-Hwan Jung [1], Han-Jun Kim [1], Dong Mun Choi [1], Young-Ung Choi [1] and Jeonghoon Han [1,*]

1 Marine Biotechnology & Bioresource Research Department, Korea Institute of Ocean Science & Technology (KIOST), Busan 49111, Republic of Korea; dw0130@kiost.ac.kr (D.-W.L.); syoh@kiost.ac.kr (S.-Y.O.); aries0324@kiost.ac.kr (Y.-H.J.); hanjunkim@kiost.ac.kr (H.-J.K.); dmchoi@kiost.ac.kr (D.M.C.); yuchoi@kiost.ac.kr (Y.-U.C.)

2 Département des Sciences, Université Sainte-Anne, Church Point, NS B0W 1M0, Canada; jordan.park@usainteanne.ca

* Correspondence: jeonghoon@kiost.ac.kr

† These authors contributed equally to this work.

Abstract: Owing to the constant wind generated by the vast ocean, energy production from offshore wind farms (OWFs) plays an important role in the expansion of renewable energy. However, areas close to large wind farms are often left unutilized, and aquaculture farmers find it difficult to efficiently utilize these unoccupied spaces due to limited information showing the feasibility of utilization of OWFs as potential scallop culture sites. To analyze whether the two scallop species Zhikong scallop (*Chlamys farreri*) and bay scallop (*Argopecten irradians*) can be grown at OWFs of Gochang and Buan, Jeollabuk-do, Republic of Korea, the growth characteristics of the two scallop species were analyzed and compared with those grown at the Tongyeong Megacosm Test Station. The results clearly showed that the growth of scallops at the OWF was significantly lower with respect to the shell lengths, height, width, and weight, compared to those grown at the megacosm station. However, scallops grown at the OWF still showed consistent growth in parallel with those grown at the megacosm test station. Yet, there was a species-specific mortality rate between the two sites. In addition, our results suggest that temperature may be a key determinant of the growth of *C. farreri* and *A. irradians*. Overall, this study contributes to establishing a foundation for the stable and continuous farming of marine bivalves (e.g., clams, oysters, mussels, and scallops) in OWF areas.

Keywords: Zhikong scallop; *Chlamys farreri*; bay scallop; *Argopecten irradians*; growth characteristics; longline aquaculture facility temperature

Citation: Lee, D.-W.; Oh, S.-Y.; Park, J.J.C.; Jung, Y.-H.; Kim, H.-J.; Choi, D.M.; Choi, Y.-U.; Han, J. Offshore Wind Farms in South Korea: A Potential Site for Scallop Culture. *J. Mar. Sci. Eng.* **2023**, *11*, 1988. https://doi.org/10.3390/jmse11101988

Academic Editors: Ana Catarina Matias and Carlos Andrade

Received: 20 September 2023
Revised: 13 October 2023
Accepted: 13 October 2023
Published: 14 October 2023

1. Introduction

Offshore wind farms (OWFs) produce sufficient energy for use as a renewable energy source; thus, they play a pivotal role in the expansion of renewable energy. In general, OWFs allow energy producers to secure large farm sites and mass-produce high-density energy using abiotic factors, including the speed and direction of sea wind [1,2]. In fact, OWFs produce energy much more efficiently and stably than onshore wind farms [3–5]. However, the development of OWFs remains challenging, largely because of their incompatibility with marine ecosystems and social acceptability among local fisheries. In addition, OWFs built in the marine environment will increase the stress on existing ecosystems that have previously been used for other purposes, such as for fisheries or shipping, or those that are yet free of human activity [6,7]. Accordingly, combining OWFs and aquaculture can be a powerful means of increasing the spatial efficiency of sea usage and supporting the livelihoods of fishers. Through this innovative approach, the scope of aquaculture activities can be expanded to enhance the efficiency of spatial utilization in marine environments [8,9]. OWFs will almost certainly gain social acceptance. Over the past few years, there have been improvements in the social acceptance of establishing OWFs, as shown in some European countries, which have generated evidence from trials to develop business models for aquaculture, tourism, fishing, and leisure. Likewise, studies on offshore aquaculture systems in the United States have identified the potential of commercial aquaculture at offshore oil fields [10]. In addition, the co-location of offshore wind power facilities with fish and shellfish farming off the coast of the United Kingdom since 2012 has been shown to positively impact the total catches and, subsequently, the socioeconomic status of fishermen [11]. Overall, OWFs reportedly bring both positive and negative impacts to the ecosystem; however, with respect to fisheries, OWFs have brought mostly positive impacts, including supporting fish communities by increasing the nursery area of key species and by the fish gathering effect to prevent fish movement toward other locations [12–14]. However, long-term impacts of anthropogenic activities including OWF establishment on the marine biodiversity still remain largely unknown [15].

The largest OWFs have been built in Gochang and Buan, Jeollabuk-do, Republic of Korea. This offshore wind power industry convergence facility project in the Southwest Sea of the Republic of Korea has been anticipated to have positive effects, such as improving fishing and increasing the income of fishermen, as well as demonstrating that the co-location of sea farm projects and OWFs is economically feasible [11]. However, limited biological evidence has shown that OWFs in this region have a positive impact on the shellfish community.

The Zhikong scallop (*Chlamys farreri*) is an epifaunal and subtidal filter-feeder species that is widely distributed across the entire coast of the Republic of Korea, northern China, and Japan [16]. This particular species is also commercially valuable and mainly inhabits pebbles and gravel at water depths of up to 10 m [17,18]. Bay scallops (*Argopecten irradians*) are distributed along the coastal areas of China, Taiwan, Korea, and Japan, where they mainly live on sandy and gravel-bottomed floors. It is another commercially important species [19]. The commercial size of *C. farreri* and *A. irradians* is as follows: lengths: 7–10 cm and weight: 50–110 g, with median growth of 2–3 cm for *C. farreri*; and lengths: 6–8 cm and weight: 40–60 g for *A. irradians* [20]. Recently, scallop farming has become an active industry worldwide; as larger quantities of scallops are consumed, the production increases accordingly [21]. In the Republic of Korea, most scallops sold on the market have been developed using aquaculture technology, to the extent that most scallops, including the Zhikong and bay scallops, are farm-raised. As such, wind-farm areas create new opportunities to develop aquaculture for shellfish farming, including scallops. However, the possibility of shellfish aquaculture, including scallops, has not been studied in OWF areas.

Here, to see whether there are differences in the growth rate of the two scallop species cultured at the OWF and megacosm station, we investigated and compared the growth characteristics (e.g., shell length, height, width, total weight, and body weight) of the *C. farreri* and *A. irradians* in the OWF area and Tongyeong Megacosm Test Station. The results obtained in this study suggest the possibility of scallop farming on OWFs.

2. Materials and Methods

2.1. Sampling of Scallops

C. farreri were purchased from Taegyeong Fisheries (Goseong-gun, Gyeongsangnam-do, Republic of Korea) in December 2022, and *A. irradians* were purchased from the same company in September 2021. The purchased scallops were kept at the Tongyeong Megacosm Test Station (Korea Institute of Ocean Science and Technology, Tongyeong-si, Gyeongsangnam-do, South Korea), and a floating shellfish farm was installed in the OWF area (Gochang-gun and Buan-gun in North Jeolla Province, Republic of Korea) at a depth of 4 m (Figure 1).

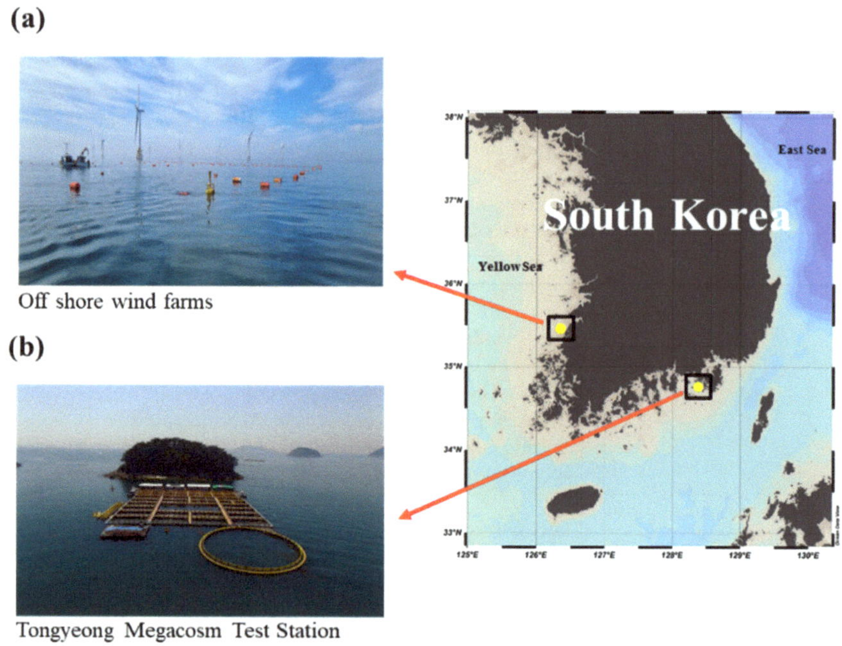

Figure 1. Map showing (**a**) the offshore wind farm (Gochang-gun and Buan-gun in North Jeolla Province, Republic of Korea) and (**b**) Tongyeong Megacosm Test Station (Korea Institute of Ocean Science and Technology, Tongyeong-si, Gyeongsangnam-do, Republic of Korea).

2.2. Deployment of the Scallop Lantern Cages in the Longline Aquaculture Facility

On 9 December 2021, 160,000 *C. farreri* (shell height: 41.63 ± 3.73 mm, total weight 8.80 ± 2.58 g) were transferred into 400 scallop lantern nets and deployed on the longline aquaculture facility within the OWF area. Further, on September 2021, 90,000 *A. irradians* (shell height 40.14 ± 2.65 mm and shell weight 10.34 ± 1.85 g) were transferred into 300 scallop lantern nets and transplanted on the longline aquaculture facility within the OWF area.

For transplantation, they were transported by vehicle to Gochang-gun, Republic of Korea, and further transported to the OWF area using two vessels. They were nurtured in the OWF's longline aquaculture facility along two 200 m long lines, and the spacing of each scallop lantern net was maintained at 100 cm. Four hundred 10-section scallop lantern nets with a diameter of 40 cm were used, with 40 shellfish per section. For proper immersion, 3 kg weights were attached to the bottoms of the scallop lantern nets (Figure 2).

2.3. Sampling and Measurement

The growth characteristics of the *C. farreri* and *A. irradians* were studied from December 2021 to August 2022 and from September to November 2021, respectively. Comparative studies were conducted simultaneously with more than 30 individuals once a month at the Tongyeong Megacosm Test Station and OWF area. The monthly growth of scallops was surveyed by measuring the shell length (mm), height (mm), width (mm), total weight (g), and body weight (g) for each species. We used a Vernier caliper (Mitutoyo, Absolute Digimatic Caliper 500–153-30, Kawasaki, Japan) to measure the lengths and a weighing scale (AND, FX-3000, Seoul, South Korea) to measure the weight of the samples. During the experimental period, cumulative mortality was measured by counting the number of dead scallops (Table 1).

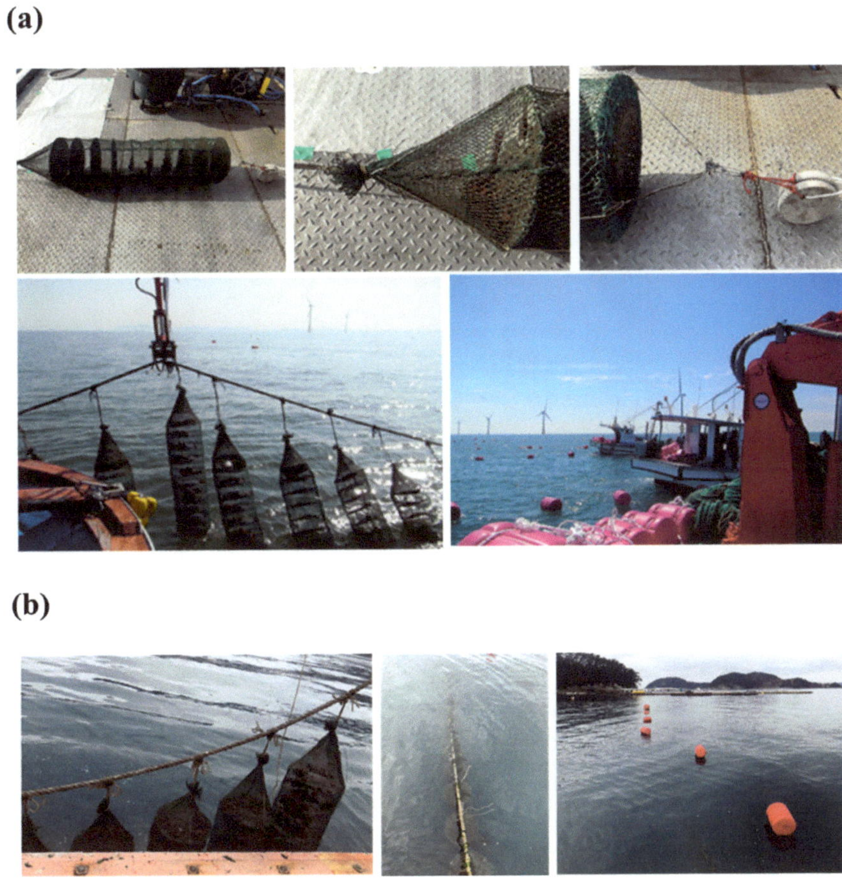

Figure 2. Deployment of the scallop lantern cages at (**a**) offshore wind farm and (**b**) Tongyeong Megacosm Test Station.

2.4. Temperature Variation of the Study Site

Surface seawater temperature data (from September 2021 to August 2022) were obtained from offshore buoys installed in Tongyeong Yeonhwa-do (34.6672 N, 128.3847 E) and Buan Wi-do (35.6584 N, 126.2610 E), which are adjacent to Tongyeong Megacosm Test Station and the OWF area (data provided by the Real-time Information System for Aquaculture environment from National Institute of Fisheries Science, South Korea) (Figure 3).

2.5. Statistical Analysis

The collected data were analyzed using one-way ANOVA followed by Scheffé's post hoc test to evaluate the effect of months in the same region. The cut-off for statistical significance was set at $p < 0.05$. An independent-sample t-test was performed to evaluate the differences in the growth characteristics of the two scallops in the two different regions for each month. Levene's test was used to assess the normal distribution and homogeneity of variance among the samples. Values are expressed as the mean ± standard deviation (SD). All data were analyzed using SPSS (version 27.0; IBM, Armonk, NY, USA).

Table 1. Cumulative mortality of the two scallops at the two different culture sites during the experimental period.

	Chalamys farreri					
	Offshore Wind Farm			Tongyeong Megacosm Test Station		
Month	**Total Individuals**	**# of Dead**	**Mortality (%)**	**Total Individuals**	**# of Dead**	**Mortality (%)**
12	33	0	0	33	0	0
2	35	0	0	35	0	0
3	35	0	0	41	3	7.32
4	32	0	0	37	1	2.7
5	40	1	2.5	42	13	30.95
6	40	1	2.5	35	10	28.57
7	35	0	0	35	11	31.43
8	40	0	0	48	17	35.42
Cumulative	290	2	0.69	306	55	17.97
	Argopecten irradians					
	Offshore Wind Farm			Tongyeong Megacosm Test Station		
Month	**Total individuals**	**# of dead**	**Mortality (%)**	**Total individuals**	**# of dead**	**Mortality (%)**
9	28		0	28	0	0
10	59	23	38.98	32	1	3.13
11	40	11	27.5	30	1	3.33
Cumulative	127	34	26.77	90	2	2.22

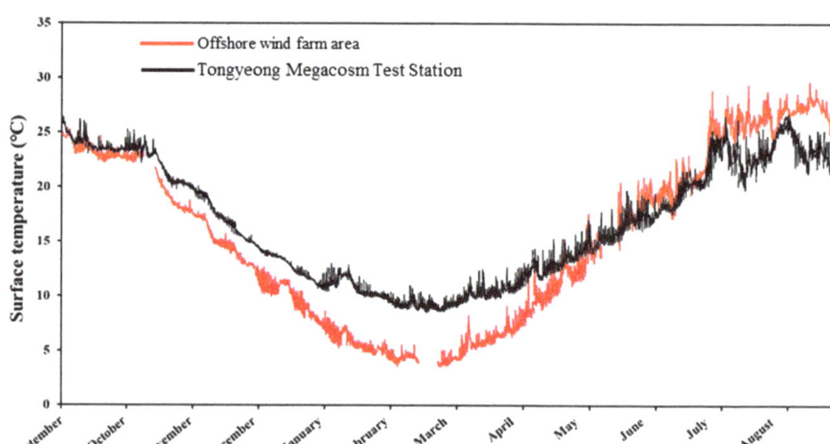

Figure 3. Monthly change in surface seawater temperatures at the sampling sites from September 2021 to August 2022.

3. Results and Discussion

In scallop aquaculture in general, scallops undergo an intermediate rearing process to reduce the costs arising from size selection of juvenile scallops and the consequent labor burden [22,23]. Early juvenile scallops typically range from 2 to 3 mm in size, which makes it difficult to accommodate them in the scallop lantern net; further, they lack resistance to changes in the environment (e.g., water temperature, buoyancy, dissolved oxygen). Therefore, it is essential to use scallops that have undergone an intermediate rearing process to produce healthy scallops and improve their survival rate. In addition, they must be at least 2–3 cm for controlled air exposure and long-distance transportation [20].

The comparative mortality analysis between the two scallop species showed species-specific mortality (Table 1). For *C. farreri*, a higher mortality was observed for those grown at Tonhyeong Megacosm Station (17.97%) compared to those grown at the OWF (0.69%) during December~August. For *A. irradians* cultured during September~November, a higher mortality was observed in those cultured at the OWF (26.77%) compared to the Tongyeong Megacosm Test Station (2.22%). Due to the short time frame and differences in sampling period, it is challenging and difficult to conclude that one species performs better than the other at a specific culture location, yet it could be suggested that a species-specific tolerance and optimum temperature range could contribute to the survival of the two scallops.

Previous studies have suggested that mortality and growth rates depend on differences in water depth, density, and temperature [24]. For example, increased mortality and poor growth rates in some scallops were observed as the water depth increased [25,26], whereas in other scallop species, such as *Plactopecten magellanicus*, the growth rate increased along with water depth [27]. In addition, scallops in longline aquaculture reportedly have the best growth rate at a water depth of 4 m [28]. Therefore, we investigated the growth rates of scallops that underwent an intermediate rearing process in the floating shellfish farm installed in the OWF area and the Tongyeong Megacosm Test Station at a depth of 4 m each.

In this study, the growth of *C. farreri* raised from a longline culture facility in the OWF area was compared with those raised at the Tongyeong Megacosm Test Station from December 2021 to August 2022 (Figure 4). Based on the specific growth rate (SGR) equation derived from the shell length and wet weight, *C. farreri* showed an SGRl of 1.46 and SGRw of 1.19 at the OWF over the 9 month period [29].

$SGR = \ln((Lf) - \ln(Li))/t \times 100$ and $SGRw = \ln((\ln Wf) - \ln(Wi))/t \times 100$, where:

Lf = final average shell length;
Wf = wet weight at the end of the experiment;
Li = initial average shell length;
Wi = wet weight at the beginning of the experiment;
ln = natural logarithm;
t = number of days of the experimental time.

Specifically, from December 2021 to August 2022, the shell length of *C. farreri* cultured in the longline culture facility at the OWF area increased from 36.21 ± 3.38 mm to 55.24 ± 4.60 mm, the shell height increased from 41.63 ± 3.73 mm to 59.66 ± 5.58 mm, the shell width increased from 11.93 ± 1.69 mm to 20.06 ± 1.81 mm, the total weight increased from 8.80 ± 2.58 g to 26.99 ± 5.20 g, and the body weight increased from 4.26 ± 1.35 g to 12.34 ± 2.80 g. Overall, *C. farreri* cultivated in the OWF area exhibited a 1.4-times-greater shell height and 4.4-times-higher total weight, after eight months (December 2021~August 2022) of cultivation. Comparatively, those Zhikong scallops (*C. farreri*) reared at the Tongyeong Megacosm Test Station during the same period showed an SGRl and SGW of 1.51 and 1.35 over the same period of those cultured at the OWF, respectively. Specifically, *C. farreri* showed an increase in the shell length from 36.21 ± 3.38 mm to 63.05 ± 5.10 mm, the shell height increased from 41.63 ± 3.73 mm to 66.45 ± 5.12 mm, the shell width increased from 11.93 ± 1.69 mm to 22.49 ± 2.55 mm, the total weight increased from 8.80 ± 2.58 g to 40.75 ± 9. 35 g, and the body weight increased from 4.26 ± 1.35 g to 15.26 ± 3.45 g. Eight months of experimentation at the Tongyeong Megacosm Test Station indicated that the growth determinants of *C. farreri* increased 1.6- and 4.6-fold for shell height and total weight, respectively. Overall, the growth rate of *C. farreri* reared at the OWF area and Tongyeong Megacosm Test Station showed that the shell height was approximately 10 mm and the total weight was approximately 17 g higher at the Tongyeong Megacosm Test Station compared to those reared in the OWF area.

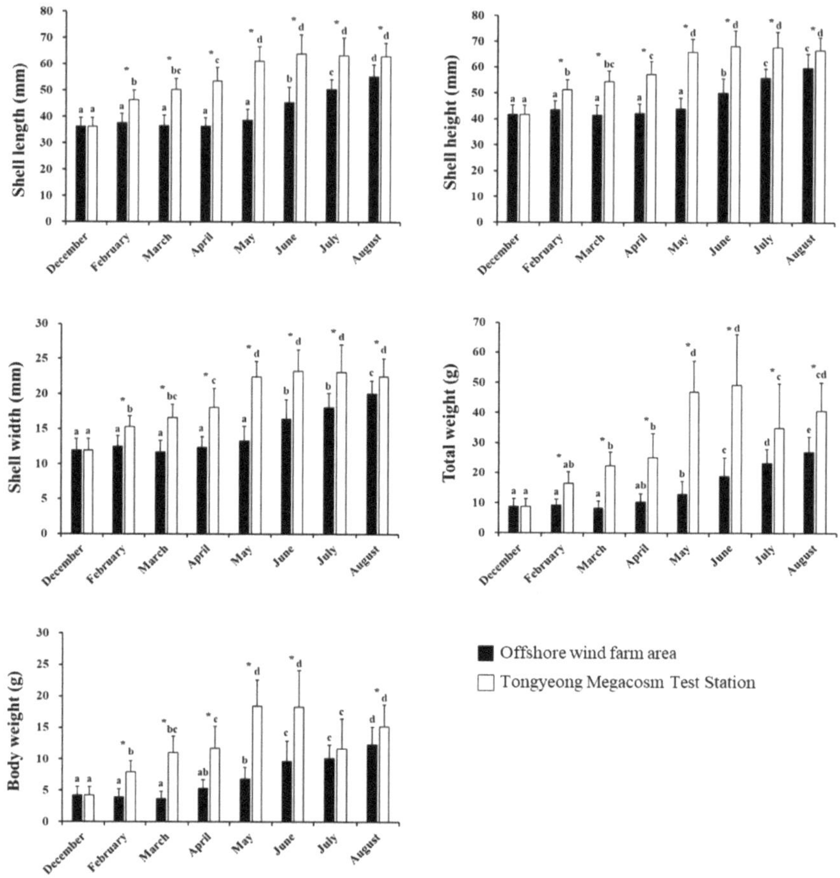

Figure 4. Bar plots depicting the comparisons of changes in size (shell length, height, width, total weight, and body weight) of *Chlamys farreri* from month to month. The letters indicate a significant difference for different months in the same region ($p < 0.05$). The symbol "*" indicates significant differences by region in the same month ($p < 0.05$). All values represent mean ± standard deviation (SD).

Generally, the water temperature for the growth of *C. farreri* ranges between 15 and 22 °C (optimal water temperature: 20 °C), with a maximum water temperature of 29 °C and minimum water temperature of 0 °C [20]. Indeed, the growth, maturation, survival, distribution and feeding, energy utilization, and metabolic activities of shellfish are known to be greatly influenced by the surrounding water temperatures [30,31]. For example, in commercial scallop *Pecten fumatus* (Reeve), the growth rate and survival of larvae significantly decrease in response to increasing temperature [32]. In the sea scallop *Placopecten magellanicus*, survival significantly increases in response to high temperatures [22,23]. Our experiments at the two different culture sites showed interesting results. In the OWF area, *C. farreri* showed a slower growth rate from December to May, but the growth rate slowly increased from June and continued significantly through August. In contrast, at the Tongyeong Megacosm Test Station, the growth rate of *C. farreri* was faster than that in the OWF area during the same months, with no significant growth changes from May to August. As previously mentioned, the growth rate is partially influenced by temperature. At the time of the growth experiment, the water temperature at the OWF area steadily decreased from December to March by up to 5 °C, and then gradually increased from June to August, reaching approximately the same temperature as the water temperature at the Tongyeong Megacosm Test Station (Figure 3).

Recently, scallop production has declined sharply due to widespread mass mortality during the summer [33,34]. Here, no mass mortality was observed in 2022, since the water temperature did not exceed 26 °C at the most; however, in the year 2021, a high-temperature period (>28 °C) resulted in the mass mortality of experimental scallops. Therefore, it was evident that the water temperature plays a crucial role in determining the growth and health of the scallops. However, the two locations naturally have different temperature ranges (i.e., the OWF area is in the west and typically has a lower temperature than the Tongyeong Macromosm station situated in the warm southern part of the Republic of Korea). Although the growth rate was not higher than that at the Tongyeong Megacosm Test Station, the results clearly show the possibility of utilizing the OWF area to cultivate *C. farreri*.

Generally, the optimum water temperature for the growth of *A. irradians* is 18–28 °C (optimal water temperature is 23 °C). The upper limit of the water temperature is 31 °C and the lower limit is 1 °C, with a growth pause below 5 °C. *A. irradians* take less than 10 months to grow into harvestable products of 5–7 cm in height. Their lifespan is approximately 12–16 months and does not exceed 24 months [20].

A. irradians was analyzed for growth differences at two sites (Tongyeong Megacosm Test Station and the longline culture facility in the OWF area) from September to November 2021 (Figure 5). Overall, *A. irradians* showed an SGRl and SGRw of 4.08 and 2.90, respectively, at the OWF over the 3 month period. Data collected from September to November 2021 at the longline culture facility within the OWF area showed increases in the shell length of the bay scallops from 34.20 ± 4.68 mm to 42.89 ± 4.90 mm, the height increased from 32.56 ± 4.49 mm to 41. 09 ± 4.84 mm, the width increased from 12.90 ± 1.78 mm to 18.18 ± 2.37 mm, the total weight increased from 6.61 ± 2.14 g to 15.51 ± 4.79 g, and the body weight increased from 2.86 ± 1.02 g to 7.00 ± 2.27 g. Three months of experimentation at the longline culture facility within the OWF area during the same period indicated that the growth determinants of the bay scallops increased 1.3- and 2.4-fold for shell length and total weight, respectively. Comparatively, those bay scallops reared at the Tongyeong Megacosm Test Station during the same period showed an SGRl and SGRw of 4.37 and 3.52. Specifically, they showed an increased shell length from 34.20 ± 4.68 mm to 54.39 ± 5.98 mm, the height increased from 32.56 ± 4.49 mm to 49.75 ± 5.42 mm, the width increased from 12.90 ± 1.78 mm to 22.63 ± 3.16 mm, the total weight increased from 6.61 ± 2.14 g to 25.65 ± 6.95 g, and the body weight increased from 2.86 ± 1.02 g to 10.21 ± 2.90 g. Three months of experimentation at the Tongyeong Megacosm Test Station indicated that the growth determinants of bay scallops increased 1.5- and 3.9-fold for shell height and total weight, respectively. Overall, the growth rate of bay scallops reared at the OWF area and the Tongyeong Megacosm Test Station showed a shell height of about 0.9 cm, with a total weight that was about 10 g higher at the Tongyeong Megacosm Test Station compared to those reared at the OWF area. Based on the preliminary experiment, the growth rates of *C. farreri* and *A. irradians* were higher at the Tongyeong Megacosm Test Station compared to the OWF area. The growth rate of *C. farreri* was found to be highest in summer (July to August) in the OWF area, whereas the growth rate was lowest during the same period at the Tongyeong Megacosm Test Station. Based on this short-term experiment of 8 months, we could see that temperature is indeed one of the key factors affecting the growth of *C. farreri*. Despite the limitation of this being a short-term experiment, the results clearly suggest the possible utilization of OWF areas as potential scallop culture locations.

In conclusion, based on the results obtained from 2021 and 2022, scallop farming in OWF areas is feasible. The culture of scallops in wind farm areas is not profitable but it can produce some protein that can supply protein to local and national markets and also may lead to an increase in the biodiversity of the marine ecosystem. However, more information on aquaculture should be obtained in parallel with the monitoring of environmental conditions. Regardless, this could be a step forward toward utilizing OWF areas for aquaculture in countries with limited fishing areas. Additionally, owing to poor accessibility and environmental conditions compared to the coast, further development of the facilities and management are required to establish stable aquaculture conditions for OWFs.

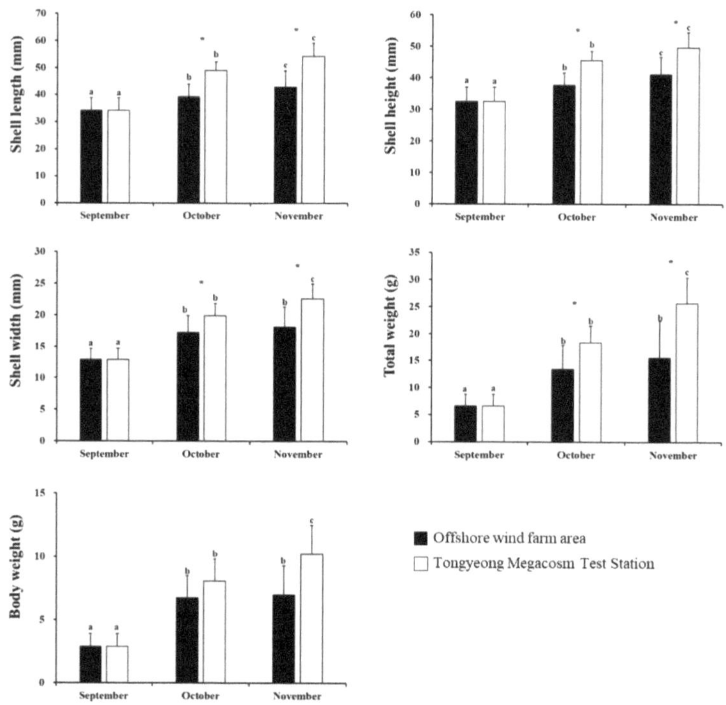

Figure 5. Bar plots depicting the comparisons of changes in size (shell length, height, width, total weight, and body weight) of *Argopecten irradians* from month to month. The letters indicate a significant difference for different months in the same region ($p < 0.05$). The symbol "*" indicates significant differences by region in the same month ($p < 0.05$). All values represent mean ± standard deviation (SD).

Author Contributions: Conceptualization, data curation, formal analysis, and writing—original draft, J.H. and J.J.C.P.; investigation, D.-W.L., Y.-H.J., H.-J.K. and D.M.C.; project administration and funding acquisition, S.-Y.O. and Y.-U.C. All authors have read and agreed to the published version of the manuscript.

Funding: This work was supported by the Korea Institute of Energy Technology Evaluation and Planning (KETEP), and the Ministry of Trade, Industry & Energy (MOTIE) of the Republic of Korea (No. 20203030020080 and No. 20203040020130).

Institutional Review Board Statement: The animal study protocol was approved by the Institutional Animal Care and Experimental Committee of the Korea Institute of Ocean Science and Technology (KIOST).

Informed Consent Statement: Not applicable.

Data Availability Statement: All data generated or analyzed during this study are available from the data repository of KIOST. Requests for material should be made to the corresponding author.

Acknowledgments: The authors thank the Korea Institute of Energy Technology Evaluation and Planning (KETEP), Ministry of Trade, Industry & Energy (MOTIE) of the Republic of Korea (No.

20203030020080 and No.20203040020130). Finally, we thank the editor and the anonymous reviewers whose comments have greatly improved the manuscript.

Conflicts of Interest: The authors declare no conflict of interest.

References

1. Kyong, N.H.; Yoon, J.E.; Jang, M.S.; Jang, D.S. An assessment of offshore wind energy resources around Korean Peninsula. *J. Korean Sol Energy Soc.* **2003**, *23*, 35–41.
2. Park, J.; Kim, B. An analysis of South Korea's energy transition policy with regards to offshore wind power development. *Renew. Sustain. Energy Rev.* **2019**, *109*, 71–84. [CrossRef]
3. Hevia-Koch, P.; Klinge Jacobsen, H.K. Comparing offshore and onshore wind development considering acceptance costs. *Energy Policy* **2019**, *125*, 9–19. [CrossRef]
4. Chen, J.; Kim, M.-H. Review of recent offshore wind turbine research and optimization methodologies in their design. *J. Mar. Sci. Eng.* **2022**, *10*, 28. [CrossRef]
5. Olczak, P.; Surma, T. Energy productivity potential of offshore wind in Poland and cooperation with onshore Wind Farm. *Appl. Sci.* **2023**, *13*, 4258. [CrossRef]
6. Rehfeldt, K.; Paschedag, U.; Bömer, J. (Eds.) *Offshore Wind Power Deployment in Germany*; Federal Ministry for the Environment; Nature Conservation and Nuclear Safety (BMU); Offshore Wind Energy Foundation: Bonn, Germany, 2007; p. 31.
7. Chaji, M.; Werner, S. Economic impacts of offshore wind farms on fishing industries: Perspectives, methods, and knowledge gaps. *Mar. Coast. Fish.* **2023**, *15*, e10237. [CrossRef]
8. Buck, B.H.; Krause, G.; Michler-Cieluch, T.; Brenner, M.; Buchholz, C.M.; Busch, J.A.; Fisch, R.; Geisen, M.; Zielinski, O. Meeting the quest for spatial efficiency: Progress and prospects of extensive aquaculture within offshore wind farms. *Helgol. Mar. Res.* **2008**, *62*, 269–281. [CrossRef]
9. Huang, C.T.; Afero, F.; Hung, C.W.; Chen, B.Y.; Nan, F.H.; Chiang, W.S.; Tang, H.J.; Kang, C.K. Economic feasibility assessment of cage aquaculture in offshore wind power generation areas in Changhua County, Taiwan. *Aquaculture* **2022**, *548*, 737611. [CrossRef]
10. Kang, K.-S.; Jeon, I.-S.; Kwak, J.-Y. Possibilities and orientation toward co-existence of offshore wind farms. *J. Wind Energy* **2016**, *7*, 5–13. [CrossRef]
11. Choi, E.J.; Kim, H.W.; Kim, J.H. An economical feasibility analysis of sea farm project using co-location in offshore wind farms. *Product. Res.* **2015**, *33*, 73–80. [CrossRef]
12. Lindeboom, H.J.; Kouwenhoven, H.J.; Bergman, M.J.N.; Bouma, S.; Brasseur, S.; Daan, R.; Fijn, R.C.; De Haan, D.; Dirksen, S.; Van Hal, R.; et al. Short term ecological effects of an offshore wind farm in the Dutch coastal zone; a compliation. *Environ. Res. Lett.* **2011**, *6*, 03510. [CrossRef]
13. Reubens, J.T.; Degraer, S.; Vincx, M. The ecology of benthopelagic fishes at offshore wind farms: A synthesis of 4 years of research. *Hydrobiologia* **2014**, *727*, 121–136. [CrossRef]
14. Stenberg, C.; Støttrup, J.G.; Van Deurs, M.; Berg, C.W.; Dinesen, G.E.; Mosegaard, H.; Grome, T.M.; Leonhard, S.B. Long-term effects of an offshore wind farm in the North Sea on fish communities. *Mar. Ecol. Prog. Ser.* **2015**, *528*, 257–265. [CrossRef]
15. Li, C.; Coolen, J.W.P.; Scherer, L.; Mogollón, J.M.; Braeckman, U.; Vanaverbeke, J.; Tukker, A.; Steubing, B. Offshore Wind Energy and Marine Biodiversity in the North Sea: Life Cycle Impact Assessment for Benthic Communities. *Environ. Sci. Technol.* **2023**, *57*, 6455–6464. [CrossRef]
16. Zhou, Y.; Yang, H.; Zhang, T.; Liu, S.; Zhang, S.; Liu, Q.; Xiang, J.; Zhang, F. Influence of filtering and biodeposition by the cultured scallop *Chlamys farreri* on benthic-pelagic coupling in a eutrophic bay in China. *Mar. Ecol. Prog. Ser.* **2006**, *317*, 127–141. [CrossRef]
17. Park, G.J.; Yoon, S.P.; Park, Y.J.; Song, H.I. Effect of stocking density on growth and survival rate of the scallop, *Chlamys farreri* (Jones & Preston, 1904) cultured in hanging culture in the west coast of Korea. *Korean J. Malacol.* **2012**, *28*, 1–6. [CrossRef]
18. Yoon, J.M. Genetic distances of scallop (*Chlamys farreri*) populations investigated by PCR procedure. *Dev. Reprod.* **2017**, *21*, 435–440. [CrossRef]
19. Fay, C.W.; Neves, R.J.; Pardue, G.B. Species Profiles. Life Histories and Environmental Requirements of Coastal Fishes and Invertebrates (Mid-Atlantic): Bay scallop. In *Virginia Polytechnic Institute and State University Blacksburg Department of Fisheries and Wildlife Science, FWS/OBS-82/11.12*; EL-82-4; U.S. Army Corps of Engineers: Washington, DC, USA, 1983; p. 17.
20. NIFS (National Institute of Fisheries Science). *Technical Manual for the Scallops Aquaculture*; NIFS (National Institute of Fisheries Science): Busan, Republic of Korea, 2019; pp. 18–23.
21. FAO. Sustainability in action (Rome). In *The State of World Fisheries and Aquaculture 2020*; FAO: Rome, Italy, 2020. [CrossRef]
22. Coleman, S.; Cleaver, C.; Morse, D.; Brady, D.C.; Kiffney, T. The coupled effects of stocking density and temperature on Sea Scallop (*Placopecten magellanicus*) growth in suspended culture. *Aquacult. Rep.* **2021**, *20*, 100684. [CrossRef]
23. Coleman, S.; Morse, D.; Christian Brayden, W.; Brady, D.C. Developing a bioeconomic framework for scallop culture optimization and product development. *Aquac. Econ. Manag.* **2021**, *27*, 25–49. [CrossRef]
24. Coleman, S.; Kiffney, T.; Tanaka, K.R.; Morse, D.; Brady, D.C. Meta-analysis of growth and mortality rates of net cultured sea scallops across the northwest Atlantic. *Aquaculture* **2022**, *546*, 737392. [CrossRef]
25. Cano, J.; Campos, M.J.; Román, G. Growth and mortality of the king scallop grown in suspended culture in Malaga, Southern Spain. *Aquacult. Int.* **2000**, *8*, 207–225. [CrossRef]

26. Yiğitkurt, S. Growth and survival performance of smooth scallop (*Flexopecten glaber* Linnaeus, 1758) at different depths in the Aegean Sea. *Mar. Sci. Technol. Bull.* **2021**, *10*, 278–285. [CrossRef]
27. Claereboudt, M.R.; Burreau, D.; Coté, J.; Himmelman, J.H. Fouling development and its effect on the growth of juveniles giant scallop (*Placopecten magallanicus*) in suspended culture. *Aquaculture* **1994**, *121*, 324–342. [CrossRef]
28. Sun, J.; Lin, C.; Li, P.; Jin, Y.; Zhou, L. The culture experiment of Scallop of *Chlamys farreri* in Nanji Islands. *J. Zhejiang Coll. Fish.* **1997**, *16*, 247–255.
29. Prato, E.; Biandolino, F.; Parlapiano, I.; Papa, L.; Denti, G.; Fanelli, G. Estimation of growth of parameters of the Black Scallop *Mimachlamys Varia* in the Gulf of Taranto (Ionian Sea, Southern Italy). *Water* **2020**, *12*, 3342. [CrossRef]
30. Mills, D. Combined effects of temperature and algal concentration on survival, growth and feeding physiology of *Pinctata maxima* (Jameson) spat. *J. Shellfish Res.* **2000**, *19*, 159–166.
31. Nan, X.; Wei, H.; Zhang, H.; Nie, H. Factors influencing the interannual variation in biomass of bottom-cultured Yesso scallop (*Patinopecten yessoensis*) in the Changhai Sea Area, China. *Front. Mar. Sci.* **2022**, *8*, 798359. [CrossRef]
32. Heasman, M.P.; O'Connor, W.A.; Frazer, A.W.J. Ontogenetic changes in optimal rearing temperatures for the commercial scallop, *Pecten fumatus* Reeve. *J. Shellfish Res.* **1996**, *15*, 627–634.
33. Xiao, J.; Ford, S.E.; Yang, H.S.; Zhang, G.F.; Zhang, F.S.; Guo, X.M. Studies on mass summer mortality of cultured Zhikong scallops (*Chlamys farreri* Jones et Preston) in China. *Aquaculture* **2005**, *250*, 602–615. [CrossRef]
34. Han, J.C.; Jo, Q.; Park, Y.C.; Park, T.G.; Lee, D.C.; Cho, K.C. A report on the mass summer mortalities of the farmed Pacific oysters, *Crassostrea gigas* and Bay scallops *Argopecten irradians* in the local waters of Goseong Bay, Korea. *Korean J. Malacol.* **2013**, *29*, 239–244. [CrossRef]

Journal of
*Marine Science
and Engineering*

MDPI

Article

Physiological Response of European Sea Bass (*Dicentrarchus labrax*) Juveniles to an Acute Stress Challenge: The Impact of Partial and Total Dietary Fishmeal Replacement by an Insect Meal

Ana Basto [1,*], Diogo Peixoto [1,2], Marina Machado [1], Benjamin Costas [1,2], Daniel Murta [3,4] and Luisa M. P. Valente [1,2]

1 CIIMAR/CIMAR-LA, Centro Interdisciplinar de Investigação Marinha e Ambiental, Universidade do Porto, 4450-208 Matosinhos, Portugal; bcostas@ciimar.up.pt
2 ICBAS, Instituto de Ciências Biomédicas Abel Salazar, Universidade do Porto, 4050-313 Porto, Portugal
3 Thunder Foods Lda., Ingredient Odisssey S.A., Rua Comendador José Júlio Eloy, Lote 7-A, 2005-332 Santarém, Portugal
4 Centro de Investigação Interdisciplinar Egas Moniz (CiiEM), Escola Superior de Saúde Egas Moniz, 2829-511 Monte de Caparica, Portugal
* Correspondence: ana.basto@ciimar.up.pt

Abstract: This study aimed to explore the effect of FM substitution by defatted *Tenebrio molitor* larvae meal (dTM) on the response of European seabass to an acute stress challenge. An FM-based diet was used as a control and two other isoproteic/isoenergetic diets were formulated to replace 50 and 100% of FM by dTM. Each diet was tested in quadruplicate groups of 15 fish (69 ± 5 g) fed until visual satiety for 16 weeks. After the feeding trial, fish were subjected to 1 min air exposure followed by 1 h of recovery before sampling. The haematological profile, plasma metabolites, and humoral immunity biomarkers, as well as hepatic oxidative stress and antioxidant capacity, were analysed. A clear response to acute stress was observed by a significant increase in haemoglobin, haematocrit, red blood cells, and almost all evaluated plasma metabolites and humoral parameters, regardless of dietary treatment. The obtained results demonstrated that partial substitution of FM by IM did not affect the stress response of seabass. However, total FM replacement increased the hepatic activity of total peroxidase and superoxide dismutase in fish fed TM100.

Keywords: aquaculture; aquafeed; sustainable protein sources; *Tenebrio molitor*

Citation: Basto, A.; Peixoto, D.; Machado, M.; Costas, B.; Murta, D.; Valente, L.M.P. Physiological Response of European Sea Bass (*Dicentrarchus labrax*) Juveniles to an Acute Stress Challenge: The Impact of Partial and Total Dietary Fishmeal Replacement by an Insect Meal. *J. Mar. Sci. Eng.* **2024**, *12*, 815. https://doi.org/10.3390/jmse12050815

Academic Editors: Dariusz Kucharczyk and Alberta Mandich

Received: 23 March 2024
Revised: 23 April 2024
Accepted: 11 May 2024
Published: 14 May 2024

1. Introduction

In 2020, more than half (i.e., 56%; 88 million tonnes) of the aquatic food available for human consumption came from aquaculture production, and this share is forecasted to increase by 22%, reaching 106 million tonnes by 2030 [1]. The expansion of the aquaculture sector necessarily implies an increase in aquafeed production, and consequently in raw materials availability. The use of fishmeal (FM) and fish oil (FO) from fish by-products (i.e., heads, viscera, skin, bones, and scales) has increased over the last years, but the production of these commodities (particularly FM) is still highly dependent on the processing of whole small pelagic fish species obtained from fisheries [2]. Thus, to guarantee the responsiveness and sustainability of the aquaculture industry over the upcoming years, it is imperative to diminish its dependence on FM and FO obtained from wild fish and promote the strategic use of these ingredients [1]. Over the past few years, the use of vegetable proteins, terrestrial processed animal proteins (PAPs), and more recently, single cell proteins as alternatives to FM has been explored extensively [3]. However, as outlined in the latest research by Glencross et al. [4], each ingredient has its own set of strengths and weaknesses, and there is not a single ingredient that can be considered perfect.

Using insects as a source of nutrients is not a novel concept, as the consumption of edible insects has been a longstanding tradition in Latin America, Africa, and Asia. [5].

However, in Europe, the interest in insect protein has increased since the inclusion of this ingredient in aquafeed was authorized by the European Union (EU) in 2017 [6]. According to Cottrell et al. [7], insect protein stands out as the most promising alternative to fishmeal when compared to other emerging protein sources like micro- and macroalgae, bacteria, and yeast. Indeed, insect meal (IM), predominantly derived from *Hermetia illucens* (HI) and *Tenebrio molitor* (TM) larvae, has demonstrated an enormous potential as an alternative protein source to partially and totally replace FM in diets for both freshwater and marine fish species, including European seabass (*Dicentrarchus labrax*), a key fish species in Mediterranean aquaculture [8]. Despite this, the majority of studies have primarily focused on fish growth, nutrient digestibility, intestinal health, and the final quality of fish intended for human consumption, overlooking the effect of the dietary inclusion of IM on fish response to stress, which in the long term may result in adverse effects on animal health and overall performance [9].

In the past few years, the expansion of aquaculture production has led to the emergence of certain practices (e.g., fish overcrowding, grading, transport, etc.) that induce stress in fish [10,11]. Stress response can be categorized into primary, secondary, or tertiary. The primary response is initiated by the activation of endocrine pathways, triggered by the central nervous system's recognition of the stressor, resulting in the release of catecholamines and cortisol. The secondary response is orchestrated by the hormones responsible for the primary response. This involves cardiovascular and respiratory changes (e.g., increased distribution of oxygen and release of energy substrates into circulation, such as glucose and lactate), as well as hydromineral dysfunction (e.g., adrenaline modifies gills' blood flow and permeability patterns). In addition, this second stage of the stress response is marked by cortisol-mediated immunosuppression. Finally, the tertiary response is an outcome of the incapacity of the animal to achieve homeostasis, resulting in suppressed immunity, metabolic dysfunctions, and impaired growth performance [12]. Moreover, stress may disturb the balance between reactive oxygen species (ROS) produced and the antioxidant scavenging ability of tissues, resulting in increased ROS levels and ultimately leading to oxidative damage. To counteract the harmful effects of ROS, the organism initiates the activation of antioxidant enzymes [13]. Nutrition plays a crucial role as a major modulator of oxidative stress. The consumption of a well-balanced diet containing antioxidant compounds can fortify the animal's redox status, contribute to allostasis, and improve flesh quality. For instance, Zarantoniello et al. [14] recently demonstrated that replacing 3 and 20% FM by full-fat spirulina-enriched black soldier fly (*H. illucens*) prepupae meal resulted in decreased lipid peroxidation of European seabass flesh. On the other hand, oxidative stress can also be triggered by feed ingredients, either through the presence of anti-nutritional factors, amino acid deficiencies, or even an excess of macronutrients [9]. Thus, to better understand the feasibility of using a novel ingredient in aquafeeds, such as IM, it is of paramount importance to assess its effects beyond growth. Specifically, evaluating its impact on immune function and the oxidative stress response becomes imperative.

In this context, the main aim of this study was to investigate, for the first time, the effects of partially and fully replacing fishmeal with defatted (d-) *Tenebrio molitor* (TM) on humoral immunity and hepatic oxidative stress in juvenile European seabass following exposure to an acute stress challenge, reflecting common handling procedures in fish farms.

2. Materials and Methods

2.1. Ingredients and Experimental Diets

Three diets were formulated and extruded by SPAROS Lda. (Olhão, Portugal) according to European seabass nutritional requirements [15]. A control diet (CTRL) was formulated to include 40% of high-quality FM and 14% FO, and two other experimental diets were obtained by replacing 50% (TM50) and 100% (TM100) of FM by dTM. All experimental diets were isoproteic, isolipidic, and isoenergetic (47% protein, 20% lipids, and 24 kJ g^{-1} on a dry matter (DM) basis, respectively) and supplemented with DL-methionine. The TM100 diet was also supplemented with L-lysine, L-threonine, and L-tryptophan.

Monocalcium phosphate was added to dTM diets. Feed ingredients and proximate composition of the dTM and experimental diets are presented in Table 1. Experimental diets are the same in this and our previous studies [16–18].

Table 1. Ingredients and proximate composition of defatted *Tenebrio molitor* larvae meal and the experimental diets.

	dTM	CTRL	TM50	TM100
Ingredients (%)				
Fishmeal [1]		40	20	-
Defatted *Tenebrio molitor* larvae meal		-	20.5	40.4
Soy protein concentrate [2]		10.5	10.5	10.5
Soybean meal [3]		13	13	13
Rapeseed meal [4]		5	5	5
Wheat meal [5]		16.2	15.2	14.3
Fish oil [6]		14.0	13.3	12.5
Vitamin and mineral premix [7]		1	1	1
Vitamin C		0.1	0.1	0.1
Vitamin E		0.1	0.1	0.1
Monocalcium phosphate		-	1.0	2.0
L-Lysine		-	-	0.2
L-Threonine		-	-	0.2
L-Tryptophan		-	-	0.1
DL-Methionine		0.1	0.2	0.3
Proximate composition (% dry matter (DM) or kJ/kg DM)				
Dry matter	97.8	93.1	92.6	92.5
Crude protein	71.0	46.9	47.3	47.2
Crude fat	12.1	19.7	19.8	19.0
Gross energy	24.3	23.2	23.5	24.0
Ash	4.8	10.2	8.1	6.3
Phosphorus	0.8	1.2	1.2	1.0

The abbreviations for the experimental diets stand for: CTRL, control diet; TM50 and TM100, diets with 50 and 100% of the fishmeal protein replaced by defatted *Tenebrio molitor* larvae meal. [1] Peruvian fishmeal super prime: 71% crude protein (CP), 11% crude fat (CF), Exalmar, Peru; [2] soy protein concentrate: 65% CP, 0.7% CF, ADM Animal Nutrition, The Netherlands; [3] soybean meal 48: dehulled solvent extracted soybean meal: 48% CP, 2% CF, Cargill, Spain; [4] rapeseed meal: 36% CP, 3% CF, PREMIX Lda., Portugal; [5] wheat meal: 10% CP, 1% CF, Casa Lanchinha, Portugal; [6] sardine oil, Sopropêche, France; [7] vitamin and mineral premix: WISIUM, ADM Portugal S.A., Portugal.

2.2. Ethical Issues

The fish trial and all procedures involving animals were approved by the CIIMAR ethical committee for Managing Animal Welfare (ORBEA), in compliance with Directive 2010/63/EU from the European Union and the Portuguese Decree-Law n° 113/2013 on "The protection of animals used for scientific purposes". Sampling procedures were performed by accredited scientists in laboratory animal science from the Portuguese Veterinary Authority (1005/92, DGAV-Portugal, following FELASA category C recommendations).

2.3. Acute Stress Challenge and Fish Sampling

Before the acute stress challenge, a growth trial was carried out for 16 weeks as described by Basto et al. [16]. Briefly, juveniles of European seabass were obtained from a commercial fish farm (Acuinuga—Acuicultura y Nutrición de Galicia S. L., Spain) and transported to the experimental facilities of CIIMAR (Portugal). After a quarantine period of 3 weeks, 12 homogeneous groups of 15 fish (69 ± 8 g) were randomly allocated to 160 L tanks of a recirculation aquaculture system (RAS) and kept at 22 ± 1 °C, 35 ± 0.5 salinity, 6 L min^{-1}, oxygen level > 90% ± 1 saturation, and an artificial 12 light/12 dark photoperiod. Water quality parameters were monitored daily as described by Basto et al. [16]. Fish were fed by automatic feeders three times a day. Each experimental diet was tested in quadruplicate for 16 weeks. After the feeding trial, and following a 48 h fasting period, 3 fish per tank (n = 12 fish) were anesthetized (ethylene glycol phenyl ether, 300 μL L^{-1})

and sampled for evaluation of innate immune status. Blood was withdrawn from the caudal vein using heparinized syringes and centrifuged ($5000\times g$ for 5 min at 4 °C), and the resulting supernatant plasma was kept at −80 °C until the metabolites and innate immunity-related parameters analysis. Then, fish were euthanized by spinal cord section, and a portion of the liver (≈150 g) was collected, instantly frozen in dry ice, and kept at −80 °C until the analysis of the activity of oxidative stress-related enzymes. To evaluate the outcome of IM dietary incorporation on oxidative stress, antioxidant capacity, and immune response to stress, other 3 fish per tank (12 fish per dietary treatment) were subjected to an acute stress challenge, consisting of 1 min of air exposure. After 1 h of recovery, the blood, plasma, and liver were sampled as previously described. All fish were individually weighted at the beginning and at the end of the experimental trial to calculate growth performance.

2.4. Haematological Parameters

The haematocrit (Ht), haemoglobin (Hb; kit ref. 1001230, Spinreact, Barcelona, Spain), and total red and white blood cell (RBC and WBC, respectively) counts were determined according to Machado et al. [19]. The mean corpuscular volume (MCV), mean corpuscular haemoglobin (MCH), and mean corpuscular haemoglobin concentration (MCHC) were then calculated as follows: MCV (μm^3) = (Ht/RBC) × 10; MCH (pg $cell^{-1}$) = (Hb/RBC) × 10; MCHC (g 100 mL^{-1}) = (Hb/Ht) × 100.

2.5. Plasma Metabolites and Innate Immunity-Related Parameters

Cortisol levels were measured using an ELISA kit (ref. RE52061, IBL International Gmbh, Hamburg, Germany) following the manufacturer's instructions and according to Azeredo, et al. [20]. Lactate and glucose were enzymatically determined using commercial kits (ref. 1,001,330 and 41,011, respectively, Spinreact, Barcelona, Spain) adapted to the microplate format. Lysozyme activity was determined according to Hutchinson and Manning [21]. Total peroxidase activity was assessed following the method described by Quade and Roth [22]. Alternative complement pathway activity (ACH50) was determined based on the lysis of rabbit blood cells according to Sunyer and Tort [23].

2.6. Hepatic Oxidative Stress and Antioxidant Capacity

Liver samples intended for the determination of lipid peroxidation (LPO) and antioxidant enzyme analysis were homogenized with phosphate buffer (0.1 M, pH 7.4) at a ratio of 1:10 (w/v), as described by Resende et al. [24]. Before enzymatic activity assessment, the hepatic soluble protein content was determined using a commercial kit (ref. 23,225, Thermo Fisher Scientific Inc., Pleasanton, CA, USA) for standardizing enzyme activity measurements. The activity of catalase (CAT), glutathione s-transferase (GST), glutathione peroxidase (GPx), and glutathione reductase (GR) was determined according to Pereira et al. [25]. Superoxide dismutase (SOD) activity and total antioxidant capacity (TAC) were determined using commercial kits (ref. 19,160 and MAK187, respectively, Sigma-Aldrich, St. Louis, MO, USA). All measurements were performed in triplicate in a Synergy HTX Multi-Mode Microplate Reader (BioTek Instruments Inc., San Diego, CA, USA).

2.7. Statistical Analysis

Data were tested for normality and homogeneity of variances by Kolmogorov–Smirnov and Levene's tests, respectively, and transformed whenever necessary. Differences were tested by a two-way ANOVA, with stress (stressed and non-stressed) and dietary treatment (CTRL, TM50, and TM100) as the main factors. A significance level of 95% ($p < 0.05$) was considered followed by a post hoc Tukey HSD test to identify significant differences amongst groups using Statistica™ 13.5.0.17 (TIBCO Software Inc., Palo Alto, CA, USA). To discriminate and classify the existing groups, a multivariate canonical discriminant analysis (DA) was performed on the dataset to evaluate linear combinations of the original variables that would best separate the groups (discriminant functions) using Addinsoft XLSTAT 2022 system software. Each discriminant function explains part of the total variance of the dataset and is loaded by variables contributing the most to that variation. Discriminatory effectiveness was assessed by Wilk's λ test, the distance between group centroids was measured by squared Mahalanobis distance, and Fisher's F statistic was applied to infer significance.

3. Results

3.1. Growth Performance

Before and after the acute stress challenge, no mortality was observed, and European seabass fed diets with partial and total FM replacement by dTM showed similar final body weight after 16 weeks of feeding (Figure 1). Therefore, the experimental diets ensured 100% survival and proper zootechnical performance, and no behavioural alterations were observed in the fish throughout the feeding experiment. Detailed results on the impact of these dietary formulations on fish growth performance are presented in Basto et al. [16].

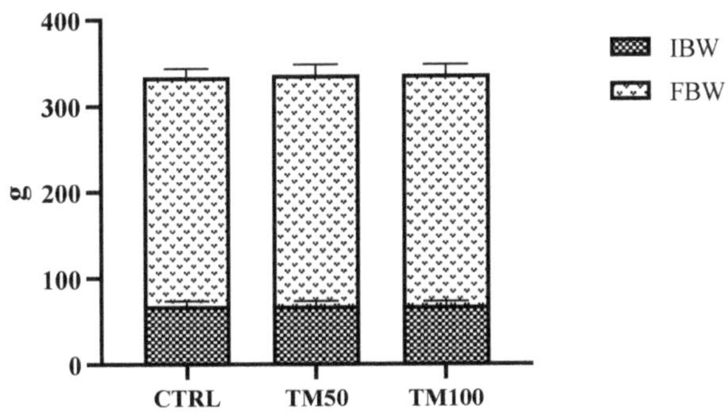

Figure 1. Final body weight (g) of European seabass fed the experimental diets for 16 weeks. The abbreviations are IBW—initial body weight and FBW—final body weight.

3.2. Haematological Profile

Partial and total FM replacement by dTM did not significantly affect haemoglobin, haematocrit, red and white blood cells, mean corpuscular volume, mean corpuscular haemoglobin, or mean corpuscular haemoglobin concentration (Figure 2). On the other hand, haemoglobin, haematocrit, red blood cells, and mean corpuscular haemoglobin concentration significantly increased after the acute stress challenge, while mean corpuscular volume and mean corpuscular haemoglobin decreased, regardless of the dietary treatment (Figure 2).

3.3. Plasma Metabolites and Innate Immunity-Related Parameters

After the acute stress challenge, and irrespective of the dietary treatment, European seabass exhibited a significant increase in plasma cortisol, glucose, and lactate levels, coupled with a significant increase in lysozyme activity and ACH50 (Figure 3). On the other hand, the levels of peroxidase in the plasma of fish fed TM100 were significantly higher than those observed in fish fed CTRL, regardless of stress condition (Figure 3).

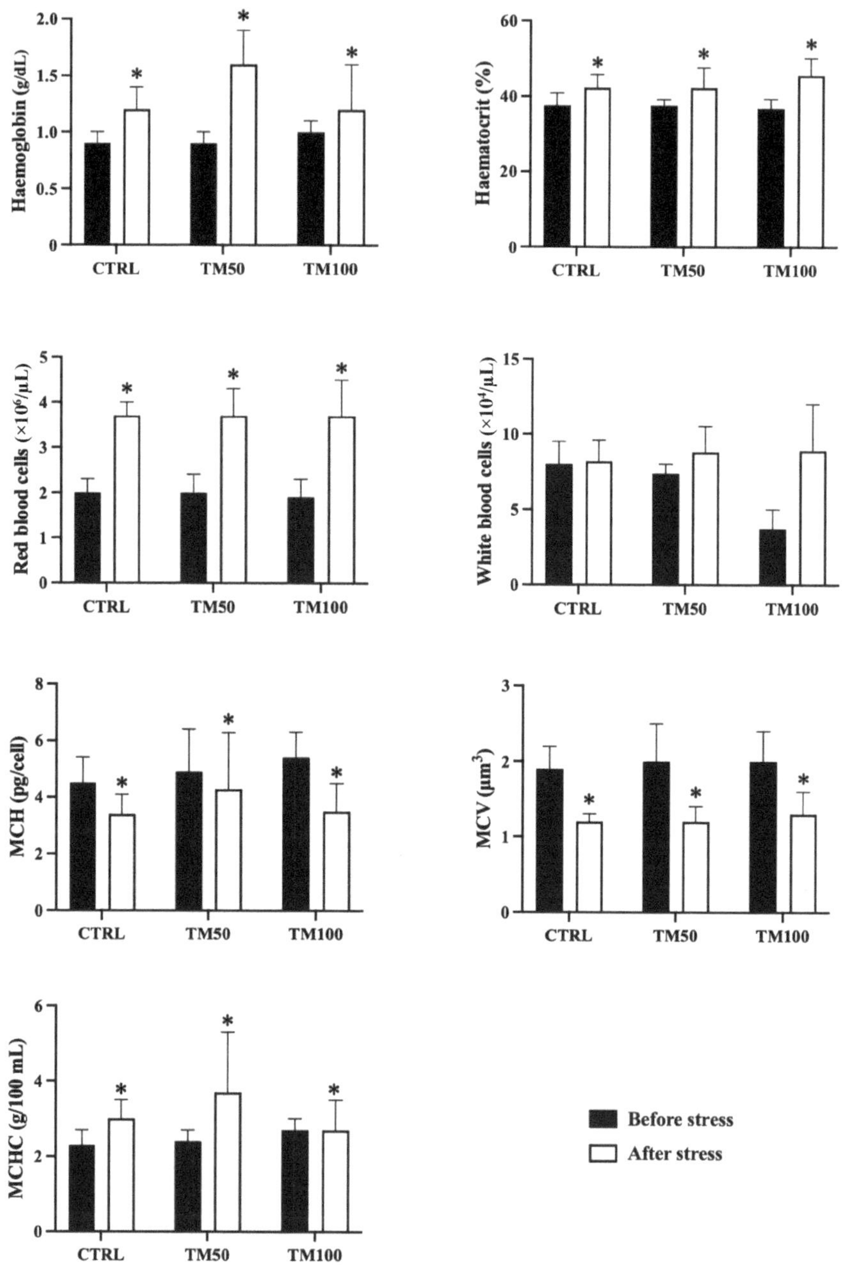

Figure 2. Haematological profile of European seabass fed the experimental diets for 16 weeks and subsequently subjected to an acute stress challenge. Bars represent mean ± standard deviation and the asterisk (*) indicates significant differences ($p < 0.05$) before and after the acute stress challenge, irrespective of dietary treatment.

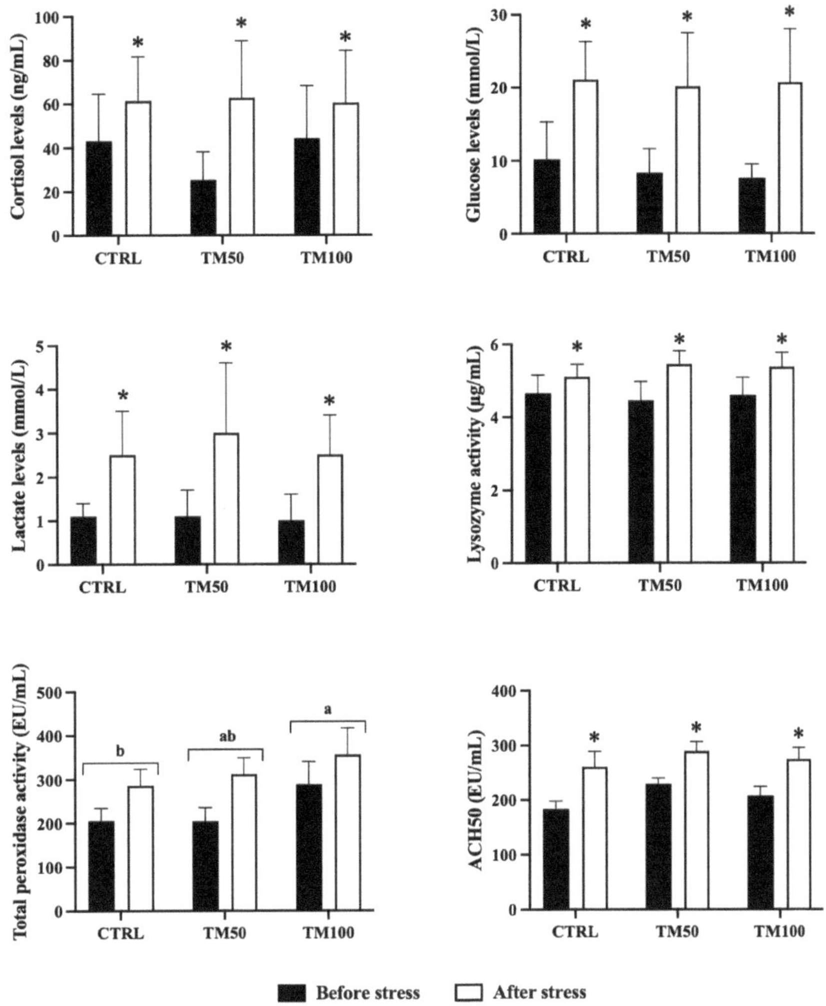

Figure 3. Plasma metabolites and innate immunity-related parameters of European seabass fed the experimental diets for 16 weeks and subsequently subjected to an acute stress challenge. Bars represent mean ± standard deviation. The asterisk (*) indicates significant differences ($p < 0.05$) before and after the acute stress challenge, irrespective of dietary treatment, and different lowercase letters indicate significant differences ($p < 0.05$), irrespective of stress condition.

3.4. Hepatic Oxidative Stress and Antioxidant Capacity

The activity of SOD in the liver of fish fed TM100 was significantly higher than those fed CTRL, irrespective of stress condition (Table 2). The biomarkers CAT, GPx, GST, GR, LPO, and TAC were not affected either by dietary treatment or stress condition (Table 2).

Table 2. Enzymatic activity of oxidative stress-related enzymes, lipid peroxidation, and total antioxidant capacity in the liver of European seabass fed the experimental diets before and after the acute stress challenge.

	Before Stress			After Stress			Two-Way ANOVA p-Value		
	CTRL	TM50	TM100	CTRL	TM50	TM100	Diet	Stress	Diet × Stress
SOD	78.5 ± 13.2 ^{b}	80.4 ± 12.8 ^{ab}	90.5 ± 1.8 ^{a}	72.2 ± 13.7 ^{b}	83.9 ± 20.3 ^{ab}	87.1 ± 1.8 ^{a}	0.02	0.46	0.45
CAT	8.1 ± 2.5	6.7 ± 3.4	8.6 ± 2.7	8.7 ± 1.3	9.5 ± 2.7	8.5 ± 2.3	0.81	0.10	0.14
GPx	196.8 ± 11.7	205.8 ± 14.4	200.2 ± 3.2	197.0 ± 5.6	191.3 ± 2.8	190.6 ± 2.7	0.67	0.06	0.06
GST	54.0 ± 9.9	50.7 ± 8.0	56.1 ± 13.6	56.1 ± 6.9	56.3 ± 1.5	61.7 ± 15.5	0.26	0.11	0.84
GR	20.5 ± 5.1	22.8 ± 6.6	23.8 ± 8.4	21.6 ± 2.0	22.3 ± 4.0	22.2 ± 4.6	0.47	0.84	0.71
LPO	10.9 ± 4.1	9.2 ± 3.9	11.0 ± 2.0	9.9 ± 3.9	13.4 ± 7.4	14.8 ± 5.3	0.23	0.06	0.14
TAC	0.21 ± 0.05	0.18 ± 0.05	0.21 ± 0.05	0.23 ± 0.05	0.20 ± 0.05	0.20 ± 0.04	0.15	0.23	0.45

Values represent mean ± standard deviation. Different lowercase letters indicate statistically significant differences (p < 0.05) for diet factor. SOD, superoxide dismutase (EU mg protein^{-1}); CAT, catalase (EU mg protein^{-1}); GPx, glutathione peroxidase (nmol mg protein^{-1}); GST, glutathione s-transferase (nmol mg protein^{-1}); GR, glutathione reductase (nmol mg protein^{-1}); LPO, lipid peroxidation (nmol g^{-1} liver); TAC, total antioxidant capacity (nmol Trolox mg^{-1} liver).

3.5. Discriminant Analysis

The overall performance of the analysis indicates good discriminatory ability (Wilks $\lambda = 0.02$, $p < 0.0001$) with the first two discriminant functions accounting for 91.69% of the total dataset variability (Figure 4; F1 83.66% and F2 8.04%). Assessing the linear functions of the variables from the analysed tissues, a clear separation by stress challenge was observed, meaning that fish fed experimental diets before the stress challenge were significantly separated from those after stress, based on the significant Mahalanobis distance of each group's multivariate mean (centroid) ($p < 0.05$). In addition, before stress, the fish fed TM100 were significantly distant from those fed CTRL. The first function (accounting for 83.66% of total dataset variability) was positively loaded mainly by the red blood cell count, haematocrit, haemoglobin, and alternative complement pathway activity, as well as cortisol, glucose, and lactate levels, while being negatively loaded by MCV.

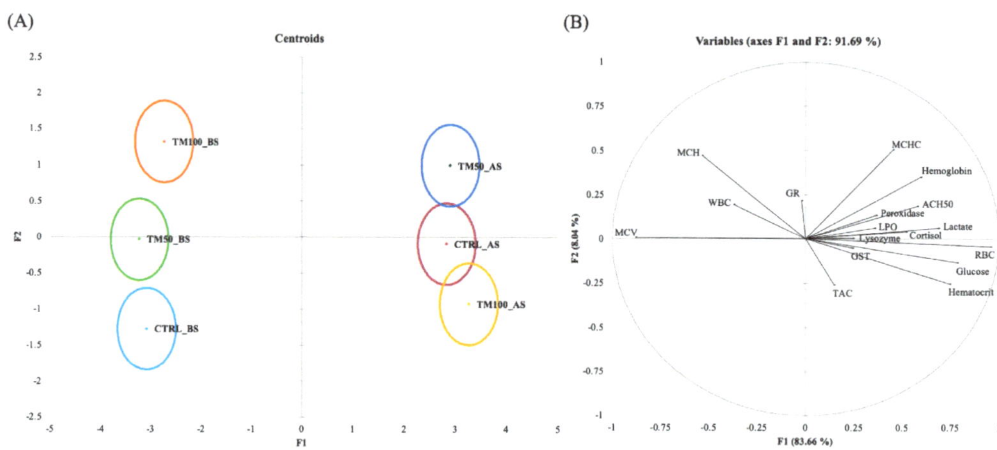

Figure 4. Canonical discriminant analysis of European seabass fed experimental diet before and after the acute stress challenge. (**A**) Canonical discriminant scores of each group. Coloured circles represent groups' centroids. (**B**) Correlation variables/factors (factor loads) for two main discriminant functions (F1 and F2).

4. Discussion

The expansion of the aquaculture sector necessarily implies an increase in aquafeed production, and consequently in raw materials availability. However, the production of conventional marine ingredients and plant protein sources has remained stable, prompting the necessity to explore alternative ingredients [2]. Additionally, the emergence of various environmental and husbandry-related stress factors also poses substantial challenges to the aquaculture industry [26]. Therefore, when a new ingredient is under consideration to be included in aquafeeds, it is increasingly important to understand if it does not compromise the animals' response to stress situations.

Among the most common stressors, handling represents a highly stressful procedure due to its frequent association with air exposure, triggering a notable physiological response [27]. Thus, in the current study, fish were exposed to an acute stress challenge consisting of air exposure, with a total duration of 1 minute, followed by 1 h of recovery before sampling. The significant increase in haemoglobin, haematocrit, red blood cells, mean corpuscular haemoglobin concentration, and cortisol, lactate, and glucose levels after the acute stress challenge validated the applied protocol, as these are typical indicators of acute stress [10,28]. Moreover, a clear separation before and after the stress challenge was observed when a discriminant analysis was applied to the dataset. Additionally, it was clearly observed that when animals were not exposed to a stressful situation, the inclusion of dTM to partially or completely replace FM did not compromise the physiological status of the animals.

After the acute stress challenge, and irrespective of the dietary treatment, European seabass displayed significantly increased values of plasma cortisol. On the other hand, the dietary inclusion of dTM did not affect the circulating levels of this hormone. This suggests that, while the primary stress response of seabass did not exhibit improvement (e.g., decreased circulating cortisol levels) with the inclusion of this protein source, it was also not compromised, even when the FM was completely replaced. To the best of the authors' knowledge, this is the first study evaluating the impact of FM substitution by an IM on the primary stress response of European seabass. In gilthead seabream (*Sparus aurata*), the inclusion of 27.6% of defatted HI to replace 75% of FM did not alter the gill cortisol levels after 17 weeks of a feeding trial [29]. Similarly, in rainbow trout (*Oncorhynchus mykiss*), 50% FM replacement by defatted HI (i.e., 15.7% dietary inclusion) also did not affect the primary stress response parameters, namely, ACTH and cortisol levels, after 18 weeks of feeding with the experimental diets [30].

The secondary stress response is characterized by (1) increased distribution of oxygen and (2) the release of energy substrates into circulation, such as glucose and lactate [10,12,31]. Thus, the observed increased haematocrit, haemoglobin, and red blood cells after air exposure, regardless of dietary treatment, may be related to a splenic contraction to release blood cells into the circulatory system as a strategy to improve the transport of oxygen [32]. In the present study, the partial and total FM replacement by dTM did not significantly affect any of the haematological parameters evaluated. The results obtained herein align with the previous findings of Abdel-Tawwab et al. [33]. Specifically, 50% of FM replacement by defatted HI (i.e., 14.8% dietary inclusion) did not demonstrate any significant impact on the values of different haematological parameters of European seabass juveniles. In Nile tilapia (*Oreochromis niloticus*), total FM replacement by full fat HI, corresponding to 10% dietary inclusion, also did not change the haematological profile of fish [34]. In contrast, in the recent study conducted by Khieokhajonkhet et al. [35], it was demonstrated that red blood cell counts and haemoglobin levels were higher in Nile tilapia fed an experimental diet completely devoid of FM and containing 21.7% giant cricket meal (*Brachytrupes portentosusas*) for 4 weeks. According to those authors, elevated levels of these haematological parameters and their inherent enhanced oxygen transport may lead to a more effective response in challenging circumstances [35]. Nevertheless, to corroborate this hypothesis, further studies would be required involving animals subjected to episodes of stress or infection.

While glucose serves as a crucial energy source in aerobic conditions, providing the fish with the necessary energy substrate to meet heightened energy requirements after stress, the limited oxygen availability in anaerobic conditions hinders the Krebs cycle. In such conditions, pyruvate undergoes conversion to lactate, which is stored in muscle. Once there is a sufficient return of oxygen availability, lactate can be transported to the liver, where it undergoes conversion back to pyruvate for gluconeogenesis. Additionally, lactate can also contribute to meeting energy demands during aerobic metabolism [12]. After the acute stress challenge, and irrespective of the dietary treatment, European seabass exhibited a significant increase in plasma glucose and lactate levels. On the other hand, the dietary inclusion of dTM did not have an impact on any of these stress biomarkers. These results align with findings from other studies that explored the utilization of different types of IM as protein sources in aquafeeds. Similarly, those studies did not observe changes in the plasmatic levels of glucose and/or lactate [36,37]. Altogether, the results herein obtained provide evidence that partial (i.e., 50%) or total FM replacement by dTM, corresponding to dietary inclusion levels of 20% and 40%, respectively, do not compromise the primary or secondary response of European seabass to stress induced by air exposure for 1 min.

Under normal conditions, organisms can neutralize reactive molecules and repair oxidative damage through their natural antioxidant defences. However, exposure to stress can trigger metabolic shifts necessary to meet the increased energy demand, leading to heightened production of ROS as by-products of cellular respiration and other metabolic processes. Oxidative stress occurs when there is an imbalance between the production of ROS and the antioxidant scavenging capacity of tissues through the action of antioxidant enzymes (e.g., SOD, CAT, GPx, GST, and GR), resulting in cellular membrane lipids damage [38,39]. In the liver of fish fed diets completely devoid of FM, the activity of SOD was increased, underscoring the significance of this adaptive response and highlighting its crucial role in antioxidant defence mechanisms. As far as we are aware, this study represents the first attempt to assess the impact of total FM substitution by an IM on the oxidative stress parameters in the liver of European seabass. However, in Atlantic salmon (*Salmo salar*), total FM replacement by dTM (i.e., 21.5% dietary inclusion) did not alter hepatic SOD activity after 12 weeks of a feeding trial [40]. On the other hand, in Siberian sturgeon (*Acipenser baerii*), the inclusion of 37.5% defatted HI to replace 50% FM led to an increase in the hepatic activity of SOD after 16 weeks of a growth trial [41]. Thus, the findings from the existing literature on partial FM substitution by IM in diets for seabass and other fish species are still inconsistent, and further research is required.

Coupled with antioxidant defences, innate immunity is considered the fish's first line of defence, with lysozyme and peroxidase used as conventional biomarkers of the innate immune response in fish. Lysozyme is involved in a defence mechanism, such as bacteriolysis, through action on the peptidoglycan layer of bacterial cellular walls. Peroxidase plays a crucial role as an enzyme that enables the conversion of the superoxide anion to generate hydrochloric acid. Moreover, peroxidase exhibits microbicidal properties by utilizing one of the oxidative radicals to generate hypochlorous acid. This process holds significant importance as it serves as an effective method for eliminating foreign microorganisms [42,43]. While the activity of lysozyme was unaffected by stress conditions or dietary treatments, the peroxidase activity was higher in fish fed TM100. Indeed, several authors state that incorporating both TM or HI in diets for various fish species can enhance the fish immune response, either before or after a bacterial challenge, by increasing lysozyme and peroxidase activities [44–46]. For example, Abdel-Latif et al. [47] observed that the lysozyme activity increased in fish that were fed diets with a 50% FM replacement by partially defatted HI (i.e., 14.8% dietary inclusion) for 8 weeks. Furthermore, fish subjected to this dietary treatment exhibited higher survival rates post-challenge with *Vibrio alginolyticus*. This immunostimulant capacity of IM might be explained by the chitin content of insects, as reviewed by Mohan et al. [48].

In conclusion, the results of this study demonstrate that total FM replacement by dTM increased total peroxidase and SOD activity in fish fed TM100, suggesting an immunostimulant and antioxidant capacity of dTM. But the partial inclusion of dTM did not compromise the physiological stress responses, antioxidant capacity, and innate immunity of European seabass juveniles. However, to better understand the impact of total FM replacement by IM on fish immune response and antioxidant capacity, it would be of high interest to deepen our knowledge of their underlying mechanisms under other challenging rearing conditions (e.g., pathogen or chronic stress exposure).

Author Contributions: Conceptualization, L.M.P.V.; Methodology, L.M.P.V. and B.C.; Validation, L.M.P.V. and B.C.; Formal Analysis, A.B., D.P. and M.M.; Investigation, A.B.; Resources, D.M. and L.M.P.V.; Data Curation, A.B. and D.P.; Writing—Original Draft Preparation, A.B.; Writing—Review and Editing, A.B., D.P., M.M., B.C., D.M. and L.M.P.V.; Supervision, L.M.P.V.; Project Administration, L.M.P.V.; Funding Acquisition, L.M.P.V. All authors have read and agreed to the published version of the manuscript.

Funding: This research was funded by the InFishMix project (PT-INNOVATION-0094) funded by Iceland, Liechtenstein, and Norway through the EEA and Norway grants. M.M. and B.C. were supported by FCT (2022.03304.CEECIND and 2020.00290.CEECIND, respectively). CIIMAR was funded by FCT, grant numbers UIDB/04423/2021 and UIDP/04423/2022.

Institutional Review Board Statement: All procedures involving animals were approved by the CIIMAR ethical committee for Managing Animal Welfare (ORBEA), in compliance with Directive 2010/63/EU from the European Union and the Portuguese Decree-Law n° 113/2013 on "The protection of animals used for scientific purposes". Sampling procedures were performed by accredited scientists in laboratory animal science from the Portuguese Veterinary Authority (1005/92, DGAV-Portugal, following FELASA category C recommendations).

Informed Consent Statement: Not applicable.

Data Availability Statement: Data are available upon request.

Conflicts of Interest: Daniel Murta is the founder and R&D advisor of Thunder Foods Lda., which is the promoter of the InFishMix project. The remaining authors declare that the research was conducted in the absence of any commercial or financial relationships that could be construed as a potential conflict of interest.

References

1. FAO. *The State of World Fisheries and Aquaculture 2022. Towards Blue Transformation*; FAO: Roma, Italy, 2022. [CrossRef]
2. Naylor, R.L.; Hardy, R.W.; Buschmann, A.H.; Bush, S.R.; Cao, L.; Klinger, D.H.; Little, D.C.; Lubchenco, J.; Shumway, S.E.; Troell, M. A 20-year retrospective review of global aquaculture. *Nature* **2021**, *591*, 551–563. [CrossRef] [PubMed]
3. Glencross, B.; Fracalossi, D.M.; Hua, K.; Izquierdo, M.; Mai, K.; Øverland, M.; Robb, D.; Roubach, R.; Schrama, J.; Small, B.; et al. Harvesting the benefits of nutritional research to address global challenges in the 21st century. *J. World Aquac. Soc.* **2023**, *54*, 343–363. [CrossRef]
4. Glencross, B.; Ling, X.; Gatlin, D.; Kaushik, S.; Øverland, M.; Newton, R.; Valente, L.M.P. A SWOT analysis of the use of marine, grain, terrestrial-animal and novel protein ingredients in aquaculture feeds. *Rev. Fish. Sci. Aquac.* **2024**, 1–39. [CrossRef]
5. Raheem, D.; Carrascosa, C.; Oluwole, O.B.; Nieuwland, M.; Saraiva, A.; Millán, R.; Raposo, A. Traditional consumption of and rearing edible insects in Africa, Asia and Europe. *Crit. Rev. Food Sci. Nutr.* **2019**, *59*, 2169–2188. [CrossRef] [PubMed]
6. European Commission. Commission Regulation (EU) 2017/893 of 24 May 2017 amending Annexes I and IV to Regulation (EC) No 999/2001 of the European Parliament and of the Council and Annexes X, XIV and XV to Commission Regulation (EU) No 142/2011 as regards the provisions on processed animal protein. *Off. J. Eur. Union* **2017**, *60*, 92–116.
7. Cottrell, R.S.; Blanchard, J.L.; Halpern, B.S.; Metian, M.; Froehlich, H.E. Global adoption of novel aquaculture feeds could substantially reduce forage fish demand by 2030. *Nat. Food* **2020**, *1*, 301–308. [CrossRef]
8. Singh, S.K.; Pawar, L.; Thomas, A.J.; Debbarma, R.; Biswas, P.; Ningombam, A.; Devi, A.G.; Waikhom, G.; Patel, A.B.; Meena, D.K.; et al. The current state of research and potential applications of insects for resource recovery and aquaculture feed. *Environ. Sci. Pollut. Res.* **2023**. [CrossRef]
9. Aragão, C.; Gonçalves, A.T.; Costas, B.; Azeredo, R.; Xavier, M.J.; Engrola, S. Alternative proteins for fish diets: Implications beyond growth. *Animals* **2022**, *12*, 1211. [CrossRef]
10. Tort, L. Stress and immune modulation in fish. *Dev. Comp. Immunol.* **2011**, *35*, 1366–1375. [CrossRef] [PubMed]
11. Martos-Sitcha, J.A.; Mancera, J.M.; Prunet, P.; Magnoni, L.J. Welfare and stressors in fish: Challenges facing aquaculture. *Front. Physiol.* **2020**, *11*, 162. [CrossRef] [PubMed]
12. Schreck, C.B.; Tort, L. The concept of stress in fish. In *Biology of Stress in Fish*; Schreck, C.B., Tort, L., Farrell, A., Brauner, C., Eds.; Academic Press: London, UK, 2016; pp. 2–34.

13. Espinosa-Diez, C.; Miguel, V.; Mennerich, D.; Kietzmann, T.; Sanchez-Perez, P.; Cadenas, S.; Lamas, S. Antioxidant responses and cellular adjustments to oxidative stress. *Redox Biol.* **2015**, *6*, 183–197. [CrossRef] [PubMed]
14. Zarantoniello, M.; de Oliveira, A.A.; Sahin, T.; Freddi, L.; Torregiani, M.; Tucciarone, I.; Chemello, G.; Cardinaletti, G.; Gatto, E.; Parisi, G.; et al. Enhancing rearing of European seabass (*Dicentrarchus labrax*) in aquaponic systems: Investigating the effects of enriched black soldier fly (*Hermetia illucens*) prepupae meal on fish welfare and quality traits. *Animals* **2023**, *13*, 1921. [CrossRef] [PubMed]
15. NRC. *Nutrient Requirements of Fish and Shrimp*, 1st ed.; Press, T.N.A., Ed.; The National Academy Press: Washington, DC, USA, 2011.
16. Basto, A.; Marques, A.; Silva, A.S.; Sá, T.; Sousa, V.; Oliveira, M.B.P.P.; Aires, T.; Valente, L.M.P. Nutritional, organoleptic and sensory quality of market-sized European sea bass (*Dicentrarchus labrax*) fed defatted *Tenebrio molitor* larvae meal as main protein source. *Aquaculture* **2022**, *566*, 739210. [CrossRef]
17. Basto, A.; Valente, L.M.P.; Soengas, J.L.; Conde-Sieira, M. Partial and total fishmeal replacement by defatted *Tenebrio molitor* larvae meal do not alter short- and mid-term regulation of food intake in European sea bass (*Dicentrarchus labrax*). *Aquaculture* **2022**, *560*, 738604. [CrossRef]
18. Basto, A.; Valente, L.M.P.; Sousa, V.; Conde-Sieira, M.; Soengas, J.L. Total fishmeal replacement by defatted *Tenebrio molitor* larvae meal induces alterations in intermediary metabolism of European sea bass (*Dicentrarchus labrax*). *J. Anim. Sci.* **2023**, *101*, skad040. [CrossRef] [PubMed]
19. Machado, M.; Azeredo, R.; Diaz-Rosales, P.; Afonso, A.; Peres, H.; Oliva-Teles, A.; Costas, B. Dietary tryptophan and methionine as modulators of European seabass (*Dicentrarchus labrax*) immune status and inflammatory response. *Fish Shellfish Immunol.* **2015**, *42*, 353–362. [CrossRef] [PubMed]
20. Azeredo, R.; Machado, M.; Afonso, A.; Fierro-Castro, C.; Reyes-Lopez, F.E.; Tort, L.; Gesto, M.; Conde-Sieira, M.; Miguez, J.M.; Soengas, J.L.; et al. Neuroendocrine and immune responses undertake different fates following tryptophan or methionine dietary treatment: Tales from a teleost model. *Front. Immunol.* **2017**, *8*, 1226. [CrossRef] [PubMed]
21. Hutchinson, T.H.; Manning, M.J. Seasonal trends in serum lysozyme activity and total protein concentration in dab (*Limanda limanda* L.) sampled from Lyme Bay, U.K. *Fish Shellfish Immunol.* **1996**, *6*, 473–482. [CrossRef]
22. Quade, M.J.; Roth, J.A. A rapid, direct assay to measure degranulation of bovine neutrophil primary granules. *Vet. Immunol. Immunopathol.* **1997**, *58*, 239–248. [CrossRef] [PubMed]
23. Sunyer, J.O.; Tort, L. Natural hemolytic and bactericidal activities of sea bream *Sparus aurata* serum are effected by the alternative complement pathway. *Vet. Immunol. Immunopathol.* **1995**, *45*, 333–345. [CrossRef] [PubMed]
24. Resende, D.; Pereira, R.; Domínguez, D.; Pereira, M.; Pereira, C.; Pintado, M.; Velasco, C. Stress response of European seabass (*Dicentrarchus labrax*) fed plant-based diets supplemented with swine blood hydrolysates. *Aquac. Rep.* **2023**, *30*, 101600. [CrossRef]
25. Pereira, R.; Costa, M.; Velasco, C.; Cunha, L.M.; Lima, R.C.; Baião, L.F.; Batista, S.; Marques, A.; Sá, T.; Campos, D.A.; et al. Comparative analysis between synthetic vitamin E and natural antioxidant sources from tomato, carrot and coriander in diets for market-sized *Dicentrarchus labrax*. *Antioxidants* **2022**, *11*, 636. [CrossRef] [PubMed]
26. Ciji, A.; Akhtar, M.S. Stress management in aquaculture: A review of dietary interventions. *Rev. Aqua.* **2021**, *13*, 2190–2247. [CrossRef]
27. Cook, K.V.; Lennox, R.J.; Hinch, S.G.; Cooke, S.J. Fish out of water: How much air is too much? *Fisheries* **2015**, *40*, 452–461. [CrossRef]
28. Guo, H.; Dixon, B. Understanding acute stress-mediated immunity in teleost fish. *Fish Shellfish. Immunol.* **2021**, *2*, 100010. [CrossRef]
29. Gai, F.; Cusimano, G.M.; Maricchiolo, G.; Caccamo, L.; Caimi, C.; Macchi, E.; Meola, M.; Perdichizzi, A.; Tartarisco, G.; Gasco, L.; et al. Defatted black soldier fly meal in diet for grow-out gilthead seabream (*Sparus aurata* L. 1758): Effects on growth performance, gill cortisol level, digestive enzyme activities, and intestinal histological structure. *Aqua. Res.* **2023**, *2023*, 3465335. [CrossRef]
30. Di Rosa, A.R.; Caccamo, L.; Pansera, L.; Oteri, M.; Chiofalo, B.; Maricchiolo, G. Influence of *Hermetia illucens* larvae meal dietary inclusion on growth performance, gut histological traits and stress parameters in *Sparus aurata*. *Animals* **2023**, *13*, 339. [CrossRef] [PubMed]
31. Witeska, M.; Kondera, E.; Ługowska, K.; Bojarski, B. Hematological methods in fish—Not only for beginners. *Aquaculture* **2022**, *547*, 737498. [CrossRef]
32. Fazio, F.; Ferrantelli, V.; Fortino, G.; Arfuso, F.; Giangrosso, G.; Faggio, C. The influence of acute handling stress on some blood parameters in cultured sea bream (*Sparus aurata* Linnaeus, 1758). *Ital. J. Food Saf.* **2015**, *4*, 4174. [CrossRef] [PubMed]
33. Abdel-Tawwab, M.; Khalil, R.H.; Metwally, A.A.; Shakweer, M.S.; Khallaf, M.A.; Abdel-Latif, H.M.R. Effects of black soldier fly (*Hermetia illucens* L.) larvae meal on growth performance, organs-somatic indices, body composition, and hemato-biochemical variables of European sea bass, *Dicentrarchus labrax*. *Aquaculture* **2020**, *522*, 735136. [CrossRef]
34. Tippayadara, N.; Dawood, M.A.O.; Krutmuang, P.; Hoseinifar, S.H.; Doan, H.V.; Paolucci, M. Replacement of fish meal by black soldier fly (*Hermetia illucens*) larvae meal: Effects on growth, haematology, and skin mucus immunity of Nile Tilapia, *Oreochromis niloticus*. *Animals* **2021**, *11*, 193. [CrossRef] [PubMed]
35. Khieokhajonkhet, A.; Thammang, S.; Aeksiri, N.; Kaneko, G.; Tatsapong, P.; Phromkunthong, W. Fish meal replacement by *Brachytrupes portentosusas* for *Oreochromis niloticus*: Effect on growth, feed utilization, fatty acid profiles, hematology, and histological changes. *Anim. Feed Sci. Technol.* **2024**, *308*, 115873. [CrossRef]

36. Magalhães, R.; Sánchez-López, A.; Leal, R.S.; Martínez-Llorens, S.; Oliva-Teles, A.; Peres, H. Black soldier fly (*Hermetia illucens*) pre-pupae meal as a fish meal replacement in diets for European seabass (*Dicentrarchus labrax*). *Aquaculture* **2017**, *476*, 79–85. [CrossRef]
37. Melenchón, F.; de Mercado, E.; Pula, H.J.; Cardenete, G.; Barroso, F.G.; Fabrikov, D.; Lourenço, H.M.; Pessoa, M.F.; Lagos, L.; Weththasinghe, P.; et al. Fishmeal dietary replacement up to 50%: A comparative study of two insect meals for rainbow trout (*Oncorhynchus mykiss*). *Animals* **2022**, *12*, 179. [CrossRef] [PubMed]
38. Song, C.; Sun, C.; Liu, B.; Xu, P. Oxidative stress in aquatic organisms. *Antioxidants* **2023**, *12*, 1223. [CrossRef]
39. Saikia, S.K.; Chowdhury, S. Oxidative stress in fish: A review. *J. Sci. Res.* **2020**, *12*, 145–160. [CrossRef]
40. Habte-Tsion, H.M.; Hawkyard, M.; Sealey, W.M.; Bradshaw, D.; Meesala, K.M.; Bouchard, D.A. Effects of fishmeal substitution with mealworm meals (*Tenebrio molitor* and *Alphitobius diaperinus*) on the growth, physiobiochemical response, digesta microbiome, and immune genes expression of Atlantic salmon (*Salmo salar*). *Aquac. Nutr.* **2024**, *2024*, 6618117. [CrossRef] [PubMed]
41. Caimi, C.; Renna, M.; Lussiana, C.; Bonaldo, A.; Gariglio, M.; Meneguz, M.; Dabbou, S.; Schiavone, A.; Gai, F.; Elia, A.C.; et al. First insights on Black Soldier Fly (*Hermetia illucens* L.) larvae meal dietary administration in Siberian sturgeon (*Acipenser baerii* Brandt) juveniles. *Aquaculture* **2020**, *515*, 734539. [CrossRef]
42. Mokhtar, D.M.; Zaccone, G.; Alesci, A.; Kuciel, M.; Hussein, M.T.; Sayed, R.K.A. Main components of fish immunity: An overview of the fish immune system. *Fishes* **2023**, *8*, 93. [CrossRef]
43. Riera Romo, M.; Perez-Martinez, D.; Castillo Ferrer, C. Innate immunity in vertebrates: An overview. *Immunology* **2016**, *148*, 125–139. [CrossRef] [PubMed]
44. Foysal, M.J.; Fotedar, R.; Tay, C.Y.; Gupta, S.K. Dietary supplementation of black soldier fly (*Hermetica illucens*) meal modulates gut microbiota, innate immune response and health status of marron (*Cherax cainii*, Austin 2002) fed poultry-by-product and fishmeal based diets. *PeerJ* **2019**, *7*, e6891. [CrossRef] [PubMed]
45. Chaklader, M.R.; Siddik, M.A.B.; Fotedar, R.; Howieson, J. Insect larvae, *Hermetia illucens* in poultry by-product meal for barramundi, *Lates calcarifer* modulates histomorphology, immunity and resistance to *Vibrio harveyi*. *Sci. Rep.* **2019**, *9*, 16703. [CrossRef] [PubMed]
46. Sankian, Z.; Khosravi, S.; Kim, Y.-O.; Lee, S.-M. Effects of dietary inclusion of yellow mealworm (*Tenebrio molitor*) meal on growth performance, feed utilization, body composition, plasma biochemical indices, selected immune parameters and antioxidant enzyme activities of mandarin fish (*Siniperca scherzeri*) juveniles. *Aquaculture* **2018**, *496*, 79–87. [CrossRef]
47. Abdel-Latif, H.M.R.; Abdel-Tawwab, M.; Khalil, R.H.; Metwally, A.A.; Shakweer, M.S.; Ghetas, H.A.; Khallaf, M.A. Black soldier fly (*Hermetia illucens*) larvae meal in diets of European seabass: Effects on antioxidative capacity, non-specific immunity, transcriptomic responses, and resistance to the challenge with *Vibrio alginolyticus*. *Fish Shellfish Immunol.* **2021**, *111*, 111–118. [CrossRef] [PubMed]
48. Mohan, K.; Rajan, D.K.; Ganesan, A.R.; Divya, D.; Johansen, J.; Zhang, S. Chitin, chitosan and chitooligosaccharides as potential growth promoters and immunostimulants in aquaculture: A comprehensive review. *Int. J. Biol. Macromol.* **2023**, *251*, 126285. [CrossRef]

Journal of
Marine Science and Engineering

MDPI

Article

Mussel Meal as a Promotor of Growth Performance for the Whiteleg Shrimp (*Litopenaeus vannamei*)

Stef Claessens [1], Cláudia Aragão [1,2,*], Flávia Banderó Hoffling [3], Isabela Pinheiro [3], Débora Machad o Fracalossi [4] and Felipe Nascimento Vieira [3,*]

1 Universidade do Algarve, 8005-139 Faro, Portugal; stefclaessens92@gmail.com
2 Centre of Marine Sciences (CCMAR), 8005-139 Faro, Portugal
3 Laboratório de Camarões Marinhos, Federal University of Santa Catarina, Barra da Lagoa 8806-600, Florianópolis, Brazil; flaviabandh@gmail.com (F.B.H.); isabelapinheiro@outlook.com (I.P.)
4 Laboratório de Nutrição de Espécies Aquícolas, Federal University of Santa Catarina, Armação 88066-260, Florianópolis, Brazil; debora.fracalossi@ufsc.br
* Correspondence: caragao@ualg.pt (C.A.); felipe.vieira@ufsc.br (F.N.V.)

Abstract: Mussel meal (species *Perna perna*) was evaluated as a potential feed additive for whiteleg shrimp (*Litopenaeus vannamei*) diets to improve growth and cold resistance. Five experimental diets (0, 1, 2, 3 and 4% of mussel meal inclusion) were tested in quadruplicate in whiteleg shrimp, using twenty polyethylene tanks of 400 L. Each tank was stocked with 40 shrimp (3.5 ± 0.5 g), filled with seawater and kept under constant aeration and a temperature of 28.4 ± 0.4 °C. After 8 weeks, the growth and feed efficiency of the shrimp were evaluated, and a thermal shock was administered. The shrimp that were fed with the addition of 3 or 4% mussel meal in their diets showed similar results as the control (0% inclusion), while the addition of 1 or 2% mussel meal in the diet resulted in a significantly higher final weight, weight gain and relative growth rate and a lower feed conversion ratio. Further, no differences were observed in thermal shock resistance and survival among the treatments. In conclusion, mussel meal can be used as a feed additive in whiteleg shrimp diets to improve growth, and quadratic regression models indicate that the best levels of inclusion range from 1.73 to 2.00%.

Keywords: feed additive; feed efficiency; pacific white shrimp; *Perna perna*; shrimp culture; shrimp nutrition; thermal shock

Citation: Claessens, S.; Aragão, C.; Hoffling, F.B.; Pinheiro, I.; Fracalossi, D.M.; Vieira, F.N. Mussel Meal as a Promotor of Growth Performance for the Whiteleg Shrimp (*Litopenaeus vannamei*). *J. Mar. Sci. Eng.* **2023**, *11*, 1670. https://doi.org/10.3390/jmse11091670

Academic Editors: Ana Catarina Matias and Carlos Andrade

Received: 26 July 2023
Revised: 18 August 2023
Accepted: 22 August 2023
Published: 25 August 2023

1. Introduction

In recent decades, global aquaculture production has experienced immense growth in terms of the total quantity, the diversity of farmed species and the production systems used [1]. One of the most important production segments of aquaculture is the group comprising marine crustaceans, which contribute to almost 13% of the total global market, with an 86% increase in total production over the past 10 years [2]. Within this group, the species *Litopenaeus vannamei*, commonly known as whiteleg shrimp or Pacific white shrimp, comprises nearly 84% of the total farmed marine shrimp production worldwide.

Whiteleg shrimp became one of the most important aquaculture species due to their fast growth, tolerance to a wide range of salinity and high (stocking) densities, low dietary protein requirements, high survival rates and high market value [3]. This species is native to the tropical marine habitats of the Eastern Pacific coast and is mostly farmed in countries in East and South Asia (83.4%) and Latin America (16.3%) [2]. The continuous increase in the production of this species requires the development of sustainable production strategies. Current trends in aquaculture production rely on finding sustainable alternative and emergent ingredients that may be used to minimize the inclusion of traditional ingredients, such as fishmeal and fish oil, in aquafeed formulations [4,5]. The search for feed additives that can improve daily production management by promoting the shrimp's robustness within this new paradigm of feed formulation is becoming popular [3,5].

Mussel meal may be a unique local alternative ingredient and/or feed additive. The nutritional characteristics of mussel meal are similar the those of fish meal, with a similar amino acid and lipid profile [6]. Additionally, mussels remove nitrogen and phosphate from water by filtering nutrient particles and microscopic organisms, converting non-food into food [7]. Undersized mussels that are not used for human consumption or are grown specifically to reduce the overload of nutrients should be used for the production of mussel meal. By using mussel meal as an alternative ingredient for aquafeeds, nitrogen and phosphate are eco-cycled, and the mussel shells may be used for poultry feed, thereby contributing to lowering the carbon footprint of production [8].

Several studies have shown that mussel meal (produced from different species) can be utilized as a viable alternative ingredient and is a good option for replacing fishmeal or as a feed additive (e.g., [9–14]). The replacement of 25 to 50% of fishmeal with mussel meal does not impair growth performance and/or feed efficiency in fish species like Artic charr (*Salvelinus alpinus*; [15]), common sole (*Solea solea*; [9]), turbot (*Scophthalmus maximus*; [16]) and Ussuri catfish (*Pseudobagrus ussuriensis*; [10]). The inclusion of mussel meal (10–30%) in soybean meal diets even improved the growth and stimulated the feeding of tiger puffer, *Takifugu rubripes* [13]. The dietary inclusion (up to 20%) of the water-soluble fraction of blue mussel, *Mytilus edulis*, increased the body weight and feed efficiency in Japanese flounder, *Paralichthys olivaceus* [17]. Furthermore, mussel meal has also been investigated for its ability to enhance palatability. In common sole, the replacement of fishmeal with mussel meal enhanced diet palatability [18]. However, mussel meal presented only a medium/low attractant potential in Senegalese sole (*Solea senegalensis*), rockfish (*Sebastes schlegeli*) and rainbow trout (*Oncorhynchus mykiss*) [19–21]. In the case of turbot, the addition of small amounts (2 to 8%) of mussel meal to a rapeseed-based diet resulted in improved palatability [11]. Additionally, also in turbot, the inclusion of mussel meal (1%) in high-plant diets increased the mRNA expression of ghrelin in the intestine [22]. In shrimp, Cavalli et al. [12] evaluated the growth and feed utilization of *Farfantepenaeus paulensis* fed with different marine protein sources (fishmeal, squid meal and mussel meal). The results indicated that growth and feed utilization were similar in shrimp fed with fishmeal- and mussel meal-based diets. Also, the dietary inclusion of mussel meal (up to 60%) did not affect the growth, feed efficiency and survival of giant tiger prawn (*Penaeus monodon*) [14].

Due to its high economic value, the areas in which whiteleg shrimp are farmed have been expanded to sub-tropical areas. Therefore, low temperatures have become some of the major constraining factors in culturing whiteleg shrimp, affecting the health of the shrimp by suppressing their immune system and disturbing physiological processes [23–25]. In Southern China, shrimp farming has been affected by winter mortality for several decades, especially in 2008 [26]. Also, in Brazil, especially in Southern Brazil, where the weather is quite unstable, cold stress plays an important role as a natural trigger for disease outbreaks [27]. Mussels have high levels of some free amino acids [28] and polyunsaturated fatty acids (PUFAs) [29] and are a rich source of vitamins and minerals [8]. In fish, several of these nutrients have been implicated in cold resistance [30,31].

This study aims to identify whether mussel meal (species *Perna perna*) can be used as a feed additive in whiteleg shrimp (*L. vannamei*) diets to improve their growth and resistance to thermal shock.

2. Materials and Methods

This experiment was carried out at the Marine Shrimp Laboratory (Laboratório de Camarões Marinhos (LCM/UFSC)) at Barra da Lagoa, Florianópolis, Brazil.

2.1. Shrimp

The shrimp were acquired from Aquatec®, a commercial laboratory located in Canguaretama, RN, Brazil. The shrimp were maintained in a 50 m^3 nursery tank and cultured in a biofloc system until they reached the initial weight required for the experiment (3.5 ± 0.5 g).

2.2. Mussel Meal Preparation

Mussels were bought in the south of Florianópolis (Brazil) at Paraíso das Ostras (27°49′00.5″ S 48°33′49.6″ W) and transported to the Marine Shrimp Laboratory. The mussel meal was prepared via the protocol developed by Dr Antón Salgado from CSIC-IMM, Vigo, Spain (personal communication, 2022). First, the mussels were cooked for 10 min, and the shells were removed. The cooked mussels were placed in a ventilated oven at 70 °C for 48 h. After, the dried mussel meat (comprising 95.8% dry matter, 56.9% crude protein, 12.7% crude fat and 11.6% ash) was minced with a commercial mincer until a powder was obtained and sieved to <600 μm.

2.3. Diets

Five diets were formulated via the animal feed formulation software Optimal Fórmula 2000® (version 19.102.009, Optimal, Campinas, SP, Brazil) to fulfil the nutritional requirements of whiteleg shrimp [32]. Based on the formulation for a commercial-like diet for this species, experimental diets with different levels of inclusion of mussel meal (*P. perna*) were formulated—0 (control), 1, 2, 3 or 4% mussel meal—at the expense of small amounts of other protein sources (Table 1). All five diets were formulated to be isoproteic, isolipidic and isoenergetic. The experimental diets were produced at the Laboratório de Nutrição de Espécies Aquícolas (LABNUTRI, Florianópolis, Brazil), where the selected ingredients were collected, weighed and sieved using a 600 μm sieve. First, the dry ingredients were mixed in a concrete mixer. After mixing the dry ingredients, soybean oil, soy lecithin and fish oil were added and mixed using a concrete mixer for 10 min. The resulting mixture was pelletized with an extruder at 70 °C and a 2.5 mm sieve. Then the pellets were placed in an oven (50 °C) and dried for approximately 24 h until the moisture content was <10%.

The diets were analyzed to obtain their proximate compositions (dry matter, ash, crude protein, crude fat and gross energy), following the procedures standardized by the Association of Official Analytical Chemists [33] and described in Teodósio et al. [34]. Following acid hydrolysis, the dietary amino acid composition (Table 2) of each diet was analyzed via ultra-performance liquid chromatography, according to the procedures described by Aragão et al. [35].

2.4. Experimental Set Up

Twenty 400 L polyethylene tanks were used, and each tank was stocked with 40 shrimp (3.5 ± 0.5 g). A 12 h light/12 h dark regime was used, and each tank was filled with seawater and kept under constant aeration and temperature. Every day, 100% of the water was exchanged to keep the water quality parameters within ideal values. During the water exchange, dead shrimp, shrimp carcasses and feces were removed. Each experimental diet (inclusions of 0, 1, 2, 3 and 4% mussel meal) was randomly assigned to four replicate (n = 4) tanks and evaluated for 8 weeks.

The shrimp were fed four times a day (at 8h00, 11h00, 14h00 and 17h00) with their respective diets following the feeding table by Van Wyk and Scarpa [36]. The experimental diets were placed in a feeding tray (0.03 m^2) underwater in the tank. After 1.5 h, the feeding tray was checked for leftovers.

Oxygen and temperature were measured twice a day (8h00 and 16h30), using a YSI Pro 20 meter that was calibrated once a month. The temperature was kept at 28.4 ± 0.4 °C and the dissolved oxygen was maintained at 6.33 ± 0.23 mg L^{-1}. Weekly, water samples were taken to measure the total ammonia nitrogen (0.98 ± 0.41 mg L^{-1}), nitrite (0.04 ± 0.09 mg L^{-1}), alkalinity (123 ± 3 mg L^{-1}), pH (8.09 ± 0.07) and salinity (30.6 ± 0.27 ‰). The total ammonia nitrogen (TAN) was measured via the indophenol with trione method [37]. Nitrite was measured via the Griess reaction method [38]. Alkalinity was measured via titration using the APHA method 2320-B [39]. pH was measured via a Tecnal pH meter (Tecnal, Piracicaba, SP, Brazil), and the salinity was measured using a YSI EcoSense® probe, model EC300A (YSI, Yellow Springs, OH, USA).

Table 1. Formulations and proximate compositions of the experimental diets containing 0 (control), 1, 2, 3 and 4% mussel meal (*P. perna*).

Ingredients (g kg^{-1})	Diets				
	0%	1%	2%	3%	4%
Mussel meal	0.0	10.0	20.0	30.0	40.0
Soybean meal	324.3	320.3	316.3	312.3	310.3
Fish meal	126.0	125.0	122.0	119.0	115.0
Poultry meal	150.0	145.0	142.0	139.0	135.0
Wheat flour	150.0	150.0	150.0	150.0	150.0
Fish oil	20.0	20.0	20.0	20.0	20.0
Soybean oil	10.0	10.0	10.0	10.0	10.0
Carboxymethyl cellulose	5.0	5.0	5.0	5.0	5.0
Soy lecithin	25.0	25.0	25.0	25.0	25.0
Vitamin C [a]	0.7	0.7	0.7	0.7	0.7
Vitamin premix [b]	5.0	5.0	5.0	5.0	5.0
Mineral premix [c]	17.0	17.0	17.0	17.0	17.0
Monocalcium phosphate	25.0	25.0	25.0	25.0	25.0
Magnesium sulphate	15.0	15.0	15.0	15.0	15.0
Kaolin	100.0	100.0	100.0	100.0	100.0
Sodium chloride	12.0	12.0	12.0	12.0	12.0
Potassium chloride	10.0	10.0	10.0	10.0	10.0
DL-Methionine	5.0	5.0	5.0	5.0	5.0
Proximate composition * (% as fed)					
Moisture	8.9	6.8	9.7	10.8	8.2
Crude protein	36.2	37.0	35.9	35.2	36.3
Crude fat	7.2	7.2	7.8	7.5	7.9
Ash	22.2	22.4	21.8	21.5	22.2
Gross energy (MJ kg^{-1})	16.1	15.7	15.2	15.1	15.5

[a] L-ascorbic acid-2-monophosphate 35%: DSM Produtos Nutricionais Brasil (São Paulo, Brazil). [b] Vitamin premix: In Vivo mix (Paulínia, Brazil)—vitamin A, 3,000,000 IU; vitamin D3, 1,000,000 IU; vitamin E, 70,000 IU; vitamin K3, 14 g; vitamin B1, 30 g; vitamin B2, 20 g; vitamin B6, 33 g; vitamin B12, 50,000 µg; pantothenic acid, 40 g; biotin, 750 mg; nicotinic acid, 70 g; folic acid, 3000 mg; excipient for 1000 g. [c] Mineral premix: InVivo mix (Paulínia, Brazil)—potassium, 6100 mg; copper, 23,330 mg; zinc, 10,000 mg; manganese, 20,000 mg; selenium, 125 mg; iodine, 1000 mg; cobalt, 50 mg; excipient for 1000 g. * All values are reported as means of duplicate analysis.

Table 2. Amino acid compositions of the experimental diets containing 0 (control), 1, 2, 3 and 4% mussel meal (*P. perna*).

Amino Acids	Diets				
(% as Fed)	0%	1%	2%	3%	4%
Arginine	2.3	2.4	2.3	2.2	2.3
Histidine	0.8	0.8	0.8	0.8	0.8
Lysine	2.1	2.2	2.1	2.0	2.1
Threonine	1.4	1.4	1.4	1.3	1.4
Isoleucine	1.7	1.8	1.7	1.6	1.7
Leucine	2.5	2.6	2.5	2.4	2.5
Valine	1.7	1.7	1.7	1.6	1.7
Methionine	1.3	1.3	1.3	1.2	1.3
Phenylalanine	1.8	1.8	1.8	1.7	1.8
Cystine	0.9	1.0	0.9	0.9	1.0
Tyrosine	1.7	1.9	1.8	1.6	1.7
Aspartic acid	3.2	3.3	3.2	3.0	3.2
Glutamic acid	5.4	5.5	5.4	5.1	5.4
Alanine	1.7	1.8	1.7	1.6	1.7
Glycine	2.2	2.2	2.2	2.1	2.1
Proline	1.9	1.9	1.9	1.8	1.8
Serine	1.6	1.7	1.6	1.6	1.6
Taurine	0.2	0.2	0.3	0.3	0.3

All values are reported as means of duplicate analyses.

2.5. Biometry

Once a week, the total biomass for each tank was determined, and the number of shrimp was counted. The weekly biometry measurements were conducted 1.5 h after the first or second round of feeding. The lid, inlet and aeration pipe of each tank were removed, and the water level was reduced to ±15 cm to make it easier and faster to catch the shrimp and thus reduce stress during the biometric analysis. The shrimp were weighed using a scale (Marte AD2000, Marte Científica, Santa Rita do Sapucai, MG, Brazil) with an accuracy of 0.01 g. After weighing, the shrimp were placed back in the tank where they were counted. With the total weight of the shrimp, the new feeding amount was calculated until the next biometry measurements were obtained.

2.6. Shrimp Growth Performance

On the final day of the experiment (the 56th day), the number of shrimp and the tank biomass were determined. The growth performance, feed conversion ratio (FCR) and survival of the shrimp were calculated using the following formulas [40–42]:

$$\begin{gathered}\text{Weight gain (\% initial weight)} = 100 \times \text{wet weight gain} \times \text{initial weight}^{-1},\\ \text{where the wet weight gain is the final weight} - \text{initial weight.}\end{gathered} \quad (1)$$

$$\text{Weekly weight gain (g week}^{-1}) = \text{wet weight gain} \times \text{number of weeks}^{-1}. \quad (2)$$

$$\text{Productivity (kg m}^{-3}) = \text{final biomass} \times \text{tank volume}^{-1}. \quad (3)$$

$$\begin{gathered}\text{Relative growth rate: RGR (\% day}^{-1}) = (e^{g} - 1) \times 100,\\ \text{where } g = (\ln Wt - \ln Wi) \times t^{-1}, Wt \text{ and } Wi \text{ are the final and initial wet weights,}\\ \text{respectively and } t \text{ is the duration of the trial in days.}\end{gathered} \quad (4)$$

$$\text{Feed conversion ratio (FCR)} = \text{feed intake} \times \text{wet weight gain}^{-1}. \quad (5)$$

$$\text{Feed intake (g shrimp}^{-1}): \text{total feed} \times \text{number of shrimp}^{-1}. \quad (6)$$

$$\text{Survival (\%)} = 100 \times \text{final number of shrimp} \times \text{initial number of shrimp}^{-1}. \quad (7)$$

2.7. Thermal Shock

On day 57 (one day after finishing the experiment), 10 shrimp from each tank were subjected to an abrupt, potentially lethal thermal stress (cold shock). The thermal shock test was applied to show the possible effects of the addition of mussel meal on the immune system of the shrimp in the face of sudden temperature changes that are common in tropical and subtropical regions [43]. The experimental design considered the parameters that occur in practice in Brazilian farms and the challenges that the shrimp may experience. The 10 shrimp (18.0 ± 1.8 g) were simultaneously transferred from tanks with seawater at a temperature of 28.4 ± 0.4 °C to a 60 L aquarium filled with ±25 L of seawater at a temperature of 10.9 ± 0.1 °C under constant aeration, where they were maintained for 1 h. After this period, the shrimp were transferred to tanks containing ±30 L of seawater at a temperature of 28.5 ± 1.0 °C, and mortality was monitored for 48 h. The seawater used in the thermal shock trial was from the same reservoir as the experiment, at a salinity of 30.5 ‰.

2.8. Statistical Analysis

The results are presented as means ± standard deviations (SDs). After analyzing the assumptions of normality (Shapiro–Wilk test) and homoscedasticity (Levene test), a one-way analysis of variance (ANOVA), followed by Tukey's multiple comparisons test, was used to identify significant differences among the treatments for growth performance parameters, feed efficiency and survival. For this statistical analysis, the tanks were considered the experimental units (n = 4). Regression analyses between the dietary mussel meal content and performance indicators were performed using a quadratic model. The mortality data after the thermal shock treatment were analyzed via a Kaplan–Meier test. All tests used a significance level of $p < 0.05$, and all results expressed as a percentage were previously arcsine transformed [44]. All statistical tests were performed using the software program SPSS 25.0 (SPSS Inc., Chicago, IL, USA).

3. Results

3.1. Growth Performance

After 56 days of the experiment, the shrimp that were fed with the 1 and the 2% mussel meal diets had significantly ($p < 0.05$) higher final weights, weight gain, average weekly weight gain and relative growth rates than the shrimp fed with the 0, 3 and 4% diets (Table 3). The productivity was the highest for the shrimp fed the 2% diet, while the lowest values were found for the shrimp fed the 0 and 4% diets. The shrimp fed with the 2% diet presented the highest feed intake and the lowest FCR, which were significantly different ($p < 0.05$) from the 0, 3 and 4% treatments, whilst the shrimp fed the 1% diet presented intermediate values (Figure 1).

Table 3. Growth performance of *L. vannamei* fed diets containing 0, 1, 2, 3 and 4% mussel meal for 56 days.

	Diets				
Growth Indicators	**0%**	**1%**	**2%**	**3%**	**4%**
Initial weight (g)	3.50 ± 0.01	3.51 ± 0.01	3.51 ± 0.03	3.50 ± 0.00	3.50 ± 0.01
Final weight (g)	18.50 ± 0.30 ^{b}	19.63 ± 0.18 ^{a}	20.34 ± 0.54 ^{a}	18.11 ± 0.31 ^{b}	18.45 ± 0.38 ^{b}
Weight gain (% initial weight)	429.1 ± 9.0 ^{b}	459.8 ± 5.0 ^{a}	479.8 ± 18.5 ^{a}	418.0 ± 9.2 ^{b}	426.7 ± 10.7 ^{b}
Weekly weight gain (g week^{-1})	1.88 ± 0.04 ^{b}	2.02 ± 0.02 ^{a}	2.10 ± 0.07 ^{a}	1.83 ± 0.04 ^{b}	1.87 ± 0.05 ^{b}
Productivity (kg m^{-3})	1.74 ± 0.10 ^{b}	1.85 ± 0.08 ^{ab}	1.93 ± 0.05 ^{a}	1.80 ± 0.01 ^{ab}	1.76 ± 0.04 ^{b}
RGR (% day^{-1})	3.02 ± 0.03 ^{b}	3.12 ± 0.02 ^{a}	3.19 ± 0.06 ^{a}	2.98 ± 0.03 ^{b}	3.01 ± 0.04 ^{b}
Survival (%)	93.8 ± 4.1	94.4 ± 4.1	95.0 ± 3.5	98.8 ± 1.3	95.6 ± 1.1

RGR: Relative growth rate. Results are means ± SDs (n = 4). Absence of letters indicate no significant differences among the treatments. Different letters indicate significant differences ($p < 0.05$) among the treatments.

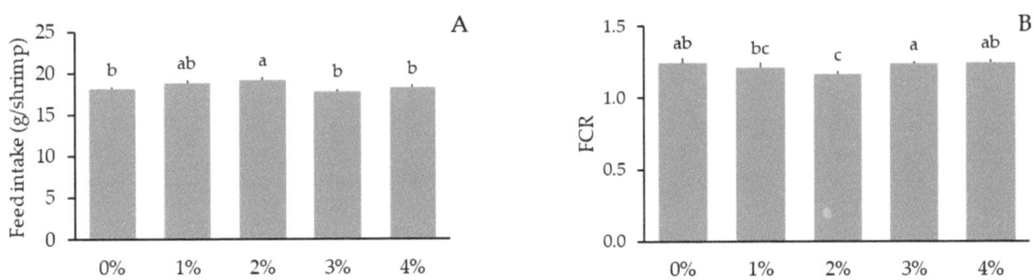

Figure 1. Feed intake (**A**) and feed conversion ratio (**B**) of *L. vannamei* fed diets containing 0, 1, 2 3 and 4% mussel meal for 56 days. Results are means ± SDs (n = 4). Different letters indicate significant differences ($p < 0.05$) among the treatments.

The quadratic regression analysis (Table S1) was significant ($p < 0.05$) for all performance indicators except for survival and feed intake ($p > 0.05$). Figure 2 shows as an example the quadratic regression of the final weight (A) and FCR (B) of the whiteleg shrimp fed diets containing different levels of mussel meal inclusion. According to the models, the best inclusion level of mussel meal is estimated to be from 1.73 to 2.00% (Table S1).

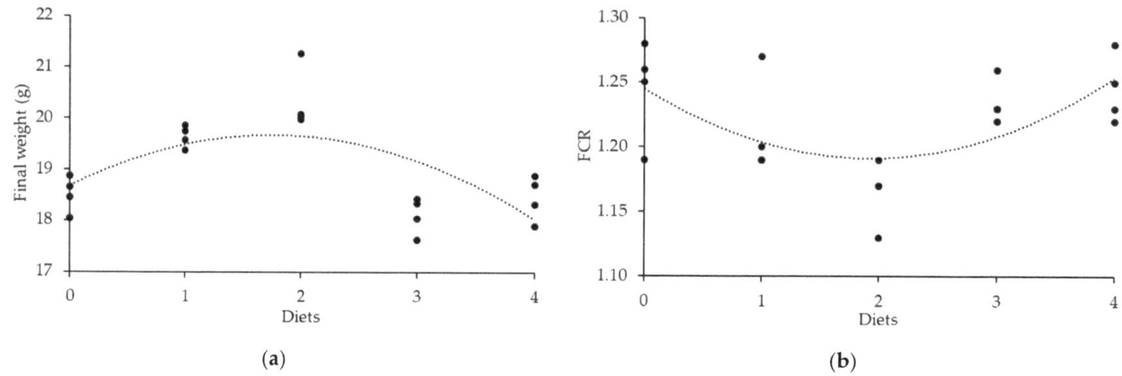

Figure 2. Quadratic regression of final weight (**a**) and feed conversion ratio (**b**) of *L. vannamei* as a function of the dietary mussel meal inclusion (0, 1, 2, 3 and 4%).

3.2. Thermal Shock

The survival of the shrimp 48 h after the thermal shock presented no significant differences among the treatments (Figure 3).

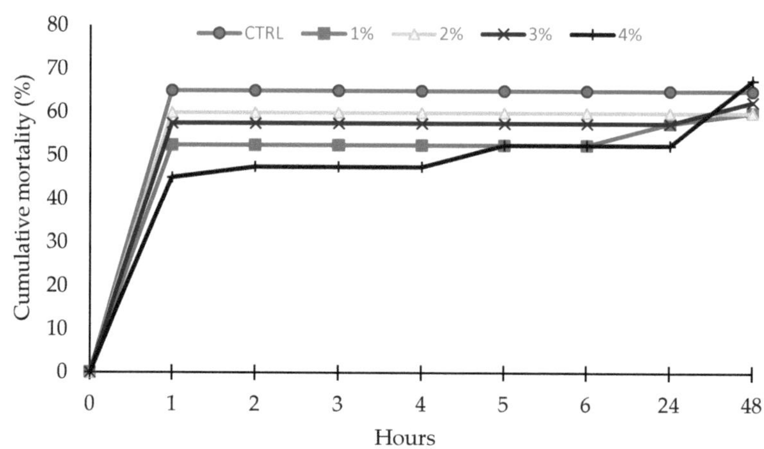

Figure 3. Cumulative mortality (%) of *L. vannamei* during 48 h after the cold shock treatment for treatment groups CTRL (0), 1, 2, 3 and 4%. No significant differences were found among the treatments.

4. Discussion

The present study investigated whether mussel meal can be used as a feed additive in whiteleg shrimp diets. Small amounts, from 1 to 4%, of mussel meal were added to commercial-like diets to see if it improved the growth of the shrimp. Secondly, this research assessed whether the inclusion of mussel meal had a positive effect on the cold resistance of the shrimp. In this study, significant differences were found in the growth performance between the five different treatments. Treatments 1 and 2% obtained significantly better results than the other treatments, with higher final weights, weight gain and RGRs. The weekly weight gain in the present study was above 2 g per week, similar to the values found by Tacon et al. [45] for whiteleg shrimp fed diets containing 35% crude protein and the inclusion of 2.5% squid meal. Nunes et al. [46], who included 1% krill meal in the shrimp diets, reported a weekly growth of 1.01 $\pm$ 0.07 g in a clear water system using the same density as the present study, 100 shrimp m^{-3}. An addition of 3 or 4% mussel meal to the diets showed similar results as the control, with a weekly gain of more or less than 1.85 g. In addition, the mortality that occurred was low and similar in all treatments and intrinsic to the possible stress caused by the management of the intensive system.

Feed costs can represent from 50 to 70% of the total cost of production in aquaculture [47–49]. Currently, the aquafeed industry uses plant ingredients such as soybean meal as the main protein source for shrimp diets since they are less expensive and more readily available compared to fishmeal. However, plant-based diets may be less attractive and palatable to farmed organisms [50,51]. Crustaceans have antennular chemoreception mechanisms to help identify attractive feeds, and amino acids and nucleotides seem to be the main triggers for these mechanisms [52,53]. In this study, despite the feed intake being significantly higher in shrimp from the 2% treatment, the tested experimental diets had very similar amino acid profiles.

The use of marine ingredients, such as krill meal, squid meal and, as in the present study, mussel meal, contain substances that act as chemoattractants and feeding stimulants for shrimp [46,52,54] and may result in increased feed consumption and decreased waste. In this study, the addition of 1 and 2% mussel meal increased feed intake, indicating that up to a certain level, this ingredient may be used as a feeding stimulant. Interestingly, besides the higher feed intake, these shrimp also grew better, resulting in a lower FCR compared to shrimp fed on the other diets. A lower feed conversion ratio was also found in shrimp fed with 3% krill meal in diets with low fishmeal content [46]. In giant tiger prawn, similar zootechnical performance was found between treatments when the prawns were fed a diet containing 20, 40 or 60% of mussel meal (*Mytella charruana*), with no adverse effects [14]. There is a limited body of recent literature exploring the inclusion of mussel meal as a potential ingredient in shrimp diets, highlighting a significant information gap that can be addressed. In this study, we identified that the inclusion of small amounts of mussel meal in the diet of whiteleg shrimp demonstrates a positive potential on zootechnical performance.

Recently, the use of marine bivalves as a potential ingredient for the aquafeed industry was suggested [55]. These resources, in addition to having high levels of protein, vitamins C and B2 and minerals such as iron and magnesium [8], are low-trophic-level filtering species and can be cultivated as tools for mitigating excesses of nutrients in marine environments [7]. For lobster farming, the industry has been testing the inclusion of mussel meal in manufactured diets to reduce reliance on mussels as fresh food [56,57]. In fish, several research studies have already revealed that mussel meal has a high potential for serving as a supplement or a fishmeal replacement in diets, resulting in similar or better growth performances. For example, in juvenile Ussuri catfish, mussel meal protein can successfully substitute 50% of fishmeal protein (~18% maximum inclusion level of mussel meal) without significant effects on growth, nutrient utilization and hepatic IGF-I gene expression [10]. In a study conducted on turbot, it was demonstrated that fishmeal can be substituted with mussel meal up to 25% (11% mussel meal inclusion) without compromising growth or negatively impacting the fish [16]. In Artic charr, similar growth performance, nutrient digestibility and intestinal barrier function were found in groups fed diets with 40% of the fishmeal replaced with mussel meal (22% mussel meal inclusion) compared to the control (0% replacement of fishmeal [15]). In juvenile common sole, a higher specific growth rate, higher feed intake and lower feed conversion ratio were found for groups fed fishmeal replacement diets (up to 75%, a 25% maximum inclusion of mussel meal [9]). Shrimp (*F. paulensis*) fed diets based on fishmeal or mussel meal (40% inclusion for both ingredients) did not present significant differences in growth performance, feed efficiency or survival [12]. Most of these studies investigated the partial or full replacement of fishmeal with mussel meal. Nevertheless, mussel meal was also tested as a feed additive, small amounts of which (2 to 8% of mussel meal inclusion) increased palatability in rapeseed-protein-based diets in turbot [11] or increased the feed intake and specific growth rate in common sole [9]. In the present work with shrimp, increases in feed intake were also found for the 1 and 2% levels of dietary mussel meal inclusion. The quadratic regression models corroborate that the best inclusion levels are from 1.73 to 2.00%, demonstrating that mussel meal can work well as a strategic nutritional additive for marine shrimp diets.

Mussel meal is rich in different compounds like minerals and unsaturated fatty acids that could improve the robustness of shrimp, as shown in previous studies. For instance, the beneficial role of a diet enriched with highly unsaturated fatty acids with respect to tolerance to handling stress and the immune response of whiteleg shrimp juveniles was reported [58]. Also, an improvement in osmoregulation capacity was monitored in whiteleg shrimp juveniles when their diets were supplemented with unsaturated fatty acids [59]. Post-larvae whiteleg shrimp fed with small amounts of brown seaweeds (1% of *Sargassum filipendula* combined with 2% of *Undaria pinnatifidia*) showed lower rates of cumulative mortality when exposed to a cold shock (13.5 ± 0.1 °C) than the control shrimp, which were fed diets without seaweed inclusion [40]. Similar results were found in post-larvae whiteleg shrimp fed with 0.5 and 2% inclusions of *S. filipendula*, showing lower rates of cumulative mortality after cold shock (12.5 ± 0.4 °C) compared with a control treatment [60]. Another study using the same seaweed at a concentration of 0.5%, demonstrated a greater resistance to cold shock in whiteleg shrimp at an even lower temperature (11.5 ± 0.1 °C; [61]). Coelho et al. [62] used the same temperature to subject whiteleg shrimp juveniles to a thermal shock challenge and observed greater survival in shrimp fed with the dietary inclusion of 2% *Ulva ohnoi*. Whiteleg shrimp fed diets enriched with 0.5 and 1% of *Aurantiochytrium* sp. flour, which is rich in docosahexaenoic acid (DHA), showed higher rates of survival for cold shock (12.5–13.0 °C) when compared to a control treatment [63]. In the current experiment, no differences were measured in survival among the treatments after the cold shock (10.9 ± 0.1 °C). This may be related to the final weight of the animals, which was higher than the final weight of the animals in the aforementioned studies. The size of the shrimp plays a significant role in their resistance to cold shock, with juvenile shrimp exhibiting greater resilience compared to post-larvae shrimp [64]. Furthermore, resistance to cold shock may be directly related to the concentration of unsaturated fatty acids in the diet, which are capable of impacting cell membrane fluidity [65,66]. In the present study, a slightly higher lipid content was observed in the diet with the inclusion of 4% mussel meal, and the lowest mortality rate during the first hours of the thermal shock challenge was observed for this treatment, although the accumulated mortality data did not show statistical differences at the end of the 48 h. Mussels are considered marine sources of fatty acids and have considerable amounts of DHA and eicosapentaenoic acid (EPA), which are long-chain unsaturated fatty acids [67,68] that improve immunity and positively impact the growth and zootechnical performance of whiteleg shrimp [69,70], although the nutritional requirement is not well defined [32]. Thus, the impact of mussel meal on the tolerance of shrimp to thermal shock and on their immune responses seem to be worth further investigation.

5. Conclusions

The current study demonstrated that shrimp fed with 1 and 2% mussel meal diets had significantly higher final weights, weight gain, relative growth rates and lower feed conversion ratios than the shrimp fed with the control, 3 and 4% mussel meal diets. The shrimp fed with the 2% mussel meal diet showed the best growth results. After 8 weeks of the experiment, the shrimp from this treatment were significantly heavier than the control shrimp by 10%. Further, no differences were observed in cold resistance and survival among the treatments. In conclusion, mussel meal can be used as a potential additive in whiteleg shrimp diets to promote growth, and inclusion levels between 1.73 and 2.00% are indicated.

Supplementary Materials: The following supporting information can be downloaded at: https://www.mdpi.com/article/10.3390/jmse11091670/s1, Table S1: Results from the regression analyses between the dietary mussel meal content and the performance indicators of *L. vannamei*.

Author Contributions: Conceptualization, F.N.V. and D.M.F.; methodology, F.N.V. and D.M.F.; validation, F.N.V. and C.A.; formal analysis: S.C., F.N.V. and C.A.; investigation, S.C., F.B.H. and I.P.; resources: F.N.V., D.M.F. and C.A.; data curation: F.N.V., C.A. and S.C.; writing—original draft preparation, S.C. and F.B.H.; writing—review and editing, C.A., F.N.V., I.P. and D.M.F.; visualization, S.C. and F.B.H.; supervision, F.N.V. and C.A.; project administration, F.N.V. and C.A.; funding acquisition, F.N.V. and C.A. All authors have read and agreed to the published version of the manuscript.

Funding: This project received funding from the European Union's Horizon 2020 Research and Innovation Programme under Grant Agreement No 818173 and Portuguese national funds from FCT—Foundation for Science and Technology to CCMAR (projects UIDB/04326/2020, UIDP/04326/2020 and LA/P/0101/2020) and to C.A. (contract DL57/2016/CP1361/CT0033). F.N.V and D.M.F. are research fellows from CNPQ (310250/2020-0). F.B.H. received a scholarship from FAPESC (48/2021). This document reflects the views only of the authors, and the European Commission cannot be held responsible for any use that may be made of the information it contains.

Institutional Review Board Statement: All applicable international and national guidelines for the care and use of animals were followed by the authors. An Institutional Review Board Statement is not necessary for studies with invertebrates other than cephalopods.

Informed Consent Statement: Not applicable.

Data Availability Statement: Data supporting reported results can be found in the publicly archived dataset generated during this study at https://doi.org/10.5281/zenodo.7722884 (accessed on 27 July 2023).

Acknowledgments: The authors would like to thank a number of people who contributed to this manuscript: Ramon Carneiro, Efrayn Wilker, Ivanilson Santos, Vitória Menoncin, Camila Miranda and Mariana Moraes for their assistance during different stages of the experiment and Priscila Costa Rezende for formulating the diets.

Conflicts of Interest: The authors declare no conflict of interest.

References

1. FAO. *The State of the World Fisheries and Aquaculture. Towards Blue Transformation*; Food and Agriculture Organization of the United Nations: Rome, Italy, 2022.
2. FAO. *World Food and Agriculture—Statistical Yearbook*; FAO: Rome, Italy, 2021.
3. Emerenciano, M.G.C.; Rombenso, A.N.; Vieira, F.d.N.; Martins, M.A.; Coman, G.J.; Truong, H.H.; Noble, T.H.; Simon, C.J. Intensification of Penaeid Shrimp Culture: An Applied Review of Advances in Production Systems, Nutrition and Breeding. *Animals* **2022**, *12*, 236. [CrossRef] [PubMed]
4. Aragão, C.; Gonçalves, A.T.; Costas, B.; Azeredo, R.; Xavier, M.J.; Engrola, S. Alternative Proteins for Fish Diets: Implications beyond Growth. *Animals* **2022**, *12*, 1211. [CrossRef] [PubMed]
5. Chen, Y.; Mitra, A.; Rahimnejad, S.; Chi, S.; Kumar, V.; Tan, B.; Niu, J.; Xie, S. Retrospect of Fish Meal Substitution in Pacific White Shrimp (*Litopenaeus Vannamei*) Feed: Alternatives, Limitations and Future Prospects. *Rev. Aquac.* 2023; *early view*. [CrossRef]
6. Jönsson, L.; Elwinger, K. Mussel Meal as a Replacement for Fish Meal in Feeds for Organic Poultry—A Pilot Short-Term Study. *Acta Agric. Scand. A Anim. Sci.* **2009**, *59*, 22–27. [CrossRef]
7. Petersen, J.K.; Hasler, B.; Timmermann, K.; Nielsen, P.; Tørring, D.B.; Larsen, M.M.; Holmer, M. Mussels as a Tool for Mitigation of Nutrients in the Marine Environment. *Mar. Pollut. Bull.* **2014**, *82*, 137–143. [CrossRef] [PubMed]
8. Suplicy, F.M. A Review of the Multiple Benefits of Mussel Farming. *Rev. Aquac.* **2020**, *12*, 204–223. [CrossRef]
9. Mongile, F.; Mandrioli, L.; Mazzoni, M.; Pirini, M.; Zaccaroni, A.; Sirri, R.; Parma, L.; Gatta, P.P.; Sarli, G.; Bonaldo, A. Dietary Inclusion of Mussel Meal Enhances Performance and Improves Feed and Protein Utilization in Common Sole (*Solea solea*, Linnaeus, 1758) Juveniles. *J. Appl. Ichthyol.* **2015**, *31*, 1077–1085. [CrossRef]
10. Luo, C.; Wang, Y.; Tao, S.; Liao, Y.; Yang, C.; Cui, C.; Yang, J.; Yang, Y.C.N.-P.D.F. Effects of Replacing Fish Meal with Mussel (*Cristaria plicata*) Meat on Growth, Digestive Ability, Antioxidant Capacity and Hepatic IGF-I Gene Expression in Juvenile Ussuri Catfish (*Pseudobagrus ussuriensis*). *Aquac. Res.* **2019**, *50*, 826–835. [CrossRef]
11. Nagel, F.; von Danwitz, A.; Schlachter, M.; Kroeckel, S.; Wagner, C.; Schulz, C. Blue Mussel Meal as Feed Attractant in Rapeseed Protein-Based Diets for Turbot (*Psetta maxima* L.). *Aquac. Res.* **2014**, *45*, 1964–1978. [CrossRef]
12. Cavalli, R.O.; Zimermann, S.; Speck, R.C. Growth and Feed Utilization of the Shrimp *Farfantepenaeus paulensis* Fed Diets Containing Different Marine Protein Sources. *Ciênc. Rural* **2004**, *34*, 891–896. [CrossRef]
13. Kikuchi, K.; Furuta, T. Use of Defatted Soybean Meal and Blue Mussel Meat as Substitute for Fish Meal in the Diet of Tiger Puffer, *Takifugu rubripes*. *J. World Aquac. Soc.* **2009**, *40*, 472–482. [CrossRef]

14. Cruz, A.C.D.; Ombac, J.E.; Siapno, K.C.; Reyes, C.R.; Aban, S.M. Growth and Survival of Giant Tiger Prawn (*Penaeus monodon*) Juveniles Fed with Formulated Diets Containing Different Levels of Charru Mussel (*Mytella charruana*) Meal. *J. Nat. Allied Sci.* **2022**, *VI*, 40–55.
15. Vidakovic, A.; Langeland, M.; Sundh, H.; Sundell, K.; Olstorpe, M.; Vielma, J.; Kiessling, A.; Lundh, T. Evaluation of Growth Performance and Intestinal Barrier Function in Arctic Charr (*Salvelinus alpinus*) Fed Yeast (*Saccharomyces cerevisiae*), Fungi (*Rhizopus oryzae*) and Blue Mussel (*Mytilus edulis*). *Aquac. Nutr.* **2016**, *22*, 1348–1360. [CrossRef]
16. Weiß, M.; Buck, B.H. Partial Replacement of Fishmeal in Diets for Turbot (*Scophthalmus maximus*, Linnaeus, 1758) Culture Using Blue Mussel (*Mytilus edulis*, Linneus, 1758) Meat. *J. Appl. Ichthyol.* **2017**, *33*, 354–360. [CrossRef]
17. Kikuchi, K.; Ueda, A.; Haruo, S.; Takeda, S. Effect of Dietary Inclusion of Blue Mussel Extract on Growth and Body Composition of Japanese Flounder *Paralichthys olivaceus*. *J. World Aquac. Soc.* **2002**, *33*, 41–47. [CrossRef]
18. Mazzoni, M.; Bonaldo, A.; Gatta, P.P.; Vallorani, C.; Latorre, R.; Canova, M.; Clavenzani, P. α-Transducin and α-Gustducin Immunoreactive Cells in the Stomach of Common Sole (*Solea solea*) Fed with Mussel Meal. *Fish Physiol. Biochem.* **2015**, *41*, 603–612.
19. Barroso, F.G.; Rodiles, A.; Vizcaino, A.J.; Martínez, T.F.; Alarcón, F.J. Evaluation of Feed Attractants in Juvenile Senegalese Sole, *Solea senegalensis*. *J. World Aquac. Soc.* **2013**, *44*, 682–693. [CrossRef]
20. Kim, H.S.; Baek, S.I.; Lee, K.W.; Jeong, H.S.; Cho, S.H. Attractiveness of Various Protein Sources to Juvenile Rockfish (*Sebastes schlegeli*, Hilgendorf 1880). *J. Appl. Aquac.* **2020**, *32*, 205–220. [CrossRef]
21. Tusche, K.; Berends, K.; Wuertz, S.; Susenbeth, A.; Schulz, C. Evaluation of Feed Attractants in Potato Protein Concentrate Based Diets for Rainbow Trout (*Oncorhynchus mykiss*). *Aquaculture* **2011**, *321*, 54–60. [CrossRef]
22. Jiang, D.; Zheng, J.; Dan, Z.; Tang, Z.; Ai, Q.; Mai, K. Effects of Five Compound Attractants in High Plant-Based Diets on Feed Intake and Growth Performance of Juvenile Turbot (*Scophthalmus maximus* L.). *Aquac. Res.* **2019**, *50*, 2350–2358. [CrossRef]
23. Zhou, M.; Wang, A.-L.; Xian, J.-A. Variation of Free Amino Acid and Carbohydrate Concentrations in White Shrimp, *Litopenaeus vannamei*: Effects of Continuous Cold Stress. *Aquaculture* **2011**, *317*, 182–186.
24. Qiu, J.; Wang, W.-N.; Wang, L.; Liu, Y.-F.; Wang, A.-L. Oxidative Stress, DNA Damage and Osmolality in the Pacific White Shrimp, *Litopenaeus vannamei* Exposed to Acute Low Temperature Stress. *Comp. Biochem. Physiol. Part C Toxicol. Pharmacol.* **2011**, *154*, 36–41. [CrossRef]
25. Fan, L.; Wang, A.; Wu, Y. Comparative Proteomic Identification of the Hemocyte Response to Cold Stress in White Shrimp, *Litopenaeus vannamei*. *J. Proteom.* **2013**, *80*, 196–206. [CrossRef]
26. Li, B.; Xian, J.-A.; Guo, H.; Wang, A.-L.; Miao, Y.-T.; Ye, J.-M.; Ye, C.-X.; Liao, S.-A. Effect of Temperature Decrease on Hemocyte Apoptosis of the White Shrimp *Litopenaeus vannamei*. *Aquac. Int.* **2014**, *22*, 761–774. [CrossRef]
27. Valenti, W.C.; Barros, H.P.; Moraes-Valenti, P.; Bueno, G.W.; Cavalli, R.O. Aquaculture in Brazil: Past, Present and Future. *Aquac. Rep.* **2021**, *19*, 100611. [CrossRef]
28. Naik, A.S.; Mora, L.; Hayes, M. Characterisation of Seasonal Mytilus Edulis By-Products and Generation of Bioactive Hydrolysates. *Appl. Sci.* **2020**, *10*, 6892. [CrossRef]
29. Carboni, S.; Kaur, G.; Pryce, A.; McKee, K.; Desbois, A.P.; Dick, J.R.; Galloway, S.D.R.; Hamilton, D.L. Mussel Consumption as a "Food First" Approach to Improve Omega-3 Status. *Nutrients* **2019**, *11*, 1381. [CrossRef]
30. Richard, N.; Silva, T.S.; Wulff, T.; Schrama, D.; Dias, J.P.; Rodrigues, P.M.L.; Conceição, L.E.C. Nutritional Mitigation of Winter Thermal Stress in Gilthead Seabream: Associated Metabolic Pathways and Potential Indicators of Nutritional State. *J. Proteom.* **2016**, *142*, 1–14. [CrossRef]
31. Abdel-Ghany, H.M.; El-Sayed, A.-F.M.; Ezzat, A.A.; Essa, M.A.; Helal, A.M. Dietary Lipid Sources Affect Cold Tolerance of Nile Tilapia (*Oreochromis niloticus*). *J. Therm. Biol.* **2019**, *79*, 50–55. [CrossRef]
32. National Research Council. *Nutrient Requirements of Fish and Shrimp*; The National Academies Press: Washington, DC, USA, 2011.
33. AOAC. *Official Methods of Analysis*, 18th ed.; Association of Official Analytical Chemists: Gaithersburgs, MD, USA, 2006.
34. Teodósio, R.; Aragão, C.; Colen, R.; Carrilho, R.; Dias, J.; Engrola, S. A Nutritional Strategy to Promote Gilthead Seabream Performance under Low Temperatures. *Aquaculture* **2021**, *537*, 736494. [CrossRef]
35. Aragão, C.; Cabano, M.; Colen, R.; Fuentes, J.; Dias, J. Alternative Formulations for Gilthead Seabream Diets: Towards a More Sustainable Production. *Aquac. Nutr.* **2020**, *26*, 444–455. [CrossRef]
36. Van Wyk, P.; Scarpa, J. Water Quality Requirements and Management. In *Farming Marine Shrimp in Recirculating Freshwater Systems*; Van Wyk, P., Davis-Hodgkins, M., Laramore, R., Main, K.L., Mountain, J.J.S., Eds.; Florida Department of Agriculture and Consumer Services: Tallahassee, FL, USA, 1999.
37. Grasshoff, K.; Ehrhardt, M.; Kremling, K. *Methods of Seawater Analysis*; Wiley-VCH: Weinhein, Germany, 1983.
38. Strickland, J.D.H.; Parsons, T.R. *A Practical Handbook of Seawater Analysis*; Fisheries Research Board of Canada: Ottawa, ON, Canada, 1972.
39. APHA. *Standard Methods for the Examination of Water and Wastewater*; American Public Health Association: Washington, DC, USA, 2005.
40. Costa Rezende, P.; Soares, M.; Guimarães, A.M.; da Rosa Coelho, J.; Seiffert, W.Q.; Dias Schleder, D.; Vieira, F.d.N. Brown Seaweeds Added in the Diet Improved the Response to Thermal Shock and Reduced *Vibrio* spp. in Pacific White Shrimp Post-Larvae Reared in a Biofloc System. *Aquac. Res.* **2021**, *52*, 2852–2861. [CrossRef]
41. Aragão, C.; Conceição, L.E.C.; Lacuisse, M.; Yúfera, M.; Dinis, M.T. Do Dietary Amino Acid Profiles Affect Performance of Larval Gilthead Seabream? *Aquat. Living Resour.* **2007**, *20*, 155–161. [CrossRef]

42. Crane, D.P.; Ogle, D.H.; Shoup, D.E. Use and Misuse of a Common Growth Metric: Guidance for Appropriately Calculating and Reporting Specific Growth Rate. *Rev. Aquac.* **2020**, *12*, 1542–1547. [CrossRef]
43. De Silva, S.S.; Soto, D. Climate Change and Aquaculture: Potential Impacts, Adaptation and Mitigation. In *Climate Change Implications for Fisheries and Aquaculture: Overview of Current Scientific Knowledge*; Cochrane, K., De Young, C., Soto, D., Bahri, T., Eds.; Food and Agriculture Organization of the United Nations: Rome, Italy, 2009; pp. 151–212.
44. Ennos, R. *Statistical and Data Handling Skills in Biology*, 2nd ed.; Pearson Prentice Hall: Harlow, UK, 2007.
45. Tacon, A.G.J.; Cody, J.J.; Conquest, L.D.; Divakaran, S.; Forster, I.P.; Decamp, O.E. Effect of Culture System on the Nutrition and Growth Performance of Pacific White Shrimp *Litopenaeus vannamei* (Boone) Fed Different Diets. *Aquac. Nutr.* **2002**, *8*, 121–137. [CrossRef]
46. Nunes, A.J.P.; Sabry-Neto, H.; Oliveira-Neto, S.; Burri, L. Feed Preference and Growth Response of Juvenile *Litopenaeus vannamei* to Supplementation of Marine Chemoattractants in a Fishmeal-Challenged Diet. *J. World Aquac. Soc.* **2019**, *50*, 1048–1063. [CrossRef]
47. Rana, K.J.; Siriwardena, S.; Hasan, M.R. *Impact of Rising Feed Ingredient Prices on Aquafeeds and Aquaculture Production*; Food and Agriculture Organization of the United Nations: Rome, Italy, 2009.
48. Asche, F.; Roll, K.H.; Sandvold, H.N.; Sørvig, A.; Zhang, D. Salmon Aquaculture: Larger Companies and Increased Production. *Aquac. Econ. Manag.* **2013**, *17*, 322–339. [CrossRef]
49. Glencross, B.; Fracalossi, D.M.; Hua, K.; Izquierdo, M.; Mai, K.; Øverland, M.; Robb, D.; Roubach, R.; Schrama, J.; Small, B.; et al. Harvesting the Benefits of Nutritional Research to Address Global Challenges in the 21st Century. *J. World Aquac. Soc.* **2023**, *54*, 343–363. [CrossRef]
50. Glencross, B.D.; Booth, M.; Allan, G.L. A Feed Is Only as Good as Its Ingredients—A Review of Ingredient Evaluation Strategies for Aquaculture Feeds. *Aquac. Nutr.* **2007**, *13*, 17–34. [CrossRef]
51. Oliva-Teles, A.; Enes, P.; Peres, H. Replacing Fishmeal and Fish Oil in Industrial Aquafeeds for Carnivorous Fish. In *Feed and Feeding Practices in Aquaculture*; Davis, D.A., Ed.; Woodhead Publishing: Oxford, UK, 2015; pp. 203–233. ISBN 978-0-08-100506-4.
52. Guillaume, J.; Cruz, R.; Cuzon, G.; Wormhoudt, V.A.; Revol, A. Growth Factors in Penaeid Shrimp Feeding. In *Actes de Colloques*; IFREMER: Tahiti, French Polynesia, 1989; pp. 327–338.
53. Derby, C.D.; Sorensen, P.W. Neural Processing, Perception, and Behavioral Responses to Natural Chemical Stimuli by Fish and Crustaceans. *J. Chem. Ecol.* **2008**, *34*, 898–914. [CrossRef]
54. Soares, R.; Peixoto, S.; Davis, R.P.; Davis, D.A. Feeding Behavior and Growth of *Litopenaeus vannamei* Fed Soybean-Based Diets with Added Feeding Effectors. *Aquaculture* **2021**, *536*, 736487. [CrossRef]
55. Albrektsen, S.; Kortet, R.; Skov, P.V.; Ytteborg, E.; Gitlesen, S.; Kleinegris, D.; Mydland, L.T.; Hansen, J.Ø.; Lock, E.J.; Mørkøre, T.; et al. Future Feed Resources in Sustainable Salmonid Production: A Review. *Rev. Aquac.* **2022**, *14*, 1790–1812. [CrossRef]
56. Simon, C.J.; Jeffs, A. Feeding and Gut Evacuation of Cultured Juvenile Spiny Lobsters, *Jasus edwardsii*. *Aquaculture* **2008**, *280*, 211–219. [CrossRef]
57. Nankervis, L.; Jones, C. Recent Advances and Future Directions in Practical Diet Formulation and Adoption in Tropical Palinurid Lobster Aquaculture. *Rev. Aquac.* **2022**, *14*, 1830–1842. [CrossRef]
58. Mercier, L.; Racotta, I.S.; Yepiz-Plascencia, G.; Muhlia-Almazán, A.; Civera, R.; Quiñones-Arreola, M.F.; Wille, M.; Sorgeloos, P.; Palacios, E. Effect of Diets Containing Different Levels of Highly Unsaturated Fatty Acids on Physiological and Immune Responses in Pacific Whiteleg Shrimp *Penaeus vannamei* (Boone) Exposed to Handling Stress. *Aquac. Res.* **2009**, *40*, 1849–1863. [CrossRef]
59. Chen, K.; Li, E.; Xu, C.; Wang, X.; Li, H.; Qin, J.G.; Chen, L. Growth and Metabolomic Responses of Pacific White Shrimp (*Litopenaeus vannamei*) to Different Dietary Fatty Acid Sources and Salinity Levels. *Aquaculture* **2019**, *499*, 329–340. [CrossRef]
60. Schleder, D.D.; da Rosa, J.R.; Guimarães, A.M.; Ramlov, F.; Maraschin, M.; Seiffert, W.Q.; do Nascimento Vieira, F.; Hayashi, L.; Andreatta, E.R. Brown Seaweeds as Feed Additive for White-Leg Shrimp: Effects on Thermal Stress Resistance, Midgut Microbiology, and Immunology. *J. Appl. Phycol.* **2017**, *29*, 2471–2477. [CrossRef]
61. Schleder, D.D.; Blank, M.; Peruch, L.G.B.; Vieira, F.d.N.; Andreatta, E.R.; Hayashi, L. Thermal Resistance of Pacific White Shrimp Fed *Sargassum filipendula*: A MALDI-TOF Mass Spectrometry Approach. *Aquaculture* **2017**, *481*, 103–111. [CrossRef]
62. Coelho, J.d.R.; Santos, U.R.A.; Rezende, P.C.; Constanza Bolívar Ramírez, N.; Vieira, F.N. Increased Resistance to Thermal Shock in Pacific White Shrimp Fed on Green Algae and Its Effect in Conjunction with Probiotics. *Int. Aquat. Res.* **2023**, *15*, 123–134. [CrossRef]
63. Rocha, J.S.; Soares, M.; Guimarães, A.M.; Schleder, D.D.; Seiffert, W.Q.; Vieira, F.N. Addition of *Aurantiochytrium* sp. Meal in the Diet Affects Immunity and Thermal Shock Resistance of the Pacific White Shrimp. *Bol. Inst. Pesca* **2021**, *47*, e631. [CrossRef]
64. Pontinha, V.A.; Vieira, F.N.; Hayashi, L. Mortality of Pacific White Shrimp Submitted to Hypothermic and Hyposalinic Stress. *Bol. Inst. Pesca* **2018**, *44*, e310.
65. Chowanski, S.; Lubawy, J.; Spochacz, M.; Ewelina, P.; Grzegorz, S.; Rosinski, G.; Slocinska, M. Cold Induced Changes in Lipid, Protein and Carbohydrate Levels in the Tropical Insect *Gromphadorhina coquereliana*. *Comp. Biochem. Physiol. A Mol. Integr. Physiol.* **2015**, *183*, 57–63. [CrossRef] [PubMed]
66. Hayward, S.A.L.; Manso, B.; Cossins, A.R. Molecular Basis of Chill Resistance Adaptations in Poikilothermic Animals. *J. Exp. Biol.* **2014**, *217*, 6–15. [CrossRef] [PubMed]
67. Taylor, A.G.; Savage, C. Fatty Acid Composition of New Zealand Green-Lipped Mussels, *Perna canaliculus*: Implications for Harvesting for n-3 Extracts. *Aquaculture* **2006**, *261*, 430–439. [CrossRef]

68. McLean, C.H.; Bulling, K.I.M.R. Differences in Lipid Profile of New Zealand Marine Species over Four Seasons. *J. Food Lipids* **2005**, *12*, 313–326. [CrossRef]
69. González-Félix, M.L.; Gatlin, D.M.; Lawrence, A.L.; Perez-Velazquez, M. Effect of Dietary Phospholipid on Essential Fatty Acid Requirements and Tissue Lipid Composition of *Litopenaeus vannamei* Juveniles. *Aquaculture* **2002**, *207*, 151–167. [CrossRef]
70. Araújo, B.C.; Mata-Sotres, J.A.; Viana, M.T.; Tinajero, A.; Braga, A. Fish Oil-Free Diets for Pacific White Shrimp *Litopenaeus vannamei*: The Effects of DHA-EPA Supplementation on Juvenile Growth Performance and Muscle Fatty Acid Profile. *Aquaculture* **2019**, *511*, 734276. [CrossRef]

Journal of
*Marine Science
and Engineering*

MDPI

Article

Alternative Feed Formulations Impact Growth Performance, Flesh Quality and Consumer Acceptance of Rainbow Trout (*Oncorhynchus mykiss*)

Gabriella do Vale Pereira [1,2,*], Luis E. C. Conceição [1,*], Filipe Soares [1], Jessica Petereit [3], Bela H. Buck [3,4], Johan Johansen [5], Jorge Dias [1] and Filippo Faccenda [6]

1 Sparos Lda, 8700-221 Olhão, Portugal; felipesoares@sparos.pt (F.S.)
2 Aquaculture Department, Federal University of Santa Catarina (UFSC), Florianópolis 88034-000, Brazil
3 Alfred Wegener Institute Helmholtz Centre for Polar and Marine Research, 27570 Bremerhaven, Germany; jessica.petereit@awi.de (J.P.); bela.h.buck@awi.de (B.H.B.)
4 Applied Marine Biology, University of Applied Sciences Bremerhaven, 27570 Bremerhaven, Germany
5 Norwegian Institute of Bioeconomy Research (NIBIO), NO-8027 Bodø, Norway; johan.johansen@nibio.no
6 Edmund Mach Foundation, Technology Transfer Centre, San Michele all'Adige, 38010 Trento, Italy
* Correspondence: gabriellapereira@sparos.pt (G.d.V.P.); luisconceicao@sparos.pt (L.E.C.C.)

Abstract: This trial aimed to assess the growth performance of trout (*Oncorhynchus mykiss*) fed novel formulations, evaluate fish welfare status, and determine flesh quality as part of the evaluation of sustainable feeds. A control diet containing fish meal and soy products (CTRL) was compared to: a diet with processed animal proteins (PAP); a diet without PAP (NoPAP); a PAP diet lower in protein (PAP−); and a NoPAP diet higher in protein (NoPAP+). Groups of 50 fish, weighing 58.84 ± 1.39 g (IBW), were allocated to 20 tanks and fed with formulated diets ad libitum over 91 days. Better growth performance was observed after the experiment in fish fed the NoPAP+ diet when compared to other diets. Protein retention was higher in CTRL diets than in PAP and PAP− diets. Protein and phosphorous digestibility were lower in fish fed PAP− diet. Diets did not influence the texture analysis. However, sensory analysis revealed higher acceptance for fish fed the NoPAP diet when compared to the PAP diet. Lysozyme was higher in the NoPAP diet than in other treatments. In addition, long-term predictions using FEEDNETICS™ software suggest some of these alternative formulations may be economically sustainable. Overall, these results support the hypothesis that the new formulations are viable options for trout farming.

Keywords: emerging ingredients; processed animal proteins; sustainable feeds; alternative diets

Citation: Vale Pereira, G.d.; Conceição, L.E.C.; Soares, F.; Petereit, J.; Buck, B.H.; Johansen, J.; Dias, J.; Faccenda, F. Alternative Feed Formulations Impact Growth Performance, Flesh Quality and Consumer Acceptance of Rainbow Trout (*Oncorhynchus mykiss*). *J. Mar. Sci. Eng.* **2023**, *11*, 1135. https://doi.org/10.3390/jmse11061135

Academic Editors: Ana Catarina Matias and Carlos Andrade

Received: 28 March 2023
Revised: 22 May 2023
Accepted: 23 May 2023
Published: 28 May 2023

1. Introduction

Over the past few decades, considerable effort has been made to find practical alternatives to reduce the aquafeed industry's reliance on marine-derived ingredients, i.e., fish meal (FM) and fish oil (FO). Sustainability concerns, i.e., responsible consumption and production and life below water goals from the United Nations (UN) Sustainable Development Goals (SDGs), have been the main driving forces behind such efforts [1,2]. Additionally, due to the unlikely increase in global production of FM and FO beyond current levels, it is expected that in the near future the overall supply will become increasingly insufficient to meet the already reduced current inclusion levels in many species and sustain the still growing aquaculture production [3–5].

So far, research undertaken since the 1960s and 1970s has led to the successful large-scale replacement of FM and FO by alternative ingredients from terrestrial plant sources. Presently, it is a common practice in aquafeed production to use ingredients derived from various terrestrial plants (e.g., soybean, canola, lupin, peas, wheat, corn, and linseed) as partial substitutes for FM and FO [6–8]. However, there are in some cases serious limitations in using these types of ingredients as total substitutes of FM and FO [3], as most plant ingredients have relatively limited nutritional value (e.g., imbalanced amino acid and fatty acid profiles) and/or biological components (e.g., anti-nutritional factors) that can negatively affect the performance, health, and well-being of some fish species [9].

Given the aforementioned achievements as well as the associated limitations, in recent years there has been a slight shift in aquafeed research, with the focus being progressively placed on expanding the range of alternative sources of ingredients.

Protein sources, such as processed animal proteins from terrestrial animals (PAPs; e.g., poultry by-product meal, blood meal, and feather meal), insect meal (e.g., black soldier fly and yellow mealworm), single-cell protein (SCP; e.g., microalgae, yeasts, bacteria, and protists), and macroalgae (e.g., Ulva, Gracilaria, and Laminaria), have been increasingly studied as potential novel alternatives to FM and terrestrial plant proteins [10,11]. Some have been shown to be well utilized by fish when incorporated at moderate levels, as is the case with PAPs that are already being used on a commercial scale [12]. Others still require further research in order to be considered viable options to be used at larger scales, as is the case with insect meal, with outcomes varying substantially with the processing methods applied, inclusion levels considered, and insect/fish species evaluated [13]. Other alternative protein sources, such as fish protein hydrolysates (FPH), have also gained attention in recent years since they can be produced from waste discarded by fish processing units (e.g., heads, skin, muscle, viscera, and bones), thus contributing to increasing the circularity of resources [14]. A detailed overview of the research carried out to date on the impact of these new protein sources on fish performance and health, along with other considerations for large-scale use, can be found in various reviews available in the literature [10,13,15,16].

Finding new alternatives to replace FO has also been a growing concern in aquafeed research, as it is critical to the sustainable growth of aquaculture. Marine fish, for example, require considerable levels of long-chain polyunsaturated fatty acids (LC-PUFAs) [17]. However, available alternatives to FO from terrestrial plant sources, such as canola, linseed, or soybean oils, have low levels of LC-PUFAs, e.g., arachidonic acid (ARA), eicosapentaenoic acid (EPA), and docosahexaenoic acid (DHA) [18–20]. This limits the incorporation of vegetable oils (VO) at high levels in aquafeeds for marine fish, as it can lead to reduced performance and adverse health issues in fish [21]. For this reason, current research has focused on developing aquafeeds that allow the use of relatively high levels of VO without impairing fish performance. Such research has been based on testing the incorporation of oils highly rich in LC-PUFAs, as is the case with oils extracted from some microalgae species [22–24]. However, due to the still relatively high cost of these types of ingredients, their incorporation is currently limited to small amounts on a commercial scale. Nevertheless, in the specific case of rainbow trout (*Oncorhynchus mykiss*), studies have demonstrated that it is possible to totally replace FO with VO without compromising growth performance or nutrient utilization efficiency [25,26]. However, the use of finfish diets rich in omega-3 fatty acids has been advised in these cases to ensure that this food product is still a viable source of omega-3 fatty acids in the human diet. This makes the search for alternative sources of LC-PUFAs a valid and relevant effort for the development of more sustainable feeding not only of marine species but of other fish species (i.e., diadromous and freshwater), including rainbow trout.

Consumer acceptance and fillet quality are also important components when developing novel aquafeeds, since consumers are at the top of the supply chain. Consumers presently are more likely to care about the sustainability of the products they are buying. Although few studies have been published on the consumer acceptance of farmed fish fed aquafeeds with alternative ingredients, the existing ones indicate that consumers are generally willing to accept the inclusion of these types of ingredients in the fish diet (e.g., insect meal) [27,28]. Other studies revealed that the level of education influences the choice of sustainable seafood consumption in Germany [29], but the choice of more sustainable aquaculture is also country-dependent and strongly impacted by misleading information from the media [30]. In terms of the impact of novel aquafeeds on the fillet quality of fish, previous research has found that, in most cases, FM and FO replacement do not lead to significant negative effects [31,32]. However, this type of sensory analysis is context-specific, and performing it in new contexts can provide a more holistic view of the overall impact of novel diets while also taking consumer perception into account [33].

Given the importance of finding practical solutions to reduce the inclusion of FM and FO in aquafeeds, the present study, carried out in the context of the EU-funded GAIN (Green Aquaculture Intensification in Europe) project (www.gain2020.com), focused on the evaluation of the impact of novel diets formulated with emerging ingredients on rainbow trout (*Oncorhynchus mykiss*) through the objectives of (1) an in vivo growth trial, carried out over 91 days under controlled conditions; (2) flesh quality and sensory analysis of fish after the in vivo trial; and (3) in silico extrapolation scenarios, performed considering commercial-like farming conditions.

2. Materials and Methods

2.1. Diets

The diet formulation framework followed to produce the test aquafeeds aimed to prioritize the use of ingredients derived from by-products (e.g., PAPs and aquaculture/fisheries by-products), as well as the inclusion of a high diversity of other novel alternative sources (e.g., micro and macroalgae, microbial meal, and insect meal), with the goal of finding more sustainable ready-to-use formulations that can be readily adopted by the aquafeed industry.

The alternative ingredients selected were defined based on circularity principles, contributing towards low waste in the agri-food value chain to optimize sustainability within the current/predictable regulatory framework. The diets were formulated according to the known nutritional requirements for rainbow trout (*Oncorhynchus mykiss*) [17] and manufactured by extrusion at SPAROS facilities. The trial comprised five experimental diets (Table 1). In particular, a control diet (CTRL), mimicking a good-quality commercial diet, contained a low inclusion level of fish meal along with traditional soy products, which was compared against four alternative formulations: a diet rich in processed terrestrial animal proteins (PAP); another one without the inclusion of PAPs (NoPAP); a diet similar to PAP but with a lower protein content (PAP−); and a diet similar to NoPAP but with a higher protein content (NoPAP+). All diets included a mix of fish protein hydrolysates produced by the GAIN project consortium based on turbot, salmon, seabream, and seabass by-products. All diets were formulated to be isolipidic, and the first three aforementioned (i.e., CTRL, PAP, and NoPAP) were also isoenergetic and isoproteic. Dietary amino acid and fatty acid contents are available in Tables S1 and S2.

Table 1. Formulation and proximate composition of the experimental diets for trout (*O. mykiss*).

Ingredients (%)	CTRL	NoPAP	NoPAP+	PAP	PAP−
Fish meal Super Prime [1]	15.00		12.50		
Krill meal [2]			5.00		
Fish protein hydrolysate [3]	2.00				
FPH-TURBOT-HEAD [4]		0.250	0.250	0.250	0.250
FPH-TURBOT-TF [5]		0.250	0.250	0.250	0.250
FPH-SALMON-HEAD [6]		0.250	0.250	0.250	0.250
FPH-SALMON-TF [7]		0.250	0.250	0.250	0.250
FPH-BREAM/BASS [8]		1.00	1.00	1.00	1.000
Feathermeal hydrolysate [9]				7.50	15.00
Porcine blood meal [10]				2.00	4.00
Poultry meal [11]				5.500	11.00
Insect meal (Black soldier fly) [12]		16.00	10.00	16.00	5.00
Fermentation biomass (*M. capsulatus*) [13]		16.00	10.00	16.00	5.00

Table 1. *Cont.*

Ingredients (%)	CTRL	NoPAP	NoPAP+	PAP	PAP−
Soy protein concentrate [14]	20.00	10.00	10.00		
Pea protein concentrate [15]		2.55	10.00		
Wheat gluten [16]	6.00	3.00	3.00		
Corn gluten meal [17]	7.00	3.50			
Soybean meal 48 [18]	10.00				
Wheat meal [19]	11.20				
Whole peas [20]		12.03		20.84	31.77
Pea starch (raw) [21]	4.00	7.00	13.35	4.00	
Vit and Min Premix-WITH I and Se [22]	1.00				
Vit and Min Premix-NO I and Se [23]		1.00	1.00	1.00	1.00
Macroalgae SHP [24]		2.00	2.00	2.00	2.00
Macroalgae SHP + Se [25]		0.05	0.05	0.05	0.05
Microalgae Se-rich [26]		0.20	0.20	0.30	0.30
Vitamin E50 [27]	0.03	0.03	0.03	0.03	0.03
Betaine HCl [28]	0.10	0.10	0.10	0.10	0.10
Antioxidant [29]	0.20	0.20	0.20	0.20	0.20
Sodium propionate [30]	0.10	0.10	0.10	0.10	0.10
Monocalcium phosphate [31]	1.04	2.26	0.70	1.80	1.70
L-Lysine HCl 99% [32]	0.50	0.60		0.45	0.65
L-Tryptophan [33]	0.04	0.01		0.04	0.17
DL-Methionine [34]					0.12
L-Taurine [35]	0.17	0.20	0.10	0.12	0.04
Yttrium oxide [36]	0.02	0.02	0.02	0.02	0.02
Fish oil [37]	5.30	2.65	2.65	2.65	2.65
Salmon oil (by-products) [38]		10.00	10.00	10.00	10.00
Algae oil (Veramaris) [39]		1.00	1.00	1.00	1.00
Rapeseed oil [40]	16.30	7.50	6.00	6.30	6.10
Total	100	100	100	100	100
Composition (feed basis)					
Crude protein, % feed	39.6	39.3	43.1	41.5	38.4
Crude fat, % feed	21.5	22.1	22.2	23.6	22.1
Fiber, % feed	1.6	1.6	1.3	1.4	2.5
Ash, % feed	6.3	6.2	5.9	6.0	5.8
Gross Energy, MJ/kg feed	23.4	22.8	23.3	22.8	23.5

[1] Sopropêche, France, Spain; [2] Aker BioMarine, Norway; [3] Sopropêche, France; [9] Coppens, The Netherlands; [10] SONAC BV, The Netherlands; [11] SAVINOR UTS, Portugal; [12] InnovaFeed, France; [13] Feedkind (Calysita); [14] ADM, The Netherlands; [15] Roquette Frères; [16] Roquette Frères; [17] Copan, Portugal; [18] CARGILL, Spain; [19] Casa Lanchinha, Portugal; [20] O Cereal, Portugal; [21] COSUCRA, Belgium; [27] DSM Nutritional Products, Switzerland; [28] ORFFA, The Netherlands; [29] Kemin Europe NV, Belgium; [30] Disproquímica, Portugal; [31] Coppens, The Netherlands; [32] Coppens, The Netherlands; [33] Ajinomoto EUROLYSINE S.A.S., France; [34] EVONIK Nutrition and Care GmbH, Germany; [35] ORFFA, The Netherlands; [36] Sigma Aldrich, USA; [37] Sopropêche, France; [38] Sopropêche, France; [39] Veramaris, The Netherlands; [40] Henry Lamotte Oils GmbH, Germany. The ingredients [4] to [8] and [24] to [26] were provided by the partners of the GAIN project consortium. [22] and [23] were manufactured at SPAROS.

All powder ingredients were admixed accordingly to the target formulation in a double-helix mixer (model 500 L, TGC Extrusion, Roullet-Saint-Estèphe, France) and ground (below 400 µm) in a micro-pulverizer hammer mill (model SH1, Hosokawa-Alpine, Augsburg, Germany). Diets (pellet size changed according to species and fish size) were manufactured with a twin-screw extruder (model BC45, Clextral, France) with a screw diameter of 55.5 mm. Extrusion conditions: feeder rate (80–85 kg/h), screw speed (247–266 rpm), water addition in barrel 1 (345–360 mL/min), temperature barrel 1 (30–35 °C), temperature barrel 3 (105–120 °C). Extruded pellets were dried in a vibrating fluid bed dryer (model DR100, TGC Extrusion, Firminy, France). After cooling, an oil blend was embedded by vacuum coating (model PG-10VCLAB, Dinnissen, Sevenum, The Netherlands). Coating conditions were: pressure (700 mbar); spraying time under vacuum (approximately 90 s); return to atmospheric pressure (120 s). Immediately after coating, diets were packed in sealed plastic buckets and shipped to the research site, where they were stored at room temperature but in a cool and aerated emplacement. Representative samples of each diet were taken for composition analysis.

2.2. Growth Trial

The growth trial was conducted at the experimental facilities of Fondazione Edmund Mach (FEM, San Michele all'Adige, Italy). A rainbow trout (*Oncorhynchus mykiss*) batch (approx. 2000 individuals), which originated from a bigger resident FEM stock, was transferred to the experimental facilities and kept in sanitary quarantine for 6 days. At the start of the trial, fish were manually sorted by weight to constitute a homogenous subsample of 1000 individuals.

Groups of 50 fish, with a mean initial body weight (IBW) of 58.84 ± 1.39 g, were randomly allocated in 20 tanks (approx. 1000 L, 700 L of water). All tanks were located indoors, supplied with well water in an independent flow-through rearing scheme, and subjected to a natural photoperiod (46°11′30.3″ N, 11°08′05.3″ E). Water quality parameters, such as temperature and dissolved oxygen, were monitored repeatedly during the trial. The mean water temperature was 13.3 ± 0.2 °C, and the dissolved oxygen was 8.0 ± 0.8 mg/L. Each diet was tested in four replicates in a randomized design. Fish were fed by hand until apparent satiation, twice a day (09.00 a.m. and 02.00 p.m.), six days per week (no feeding on Sundays), over 91 days. Administered feed per tank was registered on a daily basis. Anesthetized fish were individually weighed and measured at the start (day 0) and end (day 91) of the trial, whereas a bulk weight of all fish in each tank was carried out in the middle of the trial (day 49). At the start, 30 fish from the same source stock were sampled and stored at −20 °C for subsequent whole-body composition analysis. At the end of the trial, five fish from each tank were euthanized, from which blood samples were collected for lysozyme analysis, and six fish from each tank were sacrificed for whole-body composition analysis. In addition, twenty fish collected from the tanks fed with CTRL, PAP, and NoPAP diets were killed by a sharp blow to the head, bled, gutted, stored on ice, and sent to the lab for sensory analysis with a trained panel. The remaining individuals in each tank were assigned to flesh quality analysis (TPA and color).

2.3. Apparent Digestibility Measurements

Feces were sampled at day 91 to determine the apparent digestibility coefficients (ADC) of the diets by the indirect method [34]. All the different diets in which fish were fed throughout the experimental period contained yttrium oxide (Y_2O_3) at 0.02% as an inert marker. Feces were collected from all anesthetized fish by stripping [35], pooled per tank into a plastic container, and two subsamples (100 g each) were stored frozen at −20 °C until lab analysis.

2.4. Analytical Methods

The analysis of diets, whole fish, and feces was made with analytical duplicates, following in most cases the methodology described by [36]. Dry matter after drying at 105 °C for 24 h; total ash by combustion (550 °C during 6 h) in a muffle furnace; crude protein (N × 6.25) by a flash combustion technique followed by a gas chromatographic separation and thermal conductivity detection with a Leco N Analyzer (Model FP-528, Leco Corporation, USA); crude lipid by petroleum ether extraction (40–60 °C) using a Soxtec™ 2055 Fat Extraction System (Foss, Hillerød, Denmark), with prior acid hydrolysis with 8.3 M HCl; gross energy in an adiabatic bomb calorimeter (Werke C2000, IKA, Hohenems, Austria); total phosphorus according to ISO 27085:2009 by ICP-AES methodology) [37]; yttrium concentration in feed and feces was determined by atomic absorption spectrometry (SpectrAA 220 FS, Varian) [38]. For mineral analysis, dry samples were weighed (50–200 mg) in quartz vessels. Samples were then digested in 6 mL of nitric acid (HNO3 tracer grade, 70%) in a Discovery SP-D microwave digestion unit according to the following program: 200 °C; 4 min ramp; 3 min hold. The samples were then cooled to room temperature, and a final volume of 10 mL was achieved by adding ultrapure water. The samples were subsequently diluted 16x in ultrapure water, and standard curves were prepared in ultrapure water. Mineral quantification was performed by MP-AES (Agilent, model 4200). Blank samples containing only the decomposition acid were included to measure the matrix effects of decomposition, which were subtracted from every element in each sample.

2.5. Evaluation Criteria

Growth performance and feed utilization were evaluated according to the following equations:

Relative growth rate:

$$\mathrm{RGR}\ (\%\mathrm{BW/day}) = (e^{g} - 1) \times 100 \tag{1}$$

where 'e' is Euler's number and g = (ln (FBW) − ln (IBW))/days.

Feed conversion ratio:

$$\mathrm{FCR} = \text{crude feed intake/weight gain} \tag{2}$$

Weight gain was corrected for mortality and sampled when necessary.

Feed intake:

$$\mathrm{FI}\ (\%\mathrm{BW/day}) = (\text{crude feed intake}/(\mathrm{IBW} + \mathrm{FBW})/2/\text{days}) \times 100. \tag{3}$$

Protein efficiency ratio:

$$\mathrm{PER} = \text{wet weight gain/crude protein intake.} \tag{4}$$

The contractions IBW and FBW are assigned for initial mean body weight and final mean body weight, respectively, both expressed in grams (g).

Retention was calculated accordingly using the following equation: where NFF is the nutrient content of the final fish and NIF is the nutrient content of the initial fish, both expressed as a percentage.

$$\text{Retention}\ (\%) = 100 \times \frac{(\mathrm{FBW} \times \mathrm{NFF}) - (\mathrm{IBW} \times \mathrm{NIF})}{\text{nutrient intake}} \tag{5}$$

Apparent digestibility coefficients (ADC) of dietary nutrients and energy in the experimental diets were calculated according to the NRC [17]:

$$\mathrm{ADC}\ (\%) = 100 \times \left(1 - \frac{\%\ marker\ diet}{\%\ marker\ feces} \times \frac{\%\ nutrient\ feces}{\%\ nutrient\ diet}\right) \tag{6}$$

2.6. Lysozyme

The lysozyme activity was slightly adjusted to the protocol of [39]. The phosphate buffer consisted of 0.05 M NaH_2PO_4 + 0.05 M Na_2HPO_4 and was modified with 85% H_3PO_4 to a pH of 6.2. An amount of 30 mg of *Micrococcus luteus* (0.6 mg/mL, SIGMA M3770) was mixed with 50 mL of buffer on a daily basis, while 20 mg of lysozyme from egg whites (Lot SLCC4285, 40,382 units/mg, Sigma L6876) was mixed with 20 mL of buffer weekly. The lysozyme-buffer solution was diluted to 1000 U/mL. For the measurement, 96 flat-bottomed well plates (Brandplate 781660) were used and measured in a Berthold Tristar LB941. After calibration of standard curves, 10 μL and 5 μL of plasma were pipetted into the wells with 10 μL and 15 μL of buffer to get a total volume of 20 μL per well. Right before the measurement started, *M. luteus* was added at 130 μL to the standard curve and all sample preparations. The measurement took place at 450 nm every minute for 10 min. The plate was shaken for 5 s before the first measurement. For each measurement, the standard curve, the samples, and the blanks were measured in triplicate. Between measurements, the solutions were stored at 4 °C and the samples at −20 °C.

2.7. Fish and Fillet Quality Analysis

2.7.1. Body Indexes

The experimental pool consisted of 100 rainbow trout with an average weight of 306.1 ± 55.0 g, representing 5 fish from each tank (20 fish per treatment) collected at the end of the trial. After being killed, the fish were covered with ice and stored at 1 °C for 24 h. The day after, fish were eviscerated, and carcasses, whole viscera, liver, and mesenteric fat were weighted. Then, carcass yield (CY), hepatosomatic index (HSI), viscerosomatic index (VSI), and mesenteric fat index (MFI) were calculated as follows:

$$\text{CY (\%)} = 100 \times [\text{fish weight (g)} - \text{visceral weight (g)}/\text{fish weight (g)}] \quad (7)$$

$$\text{HSI (\%)} = 100 \times [\text{liver weight (g)}/\text{fish weight (g)}] \quad (8)$$

$$\text{VSI (\%)} = 100 \times [\text{visceral weight (g)}/\text{fish weight (g)}] \quad (9)$$

$$\text{MFI (\%)} = 100 \times [\text{mesenteric fat weight (g)}/\text{fish weight (g)}] \quad (10)$$

Fish were filleted afterwards.

2.7.2. Texture Profile Analysis (TPA)

The right fillet from each sample (100 fish) was assigned to the Texture Profile Analysis (TPA). Textural features were measured at room temperature on a sample of muscle (a section of 4 × 4 cm) withdrawn from the epaxial region of the fillet. TPA was carried out using a Zwick Roell® 109 texturometer (Ulm, Germany) equipped with a 1 kN load cell and a cylindrical probe (10 mm). Two consecutive cycles (downstroke and upstroke), with a five-second break between them, were set, and the deformation was limited to 50% of the total thickness by a crosshead speed of 100 mm/min. The raw data were collected and analyzed by the Test-Xpert II software version 3.0 (Zwick GmbH and Co. KG, Ulm, Germany). As suggested [40], the following parameters were determined:

- hardness (N), defined as the maximum force required to compress the sample;
- cohesiveness, defined as the area of work during the second compression divided by the area of work during the first compression (Area 2/Area 1);
- gumminess, calculated as hardness × cohesiveness;
- resilience (Nmm), calculated by dividing the upstroke energy of the first compression by the downstroke energy of the first compression (Area 4/Area 3);
- adhesiveness (Nmm), defined as the negative force area under the baseline between compression cycles.

2.7.3. Fillet Color

A Konica-Minolta CR-400 colorimeter was utilized for color measurement carried out according to the CIE Lab system [41]. Flesh color was measured in triplicate on the cranial, dorsal, and caudal sections of the left fillet (Figure 1), recording L* (lightness), a* (redness index), and b* (yellowness index) parameters, then chroma and hue were calculated. Chroma is an expression of the intensity of the color, and hue is an angular measurement of tint.

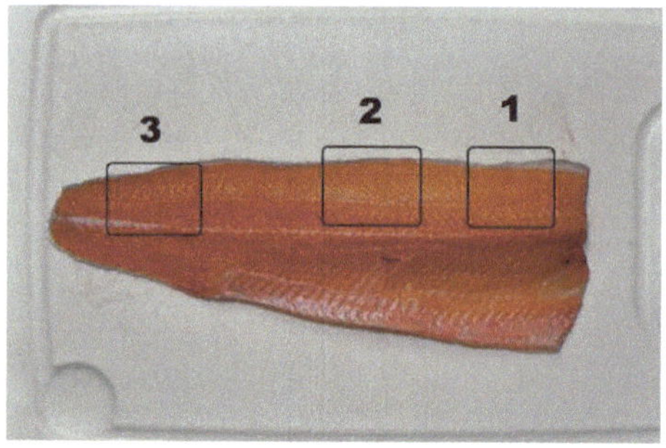

Figure 1. Sampling points for color analysis on the fillet ((**1**) cranial, (**2**) dorsal, and (**3**) caudal).

CIE Lab and CIE LCh are similar color spaces; the difference between them is the different coordinate systems used to describe the space. The relationships between their respective coordinates are, therefore, as follows:

$$L^* (LAB) \equiv L^* (LCH) \tag{11}$$

$$C^* = \sqrt{}(a^*2 + b^*2) \tag{12}$$

$$h^* = \arctan(b^*/a^*) \tag{13}$$

After that, it is possible to calculate the perceptual difference between samples. Color spaces have been developed to mathematically describe the color that a person can feel with an unaided eye, as there was a need to distinguish colors and define them as similar, identical, or completely different. ΔE is a distance vector in the color space, and his value determines the difference between two measures. Color differences (ΔE) were calculated according to the following formula:

$$\Delta E\ (\beta - \alpha) = [\ (L *\beta - L *\alpha)2 + (a *\beta - a *\alpha)2 + (b *\beta - b *\alpha)2\ ]0.5 \tag{14}$$

where α and β represent alternatively the mean color values of different treatments. ΔE lab is a single number that represents the 'distance' between two given colors, and this distance is proportional to the perceptual difference between them [42]. A standard observer perceives the differences in color presented above as follows:

- 0 < ΔE < 1—observer does not notice the difference
- 1 < ΔE < 2—only experienced observer can notice the difference
- 2 < ΔE < 3.5—unexperienced observer also notices the difference
- 3.5 < ΔE < 5—clear difference in color is noticed
- 5 < ΔE—observer notices two different colors

2.7.4. Sensory Analysis

Fifty fish collected from tanks fed the CTRL, PAP, and NoPAP diets were slaughtered and shipped packed in ice to a specialized laboratory for sensory analysis.

All tests took place at sense test—society for the study of sensory analysis of food products (Porto, Portugal), in a specific tasting room for sensory testing of food products (ISO 8589 sensory analysis—general guidance for the design of test rooms), and were performed by a panel of 100 individuals accustomed to the products being tested. Thus, the effects of physiological factors and physical conditions on human judgment were reduced. All features and products were analyzed using close to natural light (6500 K). Before each test session, a preliminary explanation was given to the tasters on the conditions and rules of the sensory test. Moreover, all rules were handed out with the proof sheets.

The trout fillets were baked in the oven for approximately 12 min at 170 °C, and a small portion was served to each taster. Each sample was served on a white dish. In the tasting booth, a knife, a fork, paper napkins, a glass of water, a spittoon, and crackers were available to the tasters. The tasters were told that the crackers should only be used between tasting sessions in order to easily remove the taste from the mouth.

After tasting, each tester answered a questionnaire in which the parameters appearance, odor, taste, and texture, as well as global acceptance of the flesh, were evaluated. The evaluation consisted of giving scores from 1 to 9, where 1 referred to "extremely disliked" and 9 referred to "extremely liked". Means and standard deviations from the scores given were calculated, and statistical analysis was performed as described in item 2.9.

2.8. Extrapolation of Trial Results Using a Nutrient-Based Model

The results obtained in the in vivo trial were extrapolated in silico using the FEEDNETICS^{TM} nutrient-based model (software; FEEDNETICS, 2022, Olhão, SPAROS) [43]. The main objective of this application was to evaluate the long-term impact of the experimental diets on the zootechnical and economic performance of rainbow trout cultivated under commercial-like farming conditions.

2.8.1. Model Validation for Trial Conditions

Before running extrapolation scenarios, the model was validated for the conditions in which the trial was conducted in order to assess whether the fish performance predicted by the model is consistent with the observations made in the trial. This gives additional confidence in the robustness of the model in predicting the nutritional effects of the experimental diets under other farming conditions. In addition, it allows us to infer if there are any potential factors not considered by the model that could affect fish performance.

The model validation was performed at the tank level, where the input data considered to run the model were based on measurements/analyses performed during the trial (i.e., daily number of fish, initial average fish weight, daily water temperature, daily amount of feed, and diet composition). The model results and robustness were evaluated qualitatively, through visual inspection of the model behavior over time in comparison with point observations made during the trial, and quantitatively, by estimating the mean absolute percentage error (*MAPE*) between the model predictions and point observations (i.e., performed in sampling days), as follows:

$$MAPE(\%) = \frac{100}{n}\sum_{i=1}^{n}\left|\frac{P_i - O_i}{O_i}\right| \tag{15}$$

where

n is the number of predicted-observed value pairs.

P_i is the predicted value.

O_i is the observed value.

2.8.2. Fish Performance under Commercial-like Farming Conditions: Extrapolation Scenarios

Extrapolation scenarios of rainbow trout fed the experimental diets were run with the model, considering hypothetical farming conditions close to what can be found in commercial settings. As a base scenario, we considered an initial average body weight of 50 g, an initial number of fish of 20,000 individuals, and a monthly mortality rate of 1% over the entire simulation period. In each scenario, the feeding rates used by the model were based on a common feeding table adjusted for rainbow trout. Data on diet proximate composition and digestibility were defined based on the analyzed/measured values, while data on diet amino acid profile and price were defined based on values estimated from the dietary ingredient composition. For each diet (i.e., CTRL, PAP, NoPAP, NoPAP+, PAP−), two extrapolation scenarios were run: one considering a constant temperature of 12 °C and the other considering a constant temperature of 16 °C. In total, 10 extrapolation scenarios were run, and their results were evaluated considering a harvest fish weight of about 400 g. Table 2 presents a brief overview of the input data used to run this model application.

Table 2. Input data used to run the extrapolation scenarios in silico with the FEEDNETICS^{TM} software.

Category	Inputs		
Farming conditions	Initial body weight: 50 g Initial number of fish: 20,000 Mortality rate: 1%/month		
Diet properties	Price * (% of change in relation to CTRL)		Proximate composition: analyzed values Digestibility: measured values Amino acid profile: estimated values Fatty acid profile: default values
	CTRL	0.0%	
	NoPAP	+22.92%	
	NoPAP+	+48.53%	
	PAP	−11.78%	
	PAP−	−6.70%	
Feeding regime	Number of meals: 2 meals/day Feeding type: feeding table Feed waste: 0%		
Temperature	Daily water temperature: constant profiles of 12 °C and 16 °C		

* Diet price data was entered in the software as €/ton, but for convenience, these values are presented here as % of change in relation to CTRL.

2.9. Statistical Analysis

The data are presented as the mean of four replicates ± standard deviation. Prior to ANOVA or ANCOVA for body indexes NP K-W, values expressed as percentages were subjected to the *arcsin* square root transformation. Statistical significance was tested at the 0.05 probability level. Statistical tests for growth performance, lysozyme, digestibility, and all analytical methods were performed using the R software (R version 4.1.0; R Core Team, 2021), and flesh quality analysis was performed using STATISTICA (version 13, TIBCO Software Inc., Palo Alto, CA, USA). Regarding the sensory panel analysis, the Wilcoxon non-parametric test was applied to compare the samples pair-wise.

3. Results

3.1. Growth Performance

The results on the growth performance of fish fed the experimental diets are presented in Table 3. Final body weight (FBW) ranged between 291.9 ± 7.3 g and 335.6 ± 6.2 g. This means an individual weight gain of over 200 g compared to their initial body weight (IBW). The NoPAP+ and PAP diets led to better growth in terms of weight gain and relative growth rate (RGR) compared to the CTRL diet, whereas no significant differences were registered between fish fed the CTRL, NoPAP, and PAP− diets. Fish fed the PAP- and NoPAP+ diets showed the worst and the best feed conversion ratio (FCR), respectively, both significantly different ($p < 0.05$) from the other dietary treatments. The verified feed intake (VFI) was significantly higher in the low-protein diet and lower in the CTRL. Protein efficiency ratio (PER) was lower in diets with PAPs. No significant differences ($p > 0.05$) were found in the whole-body composition of fish in terms of moisture, ash, crude protein, crude fat, energy, or phosphorus (Table S3).

Table 3. Growth performance and feed utilization of rainbow trout fed the experimental diets for 91 days.

Growth Parameters	CTRL	NoPAP	NoPAP+	PAP	PAP−	*p*
IBW (g)	58.2 ± 1.81	58.1 ± 1.43	58.6 ± 1.38	59.4 ± 1.40	59.9 ± 0.69	0.342
FBW (g)	292 ± 7.25 ^{c}	297 ± 11.3 ^{bc}	336 ± 6.17 ^{a}	309 ± 5.30 ^{b}	297 ± 4.45 ^{bc}	<0.001
Weight gain (g)	234 ± 5.97 ^{c}	239 ± 10.0 ^{bc}	277 ± 4.79 ^{a}	250 ± 4.70 ^{b}	237 ± 3.95 ^{bc}	<0.001
RGR (%BW/day)	1.79 ± 0.02 ^{c}	1.8 ± 0.02 ^{bc}	1.94 ± 0.01 ^{a}	1.83 ± 0.02 ^{b}	1.78 ± 0.01 ^{c}	<0.001
FCR	0.82 ± 0.01 ^{b}	0.84 ± 0.02 ^{b}	0.76 ± 0.01 ^{c}	0.83 ± 0.01 ^{b}	0.87 ± 0.01 ^{a}	<0.001
VFI (%/day)	1.46 ± 0.03 ^{c}	1.51 ± 0.02 ^{ab}	1.48 ± 0.01 ^{bc}	1.52 ± 0.01 ^{ab}	1.55 ± 0.01 ^{a}	<0.001
PER	3.09 ± 0.04 ^{a}	3.04 ± 0.06 ^{ab}	3.04 ± 0.03 ^{ab}	2.90 ± 0.05 ^{bc}	2.99 ± 0.01 ^{bc}	<0.001
Survival (%)	100 ± 0.00	98.5 ± 1.91	98.0 ± 1.63	99.5 ± 1.00	99.0 ± 2.00	0.306

Values are presented as mean ± SD. Different letters indicate significant differences (one-way ANOVA, $p < 0.05$; Tukey post hoc test, n = 4).

3.2. Nutrient Retention

The results of crude protein, crude fat, and energy retention in fish fed the experimental diets are shown in Table 4. Fish fed new-formulated diets did not present significant differences in crude fat and energy retention ($p > 0.05$). However, fish fed the PAP and PAP− diets presented significantly lower ($p < 0.05$) crude protein retention when compared to fish fed the CTRL diet.

Table 4. Nutrient and energy retention of rainbow trout fed the experimental diets for 91 days.

Retention (% of Intake)	CTRL	NoPAP	NoPAP+	PAP	PAP−	*p*
Crude Protein	48.1 ± 2.6 ^{a}	43.7 ± 0.7 ^{ab}	46.5 ± 1.3 ^{ab}	42.8 ± 1.2 ^{b}	43.1 ± 3.1 ^{b}	0.020
Crude Fat	75.6 ± 6.7	65.6 ± 6.2	71.1 ± 5.8	69.8 ± 1.4	68.6 ± 3.1	0.271
Energy	47.2 ± 2.7	44.1 ± 2.3	47.8 ± 2.2	47.5 ± 1.0	42.9 ± 1.2	0.065

Values are presented as mean ± SD. Different letters indicate significant differences (one-way ANOVA, $p < 0.05$; Tukey post hoc test, n = 4).

3.3. Apparent Digestibility Coefficients

The apparent digestibility coefficients (ADCs) of the experimental diets are presented in Table 5. Fish fed the PAP− diet presented significantly lower digestibility for crude protein and total phosphorus (Total P) when compared to fish fed the other diets. On the other hand, fish fed the NoPAP+ and PAP diets had significantly higher crude fat digestibility when compared to fish fed the CTRL diet, while significantly higher energy digestibility was observed in fish fed PAP when compared to fish fed the CTRL and PAP− diets.

Table 5. Apparent digestibility coefficients (ADCs) of rainbow trout fed the experimental diets for 91 days.

Digestibility (%)	CTRL	NoPAP	NoPAP+	PAP	PAP−	*p*
Crude protein	89.2 ± 0.1 ^{a}	88.9 ± 0.5 ^{a}	89.8 ± 0.3 ^{a}	90.3 ± 0.7 ^{a}	86.2 ± 1.3 ^{b}	<0.001
Crude fat	97.8 ± 0.5 ^{b}	98.8 ± 0.1 ^{a}	98.2 ± 0.3 ^{ab}	98.7 ± 0.2 ^{a}	98.2 ± 0.5 ^{ab}	0.01
Energy	82.3 ± 1.6 ^{b}	84.4 ± 1.1 ^{ab}	83.3 ± 1.2 ^{ab}	86.1 ± 1.2 ^{a}	81.4 ± 1.6 ^{b}	0.002
Total P	42.1 ± 4.7 ^{a}	48.1 ± 2.8 ^{a}	51.5 ± 2.6 ^{a}	63.7 ± 3.5 ^{a}	47.1 ± 4.5 ^{b}	<0.001

Values are presented as mean ± SD. Different letters indicate significant differences (one-way ANOVA, $p < 0.05$; Tukey post hoc test, n = 4).

3.4. Lysozyme

Lysozyme analysis presented different results among dietary treatments (Figure 2). Fish fed the NoPAP diet presented a significantly higher ($p < 0.05$) concentration (U/mL) of plasma lysozyme when compared to fish fed the CTRL, NoPAP+, and PAP diets.

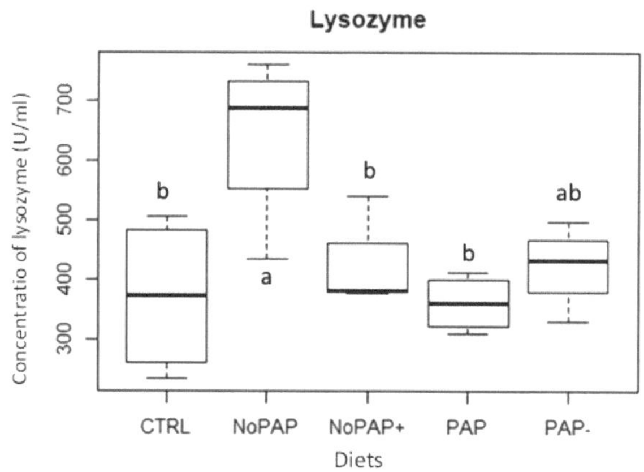

Figure 2. Lysozyme concentration (U/mL) in trout plasma after experimental period. Different letters indicate significant differences (one-way ANOVA, $p < 0.05$; Tukey post hoc test, n = 4).

3.5. Viscerosomatic Measurements

The mean weight of the liver, mesenteric fat, and gastrointestinal tract of rainbow trout fed the experimental diets is represented in Figure S1. These values, when added together, represent roughly the whole visceral weight of a fish. Using final body weight (FBW) as a predictor, an ANCOVA on organ and tissue weight was performed. The mean liver weight was significantly higher ($p < 0.05$) in fish fed the PAP diet (3.43 ± 0.64 g) in relation to fish fed the NoPAP, NoPAP+, and CTRL diets, being that the last one registered the lower mean liver weight (2.99 ± 0.71 g). A similar trend is evident in the mean gastrointestinal tract weight, where the fish fed the PAP diet had the highest value (25.7 ± 4.95 g) and were significantly different from fish fed the NoPAP and NoPAP+ diets, which registered 22.04 ± 6.18 g and 22.65 ± 3.50 g, respectively. No significant differences were observed in the quantity of mesenteric fat between dietary treatments.

The main morphological indexes are shown in Table 6. Morphological differences are mainly detected where carcass yield (CY) differs between diets. Fish fed the CTRL and NoPAP+ diets presented significantly higher ($p < 0.05$) carcass yields compared to fish fed the other diets. Moreover, it should be pointed out that significant differences ($p < 0.05$) are notable between CTRL/NoPAP+ and PAP−/PAP groups by mean ANOVA analysis followed by LSD Fisher post hoc comparisons. The viscerosomatic index (VSI) is inversely related to CY. This explains a significantly higher ($p < 0.05$) VSI for fish fed the PAP and PAP− diets compared to fish fed the CTRL and NoPAP+ diets. The same goes for hepatosomatic index (HSI), where fish fed the CTRL and NoPAP+ diets showed significantly lower values ($p < 0.05$) compared to fish fed the PAP and PAP− diets.

3.6. Fillet Quality

3.6.1. Fillet Texture

Textural features were measured 24 h after post-mortem on 100 fillets by texture profile analysis (TPA) using a Zwick Roell® 109 texturometer (Ulm, Germany). Outliers were removed from the dataset by Grubb's Test, and then a K-W ANOVA was performed. The data are presented in Table 7. The majority of variables (hardness, cohesiveness, resilience, and gumminess) did not show significant differences ($p > 0.05$) between the experimental diets, whereas statistical differences were observed in adhesiveness. Control and NoPAP+

diets obtained the highest value with 0.44 Nmm, in contrast with PAP−, which obtained the lowest value with 0.33 Nmm. Applying a multiple comparison test (mean of ranks), control differs from PAP and PAP- diets, but NoPAP+ differs only from PAP− due to the higher SD.

Table 6. Morphological indexes of carcass yield (CY), hepatosomatic index (HSI), mesenteric fat index (MFI), and viscerosomatic index (VSI) of rainbow trout fed the experimental diets for 91 days.

Indexes	CTRL	NoPAP	NoPAP+	PAP	PAP−	p
CY (%)	90.5 ± 1.3^{a}	90.3 ± 1.2^{ab}	90.5 ± 1.1^{a}	89.5 ± 0.1^{b}	89.7 ± 1.1^{b}	0.0209
HSI (%)	0.91 ± 0.1^{b}	0.99 ± 0.1^{ab}	0.94 ± 0.09^{b}	1.04 ± 0.10^{a}	1.04 ± 0.12^{a}	<0.001
MFI (%)	1.5 ± 0.5	1.5 ± 0.4	1.7 ± 0.66	1.6 ± 0.56	1.7 ± 0.47	0.72
VSI (%)	9.5 ± 1.3^{b}	9.7 ± 1.2^{ab}	9.53 ± 1.1^{b}	10.4 ± 0.9^{a}	10.32 ± 1.1^{a}	0.0209

Values are presented as mean ± SD. Different letters indicate significant differences (ANCOVA, $p < 0.05$; Fisher LSD post hoc tests, n = 20).

Table 7. Texture profile analysis of fillet of rainbow trout fed the experimental diets for 91 days.

Texture Parameters	CTRL	NoPAP	NoPAP+	PAP	PAP−	p
Hardness (N)	6.46 ± 1.79	6.93 ± 2.13	6.32 ± 1.47	6.56 ± 2.05	5.89 ± 1.88	0.604
Cohesiveness	0.22 ± 0.04	0.24 ± 0.04	0.23 ± 0.03	0.22 ± 0.04	0.21 ± 0.03	0.345
Resilience	0.04 ± 0.02	0.05 ± 0.02	0.04 ± 0.02	0.05 ± 0.02	0.04 ± 0.02	0.493
Adhesiveness (Nmm)	0.44 ± 0.10^{a}	0.39 ± 0.08^{abc}	0.44 ± 0.13^{ab}	0.36 ± 0.09^{bc}	0.33 ± 0.06^{c}	<0.001
Gumminess (N)	1.45 ± 0.51	1.66 ± 0.53	1.43 ± 0.41	1.50 ± 0.60	1.26 ± 0.46	0.233

Values are presented as mean ± SD. Different letters indicate significant differences (K-W ANOVA, $p < 0.05$; multiple comparisons of mean ranks—p value 2-tailed, n = 19–20).

3.6.2. Fillet Color

Color parameters (CIELab values) are significantly different ($p < 0.05$) among dietary treatments (Table 8). The lightness (L*) of the flesh of fish fed the CTRL and NoPAP+ diets is similar and significantly lower ($p < 0.05$) than those of the other three diets (NoPAP, PAP, and PAP−). However, it is possible to discriminate the flesh color related to the NoPAP+ from the CTRL diet by the indexes of red (a*) and yellow (b*), both higher in the former than in the latter. Moreover, the red index is rather high for fish fed the PAP− diet too, and it is significantly different ($p < 0.05$) from the CTRL and NoPAP dietary treatments. In terms of the CIE L*C*h color system, the NoPAP+ and PAP− dietary treatments registered the highest chroma value (or color saturation), but their hue differs clearly between them due to the b* index. This shows us that the flesh color of fish fed the NoPAP+ diet segregates from the other diets since it is in a different quarter of the color space (+b* quarter vs. –b* quarter).

Table 8. Color analysis of fillet of rainbow trout fed the experimental diets for 91 days.

Colour Parameters	CTRL	NoPAP	NoPAP+	PAP	PAP−	p
L*	43.19 ± 1.93^{c}	43.90 ± 2.19^{b}	42.65 ± 1.49^{c}	44.12 ± 1.73^{ab}	44.55 ± 1.49^{a}	$p < 0.001$
a*	2.18 ± 1.19^{d}	2.48 ± 1.22^{cd}	3.87 ± 1.97^{a}	2.68 ± 1.36^{bc}	3.02 ± 1.43^{b}	$p < 0.001$
b*	-0.19 ± 1.70^{bc}	-0.01 ± 1.87^{b}	0.92 ± 2.38^{a}	-0.27 ± 1.49^{bc}	-0.68 ± 1.41^{c}	$p < 0.001$
Chroma	2.79 ± 1.14^{b}	3.04 ± 1.35^{b}	4.41 ± 2.44^{a}	3.09 ± 1.33^{b}	3.48 ± 1.23^{a}	$p < 0.001$
Hue (rad)	-0.26 ± 0.67^{b}	-0.21 ± 0.61^{b}	0.01 ± 0.48^{a}	-0.22 ± 0.52^{b}	-0.35 ± 0.47^{b}	$p < 0.001$

Values are presented as mean ± sd. Different letters indicate significant differences (one-way ANOVA, $p < 0.05$; Fisher LSD post hoc tests, n = 20).

As shown in Table 9, an unexperienced observer may notice a difference in the flesh color of rainbow trout fed the NoPAP+ diet in relation to the other experimental diets. However, this analysis indicates that it would be more difficult to perceive the difference induced by the CTRL diet compared to the PAP and PAP− diets, as well as the difference induced by the NoPAP diet compared to the PAP− diet. Only the instrumental measurement can discriminate between the other matches.

Table 9. Total color differences (ΔE Lab) in CIELab color space presented as pairwise comparisons. ΔE < 1—observer does not notice the difference (white background colour), 1 < ΔE < 2—only experienced observer can notice the difference (green background colour), 2 < ΔE < 3.5—unexperienced observer also notices the difference (yellow background colour).

ΔE Lab	CTRL	NoPAP+	NoPAP	PAP	PAP−
CTRL					
NoPAP+	2.09				
NoPAP	0.79	2.09			
PAP	1.06	2.23	0.39		
PAP−	1.68	2.63	1.08	0.69	

To understand how the colors are different, we can say that a pale gray, pink, or brown color is one feature of the flesh of farmed rainbow trout fed without carotenoids. The instrumental data and the main color indexes suggest that CTRL, PAP, and NoPAP diets range close to this target, whereas a mild pink/orange pigmentation of fillets was observed in the flesh of rainbow trout fed the NoPAP+ diet compared to the other dietary treatments. In contrast with the NoPAP+ treatment, the lack of yellow (b*) in the flesh of fish fed the PAP− diet results in a shift of hue from orange to purple/blue in the eyes of an observer. To conclude, rainbow trout fed the CTRL, PAP, and NoPAP diets showed a flesh color that could be considered by consumers to be more natural in appearance. A visual comparison can be made between the pictures below (Figure 3).

3.7. Sensory Analysis

Sensory analysis of baked rainbow trout revealed few differences, resulting in high general acceptance by consumers (Table 10). Consumers considered that fish fed the NoPAP diet had better texture than fish fed the PAP diet, which consequently is reflected in the higher global acceptance of fish fed the NoPAP diet.

3.8. Extrapolation of Trial Results Using a Nutrient-Based Model

3.8.1. Model Validation for Trial Conditions

The validation results (Figure S2) indicate that the model is able to predict the growth pattern of rainbow trout fed the experimental diets over the 91-day trial period, presenting an overall MAPE of 11.2%. Nevertheless, it should be pointed out that for all tanks, the model tends to consistently underestimate fish growth by a similar magnitude. As the model calibration used in this application is generic for rainbow trout, these deviations can be associated with strain/population effects. In this case, the fish used in the in vivo trial may be slightly more efficient at retaining nutrients due to genetic differences than the generic model calibration for rainbow trout considers. Given this, and since the deviation between the predicted and observed values presents a similar magnitude between diets and tanks, we decided to run the extrapolation scenarios for all diets without any special concerns.

3.8.2. Fish Performance under Commercial-like Farming Conditions: Exploitation Scenarios

Table 11 shows the results of the extrapolation scenarios, generated in silico with the FEEDNETICS^{TM} model, of rainbow trout cultivated under commercial-like farming conditions at two constant temperatures, i.e., 12 °C and 16 °C. The results are presented for the time point at which fish reach an average harvest weight of about 400 g.

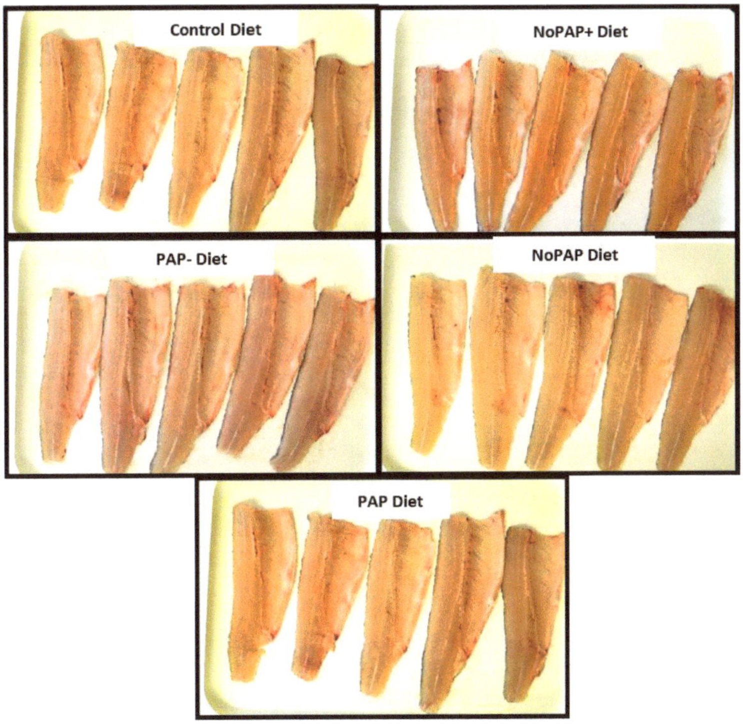

Figure 3. Each picture shows a subsample of fillets analyzed by colorimeter and CIE Lab method.

Table 10. Sensory analysis of fillet of rainbow trout fed the experimental diets for 91 days.

Sensory Analysis	CTRL	PAP	NoPAP	*p*
Appearance	7.8 ± 1.3	8.0 ± 1.0	8.0 ± 1.0	0.106
Odor	7.9 ± 1.1	7.9 ± 1.0	8.0 ± 0.9	0.932
Texture	7.9 ± 1.2 ^{ab}	7.9 ± 1.1 ^{b}	8.1 ± 0.9 ^{a}	0.009
Taste	7.0 ± 1.2	7.9 ± 1.1	8.1 ± 0.9	0.898
Global acceptance	7.9 ± 1.0 ^{ab}	7.9 ± 0.9 ^{b}	8.1 ± 0.7 ^{a}	0.043

Values are presented as mean ± SD. Different letters indicate significant differences (one-way ANOVA, $p < 0.05$; Tukey post hoc test, n = 100).

For both temperature scenarios, the model suggests better performance in terms of growth and feed conversion when fish are fed the PAP and NoPAP+ diets. Surprisingly, the NoPAP diet presented the highest FCR, even outperforming the CTRL diet at 12 °C. When looking at economic performance (economic conversion ratio—ECR) and nutrient waste indicators (total N and P waste), the figures change dramatically. In economic terms, both the CTRL and PAP diets are the most cost-efficient, with impressively lower feeding costs of up to 36% compared to the other diets. Regarding nutrient waste, for both temperature scenarios, the PAP− diet results in the lowest total N waste, while the NoPAP+ diet results in the lowest total P waste. Overall, the greatest negative impacts in terms of N and P waste are induced by the CTRL and NoPAP diets.

Table 11. Results of the extrapolation scenarios run in silico with the FEEDNETICSTM software, representing fish fed the experimental diets under commercial-like farming conditions at two constant temperatures (12 °C and 16 °C). All results are relative to the time point at which fish reach an average harvest weight of about 400 g.

		12 °C					16 °C				
Indicator	Unit	CTRL	NoPAP	NoPAP+	PAP	PAP−	CTRL	NoPAP	NoPAP+	PAP	PAP−
Time to reach 400 g	days	176 (0.0%)	179 (+1.7%)	160 (−9.1%)	158 (−10.2%)	166 (−5.7%)	139 (0.0%)	140 (+0.7%)	127 (−8.6%)	129 (−7.2%)	134 (−3.6%)
Growth rate	%/day	1.19 (0.0%)	1.17 (−1.6%)	1.31 (+9.9%)	1.33 (+11.7%)	1.27 (+6.3%)	1.52 (0.0%)	1.50 (−1.0%)	1.66 (+9.6%)	1.63 (+7.5%)	1.57 (+3.5%)
Feed conversion ratio (FCR)	g feed/g weight gain	1.19 (0.0%)	1.21 (+1.4%)	1.08 (−9.3%)	1.04 (−12.2%)	1.10 (−7.3%)	1.00 (0.0%)	1.00 (+0.5%)	0.90 (−9.4%)	0.90 (−9.4%)	0.94 (−5.3%)
Economic conversion ratio (ECR)	-	- (0.0%)	- (+25.0%)	- (+34.8%)	- (−2.3%)	- (−13.8%)	- (0.0%)	- (+22.9%)	- (+33.7%)	- (+0.6%)	- (−12.3%)
Total N waste	kg N/ton biomass gain	52.14 (0.0%)	52.60 (+0.9%)	51.14 (−1.9%)	46.09 (−11.6%)	44.51 (−4.6%)	39.82 (0.0%)	39.68 (−0.3%)	38.95 (−2.2%)	36.67 (−7.9%)	34.67 (−12.9%)
Total P waste	kg P/ton biomass gain	8.96 (0.0%)	8.25 (−7.9%)	6.53 (−27.1%)	7.28 (−18.8%)	7.75 (−3.4%)	7.63 (0.0%)	6.58 (−13.8%)	4.99 (−34.6%)	6.41 (−6.0%)	6.75 (−11.5%)

Inside parenthesis is the percentage change in relation to CTRL in each respective unit.

4. Discussion

This study is part of the GAIN project, in which a high diversity of novel alternative sources to FM, FO, and traditional vegetable ingredients were tested in trout feeds. It shows new insights for the inclusion of a combination of circular economy-driven ingredients for trout farming. Within the GAIN project, these novel formulation concepts were further tested in gilthead seabream [44], turbot [45], European seabass [46], and Atlantic salmon (in prep.). The results presented thus far support the hypothesis that using a combination of sustainable ingredients is a viable formulation concept, as for some of the tested formulations, no negative effects on fish performance or consumer acceptance were observed. Due to the complexity of the diet formulations tested in this study, it is difficult to make direct comparisons with other published studies, as most of them focused on evaluating one single ingredient at a time. Therefore, we will largely focus this discussion on other GAIN outputs.

4.1. Growth and Feed Performance

In the present study, rainbow trout kept the same growth pattern on the CTRL diet when fed diets containing low amounts of processed land animal proteins (PAP-) or without land animal proteins (NoPAP). These results corroborate findings reported by [11,47], where rainbow trout fed moderate amounts of poultry meal and insect meal showed good acceptance by fish. Additionally, trout presented better performance once fed improved versions of these diets, namely when fed NoPAP+ and PAP diets. The values of FCR, weight gain, and final body weight were generally better in fish fed the NoPAP+ diet when compared to the other diets. Thus, the quality and quantity of the novel ingredients used to replace FM and FO may have contributed to the increase in weight gain and relative growth rate (RGR) in trout fed with these two diets.

Furthermore, turbot-fed diets with the same concept of circular economy-driven ingredients showed no difference in FCR [43]. However, seabass fed NoPAP diets showed lower final body weight and similar weight gain when fed NoPAP+ compared to control [44]. On the other hand, sea bream tends to behave like turbot and do not present huge differences in growth performance parameters when fed sustainable ingredients [42]. This is also in line with the good results obtained previously with PAP-rich and alternative plant formulations for gilthead seabream [41], but a diet similar to the NoPAP diet led to poorer performance. Thus, different fish species may have different performance when fed complex alternative formulations such as those tested in the GAIN project.

In the present study, crude protein retention was lower in trout fed the PAP and PAP– diets when compared to the CTRL diet. Although diets that include proteins from land animals tend to be well accepted by fish [48], some studies have reported that the replacement of 20% of FM by poultry protein concentrate led to lower protein retention in a barramundi experiment [49]. Back to the present work, we can link protein retention with two other parameters. One is the protein efficiency ratio (PER), which was lower in the PAP-fed group, and the second is the digestibility of protein, which was lower in PAP− fed group. There are many factors that can influence the digestibility and nutritional value of land animal proteins, such as the processing methods used to treat PAPs in general, which can affect their nutritional quality, consequently leading to lower digestibility [49]. Another possible explanation may be related to the fact that these two diets, PAP and PAP−, have been formulated with higher contents of whole peas (20.8% and 31.8%, respectively). In contrast to pea protein concentrates, which can easily replace FM in salmon and seabream diets [29,50,51], whole peas are not subjected to pre-treatment before being used in fish feeds. Whole plant ingredients have a high fiber content, which is considered to be an antinutritional factor [52]. This same author reports that raw/whole pea protein is less digestible by rainbow trout when compared to a reference diet containing soy bean meal. Therefore, the combination of PAPs and high-quality whole peas could be the reason behind the lower performance of fish fed the PAP and PAP− diets in the present study. Still, we cannot conclude that the diets rich in processed land animal proteins have a negative effect on trout, as the growth performance was good and there was a complex mixture of ingredients in the diets tested. However, it may be speculated that if a higher quality of processed land animal proteins and/or a lower whole pea inclusion were used, an even higher performance may have been registered.

4.2. Welfare Indicators

It is important to mention that a lower digestibility of protein can lead to a lower digestibility of energy. This fact can be clearly seen in Table 4 with fish fed the PAP− diet. Thus, there is a clear unbalanced digestibility that might be leading to differences in some welfare indicators, such as the hepatosomatic index (HSI) and the viscerosomatic index (VSI). In the case of this study, higher values for HSI and VSI were found in fish fed both PAP and PAP- diets when compared to the CTRL and NoPAP+ diets. On the contrary, fish fed these same diets showed lower carcass yield (CY) when compared to the CTRL and NoPAP+ diets. In the case of seabass fed diets with the same concepts, the VSI was lower when compared to the control, although the HSI was not statistically different [44]. In addition, turbot fed with a mix of ingredients also presented a lower value when fed a PAP diet when compared to control fish [43]. The values that were found in this study for these indexes are in congruence with other findings from the literature [53]. Even as a welfare indicator, the discrepancy observed between treatments should not be taken as a bad or good result but only as differences reflecting disparate diet formulations, which may also represent adaptation to them.

It is commonly accepted that HSI provides an indication of the status of energy reserves and of the general metabolic activity of fish, and HSI values not exceeding 2% are generally associated with good hepatic functionality [43,44,53]. In the present study, all the HSI indexes ranged from 0.90 to 1.03%, well below this limit. Nonetheless, it must be noted that HSI values were significantly lower in all dietary groups containing a high concentration of vegetal ingredients (CTRL, NoPAP, and NoPAP+). A decrease in HSI might be attributed to reduced liver fat content. Randazzo et al. [47] reported a higher presence and distribution of lipids, fatty acids, and glycogen in the liver of rainbow trout fed graded levels of insect or poultry meal compared to fish fed a vegetable-rich diet.

Lysozyme is an important bactericidal enzyme of the humoral immune system, and it is considered a biomarker of the humoral response in fish [54]. In this study, it was possible to observe that the amount of lysozyme was markedly higher in trout fed the NoPAP diet than in trout fed the CTRL, NoPAP+, and PAP diets. The increase in serum lysozyme has been linked with the inclusion of some dietary ingredients in trout feeds, such as probiotics [55,56], fermented lupin [57], and seaweeds [58]. It is difficult to explain the increase in lysozyme due to the complexity of the formulations and the high number of ingredients used. Still, the presence of seaweed and microbial biomass may have contributed to this. However, no difference was found in the other diets tested where seaweeds and microbial biomass were added at the same level. Moreover, lysozyme levels tended to maintain similar values in seabass [44] fed diets in the same formulation context. In any case, other humoral parameters would be needed to confirm the putative effect of the NoPAP diet on trout innate immunity. Still, a diet with the NoPAP concept fed to gilthead seabream was shown to affect the expression of at least four genes (in the intestine or head kidney) involved in immune responses [42].

4.3. Fillet Analysis

Flesh color is one of the major quality attributes of farmed trout, which is generally marketed as pigmented (red) or non-pigmented (white) flesh. Salmonids depend on dietary supplements of pigments to achieve muscle color; in fact, they cannot synthesize carotenoids de novo [59]. Trout produced with diets containing no added pigments present a lower C* value and an increased L* than pigmented fish, and flesh color is expected to be white, grey, off-white, or pale pink, like the fillets of fish fed with CTRL, NoPAP, PAP, and PAP− diets. Conversely, increased dietary carotenoid levels implicate the increment of average values of a* and b* and a decrement of L* in the muscle, as measured in the fish samples from the NoPAP+ group and stated by numerous authors [60–63]. Carotenoids (β-carotene, astaxanthin) and yellow xanthophylls (lutein and zeaxanthin) are molecules widespread both in terrestrial vegetables and seaweeds [64], so the variation of color measured in NoPAP+ may be attributed to high proportions of the vegetable ingredients, particularly pea protein and starch. Color differences between fish fed NoPAP+ and other diets are confirmed by ΔE; spatial distances underline that consumers could perceive a difference in color ($2 < \Delta E < 3.5$) [42].

The TPA analysis did not show significant differences between the diets and excluded the adhesiveness parameter, but in the samples "pressure sensitive" like fish meat, harder products also create more pressure between the product and the base and thus may also create an improved bond (apparent adhesion) [65]. Anyway, NoPAP fillets achieved higher values for nearly all the measured variables (except adhesiveness). It is interesting to note that the consumer panel discriminates against No PAP compared to other samples, and this sample had the best score for textural features after cooking. The effect of cooking on salmonid flesh is not clear; some authors reported a decline in hardness [66,67] and others an increase [68,69]. Nevertheless, the reason why consumers perceived differences in textural features in line with the laboratory analysis is not clear and should be investigated [70,71].

4.4. Sensory Analysis

Despite the good performance of fish after being fed with these alternative ingredients, there are still concerns for the final consumer regarding their use in aquafeeds. These concerns range from the taste, appearance, and texture of the fish to zoonotic and risk assessment worries [72]. In the present study, we aimed to evaluate consumer acceptance of the fillet after feeding the fish for 91 days with the tested diets. Results showed that overall, people accepted the fillet of fish fed the NoPAP diet better than the group fed the PAP diet. Interestingly, these fish were the ones that presented, in general, better growth performance when compared to fish fed the PAP diets. Moreover, it is difficult to point out one single explanation for this, but it is important to highlight that PAP diets contained high quantities of land animal processed protein, such as feather meal and blood meal, that could have affected the texture attributes assessed here. On the other hand, differences in taste, appearance, and odor were not observed by the consumers participating in this analysis. In accordance with this, sensory analyses made in the laboratory were assessed by Petereit et al. [44] in seabass fed the same circularity concept diets. No differences were observed in consistency, frozen fat separation, protein separation, juice separation, or taste.

4.5. Long-Term Prediction FEEDNETICS™

The extrapolation scenarios run in FEEDNETICS™ show that the experimental diets have, in zootechnical terms, the potential to substitute current typical commercial diets (such as the CTRL diet). However, the high prices of some ingredients make some of them still an unviable option in economic terms. For example, the NoPAP and NoPAP+ diets lead to up to 38% higher feeding costs, which may compromise their use on a commercial scale. However, due to the lower cost of ingredients, the PAP and PAP− diets can lead to reduced feeding costs by up to 13.8% compared to the CTRL diet. Therefore, one of the most critical aspects of enabling the widespread adoption of alternative diet formulations, such as those presented in this paper, definitely involves reducing the costs of some novel ingredients. This challenge is fundamental and a high priority for the industry, as, when overcome, it will allow faster progress towards a more sustainable intensification of aquaculture. In addition, it may also be pointed out that the present analysis was performed taking current ingredient prices into consideration. Future price changes can rapidly make the figures presented here obsolete. Thus, this type of economic analysis is something that should be done regularly, taking into account the specificities of each farm/system.

These results put the use of novel diets under commercial farming conditions into perspective. In the first place, it is important to highlight that the relative performance of diets may change considerably with the farming conditions considered. This means that the diet that induces the best performance at a given temperature profile and/or when considering a given mortality rate may not be the same when these conditions change. In second place, it should be noted that while one diet can outperform on some important indicators (e.g., growth rate and feed conversion), it can also underperform on other, no less important ones (e.g., economic conversion and nutrient waste). The overall performance of diets is, in most cases, context-dependent, and the selection of one over the other may also differ from system to system (e.g., flow-through or RAS), as the specificities and limitations between them are distinct. Nevertheless, from the overall analysis of these results, it can be said that currently the PAP and PAP− diets seem to be viable alternatives to the CTRL diet (mimicking a typical commercial diet), as they present similar or even better performance in several aspects (i.e., zootechnical, economic, and environmental).

5. Conclusions

The findings of this research provide insights for the use of emerging ingredients in trout farming, covering important aspects in terms of zootechnical performance, welfare, and consumer acceptance. In general, it can be stated that trout accepted well the new formulations containing a wide range of emerging ingredients. Minor differences were detected in growth performance and feed conversion, but even the lowest values are still within a good performance range.

A limitation of this study is related to the fact that the formulation concepts were based on the use of a combination of different emerging ingredients rather than a classical FM replacement study with a single ingredient. Thus, the results of some indicators are hard to relate to single ingredients and consequently difficult to compare with other studies already published.

Since feeds interfere directly with the fish flesh's color, texture, and odor, sensory analysis is crucial to detecting consumer acceptance since they are strong drivers for consumer decisions. In the present study, consumers preferred the odor of fish fed the NoPAP diet. However, general acceptance did not show statistical differences. These findings suggest that the diets did not interfere with consumer acceptance.

Some emerging ingredients still have high prices or low availability in the market, making them unfeasible options for use on an industrial scale. The long-term prediction using the FEEDNETICS^{TM} tool showed that the economic conversion ratio as well as formulation costs were favorable in diet PAP−, in comparison to the CTRL diet. However, as mentioned above, future changes in ingredient prices may rapidly change the outcomes of the economic analysis presented here. Performing analysis of this kind on a regular basis is advisable. This is something that can be done easily with prediction tools based on mathematical models like FEEDNETICS^{TM}.

Clearly, feed formulations such as NoPAP and PAP, devoid of fish meal and containing a basket of alternative protein sources, are valuable options to support accelerated growth, good health, and a low feed conversion ratio in rainbow trout.

Supplementary Materials: The following supporting information can be downloaded at: https://www.mdpi.com/article/10.3390/jmse11061135/s1.

Author Contributions: All authors contributed significantly to the study's performance. More specifically: conceptualization, L.E.C.C., B.H.B., J.J. and J.D.; methodology, all authors; software, L.E.C.C. and F.S.; investigation, G.d.V.P., J.J. and J.D.; data curation, G.d.V.P. and F.F.; writing—original draft preparation, G.d.V.P., F.S., J.P. and F.F.; writing—review and editing, G.d.V.P. and L.E.C.C.; supervision, L.E.C.C. and F.F. All authors have read and agreed to the published version of the manuscript.

Funding: This research was funded by the European Union's Horizon 2020 Research and Innovation Program under grant agreement no. 773330 (GAIN, Green Aquaculture Intensification). This publication reflects the views only of the authors, and the European Commission cannot be held responsible for any use that may be made of the information contained therein. Pereira G.V. is supported by the FCT—Foundation for Science and Technology recruitment call of CEECIST/00168/2018.

Institutional Review Board Statement: All procedures involving animals were carried out in strict accordance with EU legal frameworks relating to the protection of animals used for scientific purposes (Directive 2010/63/EU). The approval was issued by the Ethics Committee of the Istituto Zooprofilattico Sperimentale delle Venezie, giving the rights to Fondazione Edmund Mach (n. 9750/2019) to perform the experiment.

Informed Consent Statement: Not applicable.

Data Availability Statement: Not applicable.

Acknowledgments: We also acknowledge the support of the fish trial team of FEM and the feed production team of Sparos. The authors thank the GAIN teams of IIM (CSIC), ANFACO, Salten Havbrukspark, and Wageningen University for providing ingredients included in the feed formulations.

Conflicts of Interest: The authors declare no conflict of interest.

References

1. Glencross, B.D.; Booth, M.; Allan, G.L. A feed is only as good as its ingredients—A review of ingredient evaluation strategies for aquaculture feeds. *Aquac. Nutr.* **2007**, *13*, 17–34. [CrossRef]
2. Naylor, R.L.; Hardy, R.W.; Bureau, D.P.; Chiu, A.; Elliott, M.; Farrell, A.P.; Forster, I.; Gatlin, D.M.; Goldburg, R.J.; Nichols, P.D. Feeding aquaculture in an era of finite resources. *Proc. Natl. Acad. Sci. USA* **2009**, *106*, 15103–15110. [CrossRef] [PubMed]
3. Hardy, R.W. Utilization of plant proteins in fish diets: Effects of global demand and supplies of fishmeal. *Aquac. Res.* **2010**, *41*, 770–776. [CrossRef]
4. Péron, G.; Mittaine, J.F.; Le Gallic, B. Where do fishmeal and fish oil products come from? An analysis of the conversion ratios in the global fishmeal industry. *Mar. Policy* **2010**, *34*, 815–820. [CrossRef]

5. Bianchi, M.C.G.; Cjopin, F.; Farme, T.; Franz, N.; Fuentevilla, C.; Garibaldi, L.; Laurenti, A.L.G. *The State of World Fisheries and Aquaculture 2020*; Sustainability in Action; FAO: Rome, Italy, 2020. [CrossRef]
6. Hardy, R.W. Aquaculture feeds and ingredients: An overview. In *New Technologies in Aquaculture*; Burnell, G., Allan, G., Eds.; Woodhead Publishing Limited: New York, NY, USA, 2009; pp. 370–386.
7. Lazzarotto, V.; Corraze, G.; Leprevost, A.; Quillet, E.; Dupont-Nivet, M.; Médale, F. Three-year breeding cycle of rainbow trout (*Oncorhynchus mykiss*) fed a plant-based diet, totally free of marine resources: Consequences for reproduction, fatty acid composition and progeny survival. *PLoS ONE* **2015**, *10*, e0117609. [CrossRef] [PubMed]
8. Ytrestøyl, T.; Aas, T.S.; Åsgård, T. Utilisation of feed resources in production of Atlantic salmon (*Salmo salar*) in Norway. *Aquaculture* **2015**, *448*, 365–374. [CrossRef]
9. Colombo, S.M. Physiological considerations in shifting carnivorous fishes to plant-based diets. In *Fish Physiology*; Benfey, T.J., Farrell, A.P., Brauner, C.J., Eds.; Academic Press: Cambridge, MA, USA, 2020; Volume 38, pp. 53–82.
10. Aragão, C.; Gonçalves, A.T.; Costas, B.; Azeredo, R.; Xavier, M.J.; Engrola, S. Alternative Proteins for Fish Diets: Implications beyond Growth. *Animals* **2022**, *12*, 1211. [CrossRef]
11. Cardinaletti, G.; Di Marco, P.; Daniso, E.; Messina, M.; Donadelli, V.; Finoia, M.G.; Petochi, T.; Fava, F.; Faccenda, F.; Contò, M.; et al. Growth and Welfare of Rainbow Trout (*Oncorhynchus mykiss*) in Response to Graded Levels of Insect and Poultry By-Product Meals in Fishmeal-Free Diets. *Animals* **2022**, *12*, 1698. [CrossRef]
12. Galkanda-Arachchige, H.S.; Wilson, A.E.; Davis, D.A. Success of fishmeal replacement through poultry by-product meal in aquaculture feed formulations: A meta-analysis. *Rev. Aquac.* **2020**, *12*, 1624–1636. [CrossRef]
13. Bandara, T. Alternative feed ingredients in aquaculture: Opportunities and challenges. *J. Entomol. Zool. Stud.* **2018**, *6*, 3087–3094.
14. Siddik, M.A.; Howieson, J.; Fotedar, R.; Partridge, G.J. Enzymatic fish protein hydrolysates in finfish aquaculture: A review. *Rev. Aquac.* **2021**, *13*, 406–430. [CrossRef]
15. Hua, K.; Cobcroft, J.; Cole, A.; Condon, K.; Jerry, D.; Mangott, A.; Praeger, C.; Vucko, M.; Zeng, C.; Zenger, K.; et al. The Future of Aquatic Protein: Implications for Protein Sources in Aquaculture Diets. *One Earth* **2019**, *1*, 316–329. [CrossRef]
16. Jones, S.W.; Karpol, A.; Friedman, S.; Maru, B.T.; Tracy, B.P. Recent advances in single cell protein use as a feed ingredient in aquaculture. *Curr. Opin. Biotechnol.* **2020**, *61*, 189–197. [CrossRef]
17. NRC National Research Council. *Nutrient Requirements of Fish and Shrimp*; The National Academies Press: Washington, DC, USA, 2011; p. 376.
18. Regost, C.; Arzel, J.; Robin, J.; Rosenlund, G.; Kaushik, S.J. Total replacement of fish oil by soybean or linseed oil with a return to fish oil in turbot (*Psetta maxima*): 1. Growth performance, flesh fatty acid profile, and lipid metabolism. *Aquaculture* **2003**, *217*, 465–482. [CrossRef]
19. Oliva-Teles, A.; Enes, P.; PerLues, H. Replacing fishmeal and fish oil in industrial aquafeeds for carnivorous fish. In *Feed and Feeding Practices in Aquaculture*; Davis, A., Ed.; Woodhead Publishing: Swaston, CA, USA, 2015; pp. 203–233.
20. Ruiz-Lopez, N.; Usher, S.; Sayanova, O.V.; Napier, J.A.; Haslam, R.P. Modifying the lipid content and composition of plant seeds: Engineering the production of LC-PUFA. *Appl. Microbiol. Biotechnol.* **2015**, *99*, 143–154. [CrossRef] [PubMed]
21. Izquierdo, M.S.; Montero, D.; Robaina, L.; Caballero, M.J.; Rosenlund, G.; Ginés, R. Alterations in fillet fatty acid profile and flesh quality in gilthead seabream (*Sparus aurata*) fed vegetable oils for a long term period. Recovery of fatty acid profiles by fish oil feeding. *Aquaculture* **2005**, *250*, 431–444. [CrossRef]
22. Carvalho, M.; Montero, D.; Rosenlund, G.; Fontanillas, R.; Ginés, R.; Izquierdo, M. Effective complete replacement of fish oil by combining poultry and microalgae oils in practical diets for gilthead sea bream (*Sparus aurata*) fingerlings. *Aquaculture* **2020**, *529*, 735696. [CrossRef]
23. Santigosa, E.; Constant, D.; Prudence, D.; Wahli, T.; Verlhac-Trichet, V. A novel marine algal oil containing both EPA and DHA is an effective source of omega-3 fatty acids for rainbow trout (*Oncorhynchus mykiss*). *J. World Aquac. Soc.* **2020**, *51*, 649–665. [CrossRef]
24. Bélanger, A.; Sarker, P.K.; Bureau, D.P.; Chouinard, Y.; Vandenberg, G.W. Apparent digestibility of macronutrients and fatty acids from microalgae (*Schizochytrium* sp.) fed to rainbow trout (*Oncorhynchus mykiss*): A potential candidate for fish oil substitution. *Animals* **2021**, *11*, 456. [CrossRef]
25. Greene, D.H.; Selivonchick, D.P. Effects of dietary vegetable, animal and marine lipids on muscle lipid and hematology of rainbow trout (*Oncorhynchus mykiss*). *Aquaculture* **1990**, *89*, 165–182. [CrossRef]
26. Richard, N.; Kaushik, S.; Larroquet, L.; Panserat, S.; Corraze, G. Replacing dietary fish oil by vegetable oils has little effect on lipogenesis, lipid transport and tissue lipid uptake in rainbow trout (*Oncorhynchus mykiss*). *Br. J. Nutr.* **2006**, *96*, 299–309. [CrossRef]
27. Mancuso, T.; Baldi, L.; Gasco, L. An empirical study on consumer acceptance of farmed fish fed on insect meals: The Italian case. *Aquac. Int.* **2016**, *24*, 1489–1507. [CrossRef]
28. Ferrer Llagostera, P.; Kallas, Z.; Reig, L.; Amores de Gea, D. The use of insect meal as a sustainable feeding alternative in aquaculture: Current situation, Spanish consumers' perceptions and willingness to pay. *J. Clean. Prod.* **2019**, *229*, 10–21. [CrossRef]
29. Hoerterer, C.; Petereit, J.; Krause, G. Informed choice: The role of knowledge in the willingness to consume aquaculture products of different groups in Germany. *Aquaculture* **2022**, *556*, 738319. [CrossRef]
30. Petereit, J.; Hoerterer, C.; Krause, G. Country-specific food culture and scientific knowledge transfer events—Do they influence the purchasing behaviour of seafood products? *Aquaculture* **2022**, *560*, 738950. [CrossRef]

31. Sanchez-Lozano, N.B.; Martinez-Llorens, S.; Tomas-Vidal, A.; Cerda, M.J. Effect of high-level fish meal replacement by pea and rice concentrate protein on growth, nutrient utilization and fillet quality in gilthead seabream (*Sparus aurata* L.). *Aquaculture* **2010**, *298*, 83–89. [CrossRef]
32. Matos, E.; Gonçalves, A.; Bandarra, N.; Colen, R.; Nunes, M.L.; Valente, L.M.P.; Dinis, M.T.; Dias, J. Plant proteins and vegetable oil do not have detrimental effects on post-mortem muscle instrumental texture, sensory properties and nutritional value of gilthead seabream. *Aquaculture* **2012**, *358*, 205–212. [CrossRef]
33. Calanche, J.B.; Beltran, J.A.; Hernandez, A.A.J. Aquaculture and sensometrics: The need to evaluate sensory attributes and the consumers' preferences. *Rev. Aquac.* **2020**, *12*, 805–821. [CrossRef]
34. De La Noüe, J.; Choubert, G. Digestibility in Rainbow Trout: Comparison of the Direct and Indirect Methods of Measurement, *Prog. Fish Cult.* **1986**, *48*, 190–195. [CrossRef]
35. Austreng, E. Digestibility determination in fish using chromic oxide marking and analysis of contents from different segments of the gastrointestinal tract. *Aquaculture* **1978**, *13*, 265–272. [CrossRef]
36. AOAC. *Official Methods of Analysis*, 18th ed.; Association of Official Analytical Chemists: Gaithersburgs, MD, USA, 2006.
37. *ISO 27085:2009*; Animal Feeding Stuffs. Determination of Calcium, Sodium, Phosphorus, Magnesium, Potassium, Iron, Zinc, copper, Manganese, Cobalt, Molybdenum, Arsenic, Lead and Cadmium by ICP-AES. International Organisation for Standardization: Geneva, Switzerland, 2009; p. 24.
38. Reis, P.A.; Valente, L.M.P.; Almeida, C.M.R. A fast and simple methodology for determination of yttrium as an inert marker in digestibility studies. *Food Chem.* **2008**, *108*, 1094–1098. [CrossRef]
39. Milla, S.; Mathieu, C.; Wang, N.; Lambert, S.; Nadzialek, S.; Massart, S.; Henrotte, E.; Douxfils, J.; Mélard, C.; Mandiki, S.N.; et al. Spleen immune status is affected after acute handling stress but not regulated by cortisol in Eurasian perch, *Perca fluviatilis*. *Fish Shellfish. Immunol.* **2010**, *28*, 931–941. [CrossRef]
40. Bourne, M.C. Texture profile analysis. *Food Technol.* **1978**, *32*, 62–66.
41. C.I.E. Compagnie Internationale de l'Eclairage. International. In *Commission on Illumination, Recommendations on Uniform Colorspaces*; Supplement No. 2 to CIE Publication No. 15.2; Bureaucentral: Paris, France, 1978.
42. Mokrzycki, W.; Tatol, M. Color difference ΔE: A survey. *Mach. Graph. Vis.* **2011**, *20*, 383–411.
43. Soares, F.M.R.C.; Nobre, A.M.D.; Raposo, A.I.G.; Mendes, R.C.P.; Engrola, S.A.D.; Rema, P.J.A.P.; Conceição, L.E.C.; Silva, T.S. Development and Application of a Mechanistic Nutrient-Based Model for Precision Fish Farming. *J. Mar. Sci. Eng.* **2023**, *11*, 472. [CrossRef]
44. Piazzon, M.C.; Naya-Català, F.; Pereira, G.V.; Estensoro, I.; Del Pozo, R.; Calduch-Giner, J.A.; Nuez-Ortín, W.G.; Palenzuela, O.; Sitjà-Bobadilla, A.; Dias, J.; et al. A novel fish meal-free diet formulation supports proper growth and does not impair intestinal parasite susceptibility in gilthead sea bream (*Sparus aurata*) with a reshape of gut microbiota and tissue-specific gene expression patterns. *Aquaculture* **2022**, *558*, 738362. [CrossRef]
45. Hoerterer, C.; Petereit, J.; Lannig, G.; Johansen, J.; Pereira, G.V.; Conceição, L.E.C.; Pastres, R.; Buck, B.H. Sustainable fish feeds: Potential of emerging protein sources in diets for juvenile turbot (*Scophthalmus maximus*) in RAS. *Aquac. Int.* **2022**, *30*, 1481–1504. [CrossRef]
46. Petereit, J.; Hoerterer, C.; Bischoff-Lang, A.A.; Conceição, L.E.C.; Pereira, G.; Johansen, J.; Pastres, R.; Buck, B.H. Adult European Seabass (*Dicentrarchus labrax*) Perform Well on Alternative Circular-Economy-Driven Feed Formulations. *Sustainability* **2022**, *14*, 7279. [CrossRef]
47. Randazzo, B.; Zarantoniello, M.; Gioacchini, G.; Cardinaletti, G.; Belloni, A.; Giorgini, E.; Faccenda, F.; Cerri, R.; Tibaldi, E.; Olivotto, I. Physiological response of rainbow trout (*Oncorhynchus mykiss*) to graded levels of Hermetia illucens or poultry by-product meals as single or combined substitute ingredients to dietary plant proteins. *Aquaculture* **2021**, *538*, 736550. [CrossRef]
48. Moutinho, S.; Martínez-Llorens, S.; Tomás-Vidal, A.; Jover-Cerdá, M.; Oliva-Teles, A.; Peres, H. Meat and bone meal as partial replacement for fish meal in diets for gilthead seabream (*Sparus aurata*) juveniles: Growth, feed efficiency, amino acid utilization, and economic efficiency. *Aquaculture* **2017**, *468*, 271–277. [CrossRef]
49. Simon, C.J.; Salini, M.J.; Irvin, S.; Blyth, D.; Bourne, N.; Smullen, R. The effect of poultry protein concentrate and phosphorus supplementation on growth, digestibility and nutrient retention efficiency in barramundi *Lates calcarifer*. *Aquaculture* **2019**, *498*, 305–314. [CrossRef]
50. Øverland, M.; Sørensen, M.; Storebakken, T.; Penn, M.; Krogdahl, Å.; Skrede, A. Pea protein concentrate substituting fish meal or soybean meal in diets for Atlantic salmon (*Salmo salar*)—Effect on growth performance, nutrient digestibility, carcass composition, gut health, and physical feed quality. *Aquaculture* **2009**, *288*, 305–311. [CrossRef]
51. Hansen, J.Ø.; Shearer, K.D.; Øverland, M.; Penn, M.H.; Krogdahl, Å.; Mydland, L.T.; Storebakken, T. Replacement of LT fish meal with a mixture of partially deshelled krill meal and pea protein concentrates in diets for Atlantic salmon (*Salmo salar*). *Aquaculture* **2011**, *315*, 275–282. [CrossRef]
52. Thiessen, D.L.; Campbell, G.L.; Addizi, P.D. Digestibility and growth performance of juvenile rainbow trout (*Oncohychus mykiss*) fed with pea and canola products. *Aqua. Nutr.* **2003**, *9*, 67–75. [CrossRef]
53. Sallam, A.E.; El-feky, M.M.; Ahmed, M.S.; Mansour, A.T. Potential use of whey protein as a partial substitute of fishmeal on growth performance, non-specific immunity and gut histological status of juvenile European seabass, *Dicentrarchus labrax*. *Aquac. Res.* **2021**, *53*, 1527–1541. [CrossRef]

54. Girón-Pérez, M.I.; Velázquez-Fernández, J.; Díaz-Resendiz, K.; Diaz-Salas, F.; Canto-Montero, C.; Medina-Diaz, I.; Zaitseva, G. Immunologic parameters evaluations in Nile tilapia (*Oreochromis niloticus*) exposed to sublethal concentrations of diazinon. *Fish Shellfish Immunol.* **2009**, *27*, 383–385. [CrossRef]
55. Mohammadian, T.; Nasirpour, M.; Tabandeh, M.R.; Heidary, A.A.; Ghanei-Motlagh, R.; Hosseini, S.S. Administrations of autochthonous probiotics altered juvenile rainbow trout *Oncorhynchus mykiss* health status, growth performance and resistance to *Lactococcus garvieae*, an experimental infection. *Fish Shellfish Immunol.* **2019**, *86*, 269–279. [CrossRef]
56. Ramos, M.A.; Gonçalves, J.F.M.; Batista, S.; Costas, B.; Pires, M.A.; Rema, P.; Ozório, R.O.A. Growth, immune responses and intestinal morphology of rainbow trout (*Oncorhynchus mykiss*) supplemented with commercial probiotics. *Fish Shellfish Immunol.* **2015**, *45*, 19–26. [CrossRef]
57. Davies, S.J.; El-Haroun, E.R.; Hassaan, M.S.; Bowyer, P.H. A Solid-State Fermentation (SSF) supplement improved performance, digestive function and gut ultrastructure of rainbow trout (*Oncorhynchus mykiss*) fed plant protein diets containing yellow lupin meal. *Aquaculture* **2021**, *545*, 737177. [CrossRef]
58. Vazirzadeh, A.; Marhamati, A.; Rabiee, R.; Faggio, C. Immunomodulation, antioxidant enhancement and immune genes up-regulation in rainbow trout (*Oncorhynchus mykiss*) fed on seaweeds included diets. *Fish Shellfish Immunol.* **2020**, *106*, 852–858. [CrossRef]
59. Barbosa, M.J.; Morais, R.; Choubert, G. Effect of carotenoid source and dietary lipid content on blood astaxanthin concentration in rainbow trout (*Oncorhynchus mykiss*). *Aquaculture* **1999**, *176*, 331–341. [CrossRef]
60. Skrede, G.; Storebakken, T. Instrumental colour analysis of farmed and wild Atlantic salmon when raw, baked and smoked. *Aquaculture* **1986**, *53*, 279–286. [CrossRef]
61. Choubert, G.; Blanc, J.-M.; Courvalin, C. Muscle carotenoid content and colour of farmed rainbow trout fed astaxanthin or canthaxanthin as affected by cooking and smoke-curing procedures. *Int. J. Food Sci.* **1992**, *27*, 277–284. [CrossRef]
62. Christiansen, R.; Struksnæs, G.; Estermann, R.; Torrissen, O.J. Assessment of flesh colour in Atlantic salmon, *Salmo salar* L. *Aquac. Res.* **1995**, *26*, 311–321. [CrossRef]
63. Wathne, E.; Bjerkeng, B.; Storebakken, T.; Vassvik, V.; Odland, A.B. Pigmentation of Atlantic salmon (*Salmo salar*) fed astaxanthin in all meals or in alternating meals. *Aquaculture* **1998**, *159*, 217–231. [CrossRef]
64. Wells, M.L.; Potin, P.; Craigie, J.S.; Raven, J.A.; Merchant, S.S.; Helliwell, K.E.; Smith, A.G.; Camire, M.E.; Brawley, S.H. Algae as nutritional and functional food sources: Revisiting our understanding. *J. Appl. Phycol.* **2017**, *29*, 949–982. [CrossRef]
65. Texture Profile Analysis Texture Technologies Corp. and Stable Micro Systems, Ltd. Available online: https://texturetechnologies.com/resources/texture-profile-analysis (accessed on 15 May 2023).
66. Mørkøre, T.; Hansen, A.Å.; Unander, E.; Einen, O. Composition, liquid leakage, and mechanical properties of farmed rainbow trout: Variation between fillet sections and the impact of ice and frozen storage. *J. Food Sci.* **2006**, *67*, 1933–1938. [CrossRef]
67. Aussanasuwannakul, A.; Kenney, P.B.; Weber, G.M.; Yao, J.; Slider, S.D.; Manor, M.L.; Salem, M. Effect of sexual maturation on growth, fillet composition, and texture of female rainbow trout (*Oncorhynchus mykiss*) on a high nutritional plane. *Aquaculture* **2010**, *317*, 79–88. [CrossRef]
68. Ginés, R.; Valdimarsdottir, T.; Sveinsdottir, K.; Thorarensen, H. Effects of rearing temperature and strain on sensory characteristics, texture, colour and fat of Arctic charr (*Salvelinus alpinus*). *Food Qual. Prefer.* **2004**, *15*, 177–185. [CrossRef]
69. Martelli, R.; Franci, O.; Lupi, P.; Faccenda, F.; Parisi, G. Physico-Chemical Traits of Raw and Cooked Fillets of Rainbow Trout (*Oncorhynchus mykiss*) from Different Strains and Farms. *Ital. J. Anim. Sci.* **2014**, *13*, 3417. [CrossRef]
70. Husein, Y.; Secci, G.; Dinnella, C.; Parisi, G.; Fusi, R.; Monteleone, E.; Zanoni, B. Enhanced utilisation of nonmarketable fish: Physical, nutritional and sensory properties of 'clean label' fish burgers. *Int. J. Food Sci. Technol.* **2019**, *54*, 593–601. [CrossRef]
71. Bruni, L.; Secci, G.; Husein, Y.; Faccenda, F.; Medeiros, A.C.L.; Parisi, G. Is it possible to cut down fishmeal and soybean meal use in aquafeed limiting the negative effects on rainbow trout (*Oncorhynchus mykiss*) fillet quality and consumer acceptance? *Aquaculture* **2021**, *543*, 736996. [CrossRef]
72. Morales, J.B.C.; Tomás-Vidal, A.; Cusiyunca, P.E.R.; Martínez-Llorens, S.; Marquina, P.L.; Jover-Cerdá, M.; Beltrán, J.A. An Approach to the Spanish Consumer's Perception of the Sensory Quality of Environmentally Friendly Seabass. *Foods* **2021**, *10*, 2694. [CrossRef] [PubMed]

Journal of
Marine Science
and Engineering

MDPI

Article

Anti-Inflammatory Activity of Mycobiont Extract of *Parmotrema austrosinense* (Zahlbr.) Hale in a Zebrafish Model

Kalidoss Rajendran [1], Ponmurugan Karuppiah [2,*], Ponmurugan Ponnusamy [3], Mohammed Rafi Shaik [4], Mujeeb Khan [4], Tae Hwan Oh [5] and Baji Shaik [5,*]

1 Department of Education, St. Joseph University in Tanzania, Dar es Salaam P.O. Box 11007, Tanzania; doss_rk@yahoo.com
2 Department of Botany and Microbiology, College of Science, King Saud University, P.O. Box 2455, Riyadh 11451, Saudi Arabia
3 Biomedical Research Laboratory, Department of Botany, Bharathiar University, Coimbatore 641046, Tamil Nadu, India; drponmurugan@gmail.com
4 Department of Chemistry, College of Science, King Saud University, P.O. Box 2455, Riyadh 11451, Saudi Arabia; mrshaik@ksu.edu.sa (M.R.S.); kmujeeb@ksu.edu.sa (M.K.)
5 School of Chemical Engineering, Yeungnam University, Gyeongsan 38541, Republic of Korea; taehwanoh@ynu.ac.kr
* Correspondence: pkaruppiah@ksu.edu.sa (P.K.); shaikbaji@yu.ac.kr (B.S.)

Citation: Rajendran, K.; Karuppiah, P.; Ponnusamy, P.; Shaik, M.R.; Khan, M.; Oh, T.H.; Shaik, B. Anti-Inflammatory Activity of Mycobiont Extract of *Parmotrema austrosinense* (Zahlbr.) Hale in a Zebrafish Model. *J. Mar. Sci. Eng.* **2023**, *11*, 1081. https://doi.org/10.3390/jmse11051081

Academic Editors: Ana Catarina Matias and Carlos Andrade

Received: 14 April 2023
Revised: 10 May 2023
Accepted: 18 May 2023
Published: 20 May 2023

Abstract: Thousands of different kinds of lichen metabolites are being examined for their biological activities, including anticancer properties. In this context, the present study aims to assess the anti-inflammatory activity of the acetone extract of *Parmotrema austrosinense* mycobiont. A lipid peroxidation assay was performed with the acetone extracts of *P. austrosinense* mycobiont, which was further used to evaluate its anti-inflammatory efficacy using a zebrafish model. Furthermore, the histopathological study was also carried out with muscle tissues and amplification of its inflammation marker. The results revealed that the lichen compound (i.e., lecanoric acid) in the acetone extract of *P. austrosinense* possesses anti-inflammatory activity. Histopathology studies confirmed the decreased numbers of neutrophil cells in the 2,4,6-trinitrobenzene sulfonic acid (TNBS)-induced zebrafishes, as confirmed by changes in the fishes' weight before and after the sample treatment, prompted by TNBS inflammation. The present results also demonstrated a dose-dependent decrease in the lipid peroxidation (LPO) levels in the muscle tissues of zebrafishes. Gene amplification studies suggested that the lichen compound might perform dose-dependent downregulation of the inflammatory gene marker of the tumor necrosis factor (TNF)-α gene; this further confirms that the extract should possess anti-inflammatory activity. As per the literature, this study is one of the most complete, comprehensive in vivo anti-inflammatory analyses in which inflammation was induced in zebrafish by using 2,4,6-trinitrobenzene sulfonic acid (TNBS). Particularly, this study successfully identified a bioactive compound isolated from the lichen *P. austrosinense*, and which exhibited decent anti-inflammatory activity.

Keywords: lichen; anti-inflammatory activity; lipid peroxidation assay; zebrafish

1. Introduction

Phenolic compounds of lichen extracts used for biological activities are all derivatives of the orcinol type [1,2]. They are aromatic compounds containing phenolic carboxylic acids. All of these can be broadly classified into two types: depsides and depsidones [3]. Some of the depsides used in the research laboratory include lecanoric acid and erythrin. Lecanoric acid is distributed in *Parmotrema austrosinense* and *P. tinctorum* [4], while erythrin is present in *Roccella montagnei* [5]. There are wide varieties of depsidones that are used for their biological activities, such as salazinic acid, protocetraric acid, etc. [6,7]. Salazinic acid is widely distributed in *Parmotrema reticulatum* and *Parmelinella wallichiana*, whereas protocetraric acid has been found in *Flavoparmelia caperata* [8]. The most commonly used lichen compound in the pharmacological research laboratory setting is usnic acid, which is a novel compound derived from dibenzofuran derivatives [9].

Several methods can be used to determine the anticancer abilities of lichen substances—mainly computational, in vivo, and in vitro methods [10]. Some species contain more than one metabolite. The yield of the metabolites from various lichen species must be ascertained before their clinical use. This accounts for 22% of the total dry weight of *Parmotrema tinctorum*. A large number of lichen compounds—such as anziaic acid, bonimic acid, divaricatic acid, lobaric acid, olivetoric acid, protolichesterinic acid, ramalinolic acid, spherophorin, and sekikaic acid—are associated with antibacterial action against *Bacillus subtilis*, *Mycobacterium tuberculosis*, *Mycobacterium smegmatis*, *Pneumococcus* sp., *Salmonella typhimurium*, *Staphylococcus aureus*, and *Streptococcus pyogenes* [11,12]. On the other hand, lichen substances such as caperatic acid, rangiformic acid, zeorin, atranorin, thamnolic acid, salazinic acid, psoromic acid, fumarprotocetraric acid, pannarin, and endocrocin are not effective against several test strains [13].

In India, the bioprospecting of lichen started very recently. The ethnobotanical uses of lichens by various ethnic groups in India have been documented. Over 80% of the world's population—especially in underdeveloped nations—still relies on herbal remedies as their primary form of healthcare today [14]. This has drawn substantial attention from the scientific community to investigate the mycobionts and photobionts (and their related biological activities), as a result of the influence of lichens in bioprospecting studies and their application [15]. The *Parmotrema* spp. thallus contains 23.5% lecanoric acid and 37.5% usnic acid. The medullary layer also contains protocetraric acid, physodalic acid, and physodic acid. These lichens have been identified and used throughout human history [6].

Considering their importance, the lichen compounds of specific species play a significant role in pharmacopeias due to their antimicrobial, antiviral, anti-inflammatory, antioxidant, anti-analgesic, antipyretic, antiproliferative, and anticancer activities. Moreover, they are receiving wide attention in the field of pharmacy and have already contributed to several advances in medicinal applications, including anti-snake-venom properties [16,17]. A considerable number of lichen species can cure jaundice or coughs and control rabies. The most important species are *Cladonia*, *Peltigera*, *Roccella*, and *Usnea*. Some species, such as *Parmelia sulcata* and *Lobaria pulmonaria*, can cure cranial and pulmonary diseases. *Xanthoria paritinais* has been used in the treatment of jaundice. *Cetraria islandica* displays medicinal properties for bladder, kidney, and lung diseases and heals oral inflammation. It can also control the proliferation of cancer cell lines [18]. More than a thousand different secondary metabolites of lichens have had their structures determined so far, although many of them still need to be described [19]. Cytotoxic ability, which is imparted by the presence of polyphenolic compounds in the fungal cells, is the term used to describe the most diverse biological activities of this group of organisms. Moreover, various secondary metabolites of lichens are known to exhibit anti-inflammatory properties, which are less frequently reported. Inflammation is a defense mechanism that controls and repairs the damage caused in the body, accompanying a number of diseases. In the event of inflammation, biochemical changes leading to redness, swelling, itching, pain, and increased temperature occur [20]. Particularly, depsides and depsidones have been reported to exhibit decent anti-inflammatory properties. For instance, the depside atranorin, which is present in the thalli of different species of lichen, has anti-inflammatory properties [21]. Bugni et al. studied the effects of a cyclooxygenase inhibitor [22]. The obtained results indicated that the tested depside inhibited cyclooxygenase-1 (COX-1) in a dose-dependent manner and, at the concentration of 17 μg/mL (45 μM), caused a decrease in enzyme activity by 50%. In addition, lecanoric acid is another depside produced by various lichen species, and it is also known to possess anti-inflammatory properties [23]. In this context, the present study is aimed at the isolation and characterization of bioactive compounds from the lichen mycobiont *P. austrosinense*, with a major objective of evaluating the anti-inflammatory effect of lecanoric acid in a zebrafish model.

2. Materials and Methods

2.1. Chemicals

All of the chemicals used in this study were procured from SD Fine Chem Limited, Chennai, Tamil Nadu, India. All of the chemicals were of analytical grade.

2.2. Collection and Identification of *P. austrosinense*

The Tamil Nadu Western Ghats (10°10′11″ N 77°03′40″ E), India, at an elevation of 2695 m (8842 ft) above sea level, is an ideal location for the present study, owing to the presence of wildlife reserves and national parks. The major reason behind the selection of this area was the availability of ethnobotanical samples obtained through prior research conducted in the region. Additionally, the region is readily accessible [24,25]. To authenticate a particular lichen sample, the representative samples were compared with the preserved samples at the National Botanical Research Institute (NBRI), Lucknow, India. The identification manual of Awasthi [13,26] was closely monitored. Thus, in addition to microscopic examination, representative samples were subjected to spot tests, chemical profiling, and thin-layer chromatography (TLC) testing to determine the lichen species. To study the morphological, structural, vegetative, and reproductive features of the lichen thalli, stereomicroscopy was used, except for anatomical sections, for which a compound light microscope was used. The lichen compounds were extracted from the thallus by the Soxhlet method, using acetone as the solvent [27].

2.3. Extraction of Bioactive Compound

Lecanoric acid was produced and extracted by the method described by Verma and Behera [28]. Briefly, Modified Bold's Basal Salt (MBBS) medium was prepared with 1% potassium nitrate and 4% sucrose in a flask and inoculated with the mycobiont isolate, and it was incubated at 25 °C for 4–8 weeks in a shaking incubator at 150 rpm agitation. During the subsequent incubation, part of the flask's contents was used for extraction of the bioactive compound with acetone through the Soxhlet apparatus. Later, the crude culture extract was concentrated with a rotary vacuum evaporator, and it was stored at −20 °C until further use. The crude extract was investigated in the presence of bioactive compounds using TLC and microcrystallization methods.

2.4. Zebrafish Collection and Maintenance

Healthy adult zebrafishes that were free from illness or deformities were selected for the present study. The average weight and length of the zebrafishes ranged from 0.18 to 0.22 g and from 2.55 to 2.95 cm, respectively. All of the fishes were habituated in a laboratory setting before the beginning of the experiments, in accordance with the Organization for Economic Co-operation and Development (OECD) recommendations [29]. For the treatment with the test sample, the zebrafishes were divided into eight different groups, each containing 12 zebrafishes (control, TNBS-induced group, standard, 200, 400, 600, 800, and 1000 μg/mL), and all maintained in different tanks [30].

2.5. Determination of LC_{50}

The zebrafishes of interest were weighed, and their weight was used to determine how much of the sample was to be injected in a 20 μL volume of phosphate-buffered saline (PBS) at pH 7.4 [31]. In this experiment, 30 G caliber injections were applied. The doses tested were 20, 40, 60, 80, and 100 μg/mL. A foundation plane was prepared beforehand to hold the fish: a soft sponge with a 1.5 cm section cut out from the center, in a Petri dish with ice water (12 °C). Prior to anesthesia, test solutions were placed into the injection. By putting ice chips into the fish water until the temperature hit 12 °C, each fish was gradually put to sleep. Holding the fish in the sponge with the abdomen facing up, it was transferred to the base plane when the fish remained still in response to any external stimuli. Without waiting, the solutions were injected intraperitoneally, and the fish were then moved to normal water (35 °C) to recover and resume their normal behaviors [32]. The fish were monitored for 7 days following the injection [33]. From day 0 to day 7, the number of animals that died in each concentration was recorded, and the LC_{50} value was calculated.

2.6. Induction of Inflammation and Efficacy Analysis

The 2,4,6-trinitrobenzene sulfonic (TNBS) acid injection was administered based on the weight of fish, at 1 μL per 1 mg of body weight. The 2,4,6-trinitrobenzene sulfonic acid (TNBS) is a kind of hapten. It binds to a protein (tissue), turns into an antigen, and provokes immunological reactions. TNBS-induced colitis is a delayed hypersensitivity reaction to haptenized proteins, and it disrupts the epithelial barrier. For testing the efficacy, the zebrafish were divided into eight different groups: control, TNBS-induced, standard, and five different concentrations. The acetone extracts were administered at final doses of 20, 40, 60, 80, and 100 g/mL after the TNBS injections, which were timed to elicit the best inflammation in adult zebrafishes within 4 h. The zebrafish weight and the volume of TNBS to be injected were determined, and the injection was adjusted to a final volume of 20 μL in PBS. Following the same procedure as mentioned above, TNBS was injected and weight was measured before and after treatment. The zebrafishes were killed immediately to finish the determination of the biological effects upon completion of the experiment after 4 h of treatment. The experimental zebrafish model was performed in triplicate.

2.7. Histopathology Study of Muscle Tissue

After 4 h, the zebrafishes were decapitated, and their abdominal region was dissected and fixed in Dietrich's fixative before being transferred for histopathology study. The remaining tissues were immediately stored at 4 °C for extraction of DNA. The tissues were taken out of the fixative, washed with saline water, and dehydrated with 90% alcohol. Then, the tissues were cleaned up with xylene. The sample tissues were embedded and impregnated using paraffin wax. Furthermore, the embedded block containing the tissue was trimmed for easy cutting into fine sections. Later, the block containing the tissue was cut to a 5–8 μm thickness with the help of a microtone, and the cut sections adhered to the slides firmly and in the right position. Before attaching the portions to the slide, it was coated with egg albumin. The slides were stained using azan, hematoxylin, and eosin. For pathological observations, the stained sections were examined under a light microscope, and microphotographs were obtained [33].

2.8. Genomic DNA Isolation

At this stage, 100 mg of muscle tissue was placed in a mortar and pestle and thoroughly crushed after freezing with liquid nitrogen. The pulverized tissues were dissolved in 1.2 mL of digestion buffer and incubated for 12 to 18 h at 55 °C in a water bath. Equal amounts of phenol, chloroform, and isoamyl alcohol were added, gently mixed, and centrifuged at 3000 rpm for 5 min at 37 °C. Without disturbing the middle or bottom layers, the aqueous phase was carefully removed and transferred to a new tube. Half of the volume of 7.5 M ammonium acetate and twice the volume of 100% ethanol were then added, thoroughly mixed, and centrifuged at 3000 rpm for 5 min at ambient temperature. Then, 70% ethanol was used to rinse the particles, and the ethanol was then thrown away. The pellet was kept to dry in the air. For further processing, the pellet was dissolved in 30 L of nuclease-free water after it had dried. The standard procedure recommended by Lee et al. [34] for agarose gel electrophoresis was followed. A UV illuminator was used to expose the gel to UV light. Bands of orange luminous color represented DNA, and the gel was discarded after taking a picture of the DNA bands.

2.9. Quantification of Lipid Peroxidation

Lipid peroxidation was determined using the thiobarbituric-acid-reactive substances (TBARS) assay. Accordingly, the tissues treated with various concentrations of the test sample were acquired for assay. The tissue was homogenized with ice-cold KCl buffer; then, the homogenized tissue sample was collected in an Eppendorf tube and centrifuged at 5000× *g* for 5 min, after which the pellet was rinsed and redispersed in 0.2 mL of 8.1% sodium dodecyl sulfate (SDS). Next, 1.5 mL of 20% acetic acid was added to this suspension, and the mixture was incubated for 10 min. The solution was then combined with 1.5 mL of 0.8% Tris-borate EDTA (TBA) buffer and 0.7 mL of distilled water. Later, after one hour at 95 °C, the reaction mixture was cooled to 25 °C. After adding around 5 mL of a 15:1 combination of butanol and pyridine, the reaction liquid was once more centrifuged at 5000× *g* for 15 min to remove debris. The absorbance was recorded at 532 nm. The untreated fish were taken as a control. The standard was performed with malondialdehyde (MDA). The sample values were compared with the standard, and the concentration was expressed as μM.

2.10. Amplification of Inflammatory Markers

In the present study, nitric oxide synthase (iNOS), tumor necrosis factor (TNF)-α, and β-actin primers were employed. All of the inflammatory markers (i.e., TNF-α and iNOS) were designed using the Basic Local Alignment Search Tool (BLAST) (Table 1). The primers were as follows:

Table 1. The primers used for inflammatory marker genes amplification.

Name of Primers	Sequences
iNOS	Forward—5′ACACTTCGAAAAGCAAGATGG 3′ Reverse—5′ ACGGGCATCGAAAAGCTGTA 3′
TNF-α	Forward—5′ TCATTTTGGCTGTGGGCCTT 3′ Reverse—5′ GGCGGTTCAAAATCTCACTCAC 3′
β-Actin	Forward—5′ CCATCGGCAATGAGCGTTTC3′ Reverse—5′ CATCCTGAGTCAATGCGCCA 3′

In addition to the inflammatory indicators used here, β-actin was chosen as the housekeeping control gene.

PCR kits were procured from Takara Bio Inc., Japan © (EMERALD PCR master mix), and the procedure was carried out as follows: EMERALD master mix—12.5 μL, DNA template—2.5 μL, forward and reverse primer—1 μL, distilled H_2O—8 μL. The above cocktail for each sample was amplified in the following order and conditions: initial denaturation at 95 °C for 2 min, denaturation at 95 °C for 1 min, annealing respective to the primer chosen for 30 secs, extension at 72 °C for 1 min, and a final extension at 72 °C for 5 min. Amplification was carried out in 34 cycles, the resulting PCR product was resolved in 1% agarose gel electrophoresis, and the results were interpreted.

2.11. Statistical Analysis

All of the experiments were performed in triplicate. The final results were expressed as the mean and standard deviation (mean ± SD). All of the data were checked for normal distribution and homogeneity by one-way ANNOVA. Then, the data were subjected to Tukey's test, which enabled us to identify the specific group means that were different and compare them with the values of honest significant difference.

3. Results

3.1. Determination of LC_{50}

In order to investigate the effect of dosage on the behavioral changes in the fish, several experiments were performed using varied dosages. The LC_{50} results demonstrated that, up to 10 μg/mL dosage, no mortality was recorded, nor were any behavioral changes were observed in the fish. Later, the dosages were gradually increased up to 50 μg/mL, i.e., different dosage ranges (including 10, 20, 30, 40, and 50 μg/mL) were used, and the fishes were closely observed at these dosages for up to 7 days. Notably, at these dosages—between 10 and 50 μg/mL—the fish survived, and no abnormal behavior was detected in the fish.

Encouraged by these results, the dosage was further increased, and higher dosages of 200, 400, 600, 800, and 1000 µg/mL were applied for total of 7 days. There was no evident effect in the entire dosage range, and the fish survived normally even up to 800 µg/mL. Furthermore, no prominent toxic effects or abnormal behavior were observed throughout the treatment period—even in the acute phase.

3.2. Anti-Inflammatory Activity

The inflammation induced by TNBS was dependent on the fishes' weight. The samples in the range of 200, 400, 600, 800, and 1000 µg/mL were given by gavage. The body weight of the fishes was noted. Figure 1 and Table 2 shows the initial weight of the fishes, their weight before treatment, and their weight after treatment (4 h).

Figure 1. The image represents the fishes without TNBS induction, with TNBS induction, and having recovered from inflammation after treatment with the sample. (**a**) Fish without TNBS induction. (**b**) Fish with TNBS induction (1 h later). (**c**) Fish after TNBS induction and treatment with the compound (4 h later).

Table 2. Anti-inflammatory activity of mycobiont extract towards zebrafish at different stages.

Dosage (µg/mL)	$ Weight before TNBS Injection (g)	Weight before Sample Treatment (g)	Weight after Sample Treatment (g)
Control	0.62 ± 0.01^{c}	0.64 ± 0.00^{d}	0.64 ± 0.00^{g}
TNBS-induced group	0.65 ± 0.01^{b}	0.74 ± 0.01^{bc}	0.77 ± 0.01^{b}
Standard	0.61 ± 0.01^{c}	0.72 ± 0.01^{c}	0.67 ± 0.01^{ef}
200	0.61 ± 0.00^{c}	0.63 ± 0.01^{d}	0.68 ± 0.00^{e}
400	0.70 ± 0.01^{a}	0.72 ± 0.01^{c}	0.79 ± 0.01^{a}
600	0.65 ± 0.01^{b}	0.75 ± 0.01^{b}	0.71 ± 0.00^{d}
800	0.69 ± 0.01^{a}	0.83 ± 0.01^{a}	0.74 ± 0.01^{c}
1000	0.62 ± 0.01^{c}	0.72 ± 0.01^{c}	0.66 ± 0.01^{fg}
F7,16	21.240 ***	68.146 ***	80.742 ***

$ Each value is a mean of five replicates ± SD. This means that a column followed by the same superscript ($p > 0.05$) differs significantly according to Tukey's range test. *** Significant at $p < 0.001$.

Transverse cuts were made in the abdominal muscular tissues, and slices were then stained with hematoxylin and eosin. Figure 2 denotes the control, TNBS-induced group, and treated group with standard and sample (200 µg/mL, 600 µg/mL, and 1000 µg/mL). The reference image depicts muscle tissue as it would typically appear, with intact nuclei and cytoplasm (cf. Figure 2a). The control muscle section revealed normal histological elements, such as fiber bundles, connective tissue, and the arrangement of muscle bundles. However, the fish in the TNBS-induced group displayed aberrant muscle tissue arrangements (cf. Figure 2b). The standard treatment showed recovery of tissues from the inflammation (cf. Figure 2c), where the sample-treated images showed the recovery process of muscle tissues in a dose-dependent manner (cf. Figure 2d–f). The progression of recovery of the muscle tissue arrangements was observed in the increased dose ranges. All of the morphological changes are indicated by black arrows, first indicating the neutrophil cells (Figure 2a). Later, an increased number of neutrophil cells was observed in the TNBS-induced group (Figure 2b), which recovered while treated with the samples and also showed recovered muscle tissues (Figure 2d–f). Overall, from the results, it was established that the mycobiont extracts possessed anti-inflammatory properties.

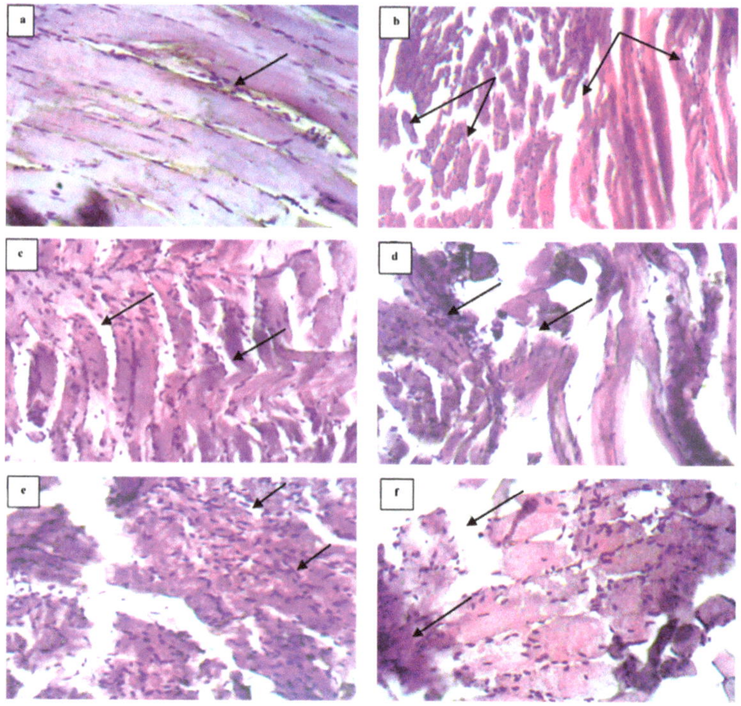

Figure 2. Histopathology of zebrafish muscle tissues: (**a**) control group; (**b**) induced group; (**c**) standard drug; (**d**) treated group (200 µg/mL); (**e**) treated group (600 µg/mL); (**f**) treated group (1000 µg/mL).

3.3. Lipid Peroxidation Assay

Figure 3 depicts the differences in lipid peroxidation (LPO) levels between the different groups. The control fish without TNBS induction showed 0.175 µM LPO release. The concentration of LPO increased in the induced group, whereas in the standard treatment the LPO level decreased. The samples showed decreasing LPO release in the range from 20 to 100 µg/mL. The figure shows the prominent levels of difference between the different groups. The graph depicts the evident change in the LPO when the inflammation was treated with the varied sample range, proving that there were significant differences in the LPO levels. The results showed that the lipid peroxidation effect ranged between 0.55 µMol (induced group) and 0.18 µMol (100 µg/mL group), whereas the variation between treatments ($F_{7,39} = 449.289$, $p < 0.001$) was highly significant. The lipid peroxidation assay was significantly related to the extent of the mycobiont extracts of the lichen. This confirms that the samples acted on the recovery at the site of inflammation induced by TNBS.

3.4. Amplification Studies of the Inflammatory Markers

The PCR amplification of DNA extracted from the control group, induced group, and treated group (standard and sample)—gene expression of (a) β actin, (b) iNOS, and (c) TNF-α—is shown in Figure 4, which reveals that the samples evoked significant downregulation of the inflammatory gene marker TNF-α in a dose-dependent manner compared to TNBS-induced group. Among the treated groups, there was a marked decrease in the expression of TNF-α at 100 µg/mL. The standard treatment showed a low expression level compared to the sample-treated groups. There were no remarkable changes in iNOS expression levels between the groups. β-Actin was used as an internal loading control. The gene expression studies confirmed the downregulation of the inflammatory marker TNF-α, which is an indicator of anti-inflammatory activity.

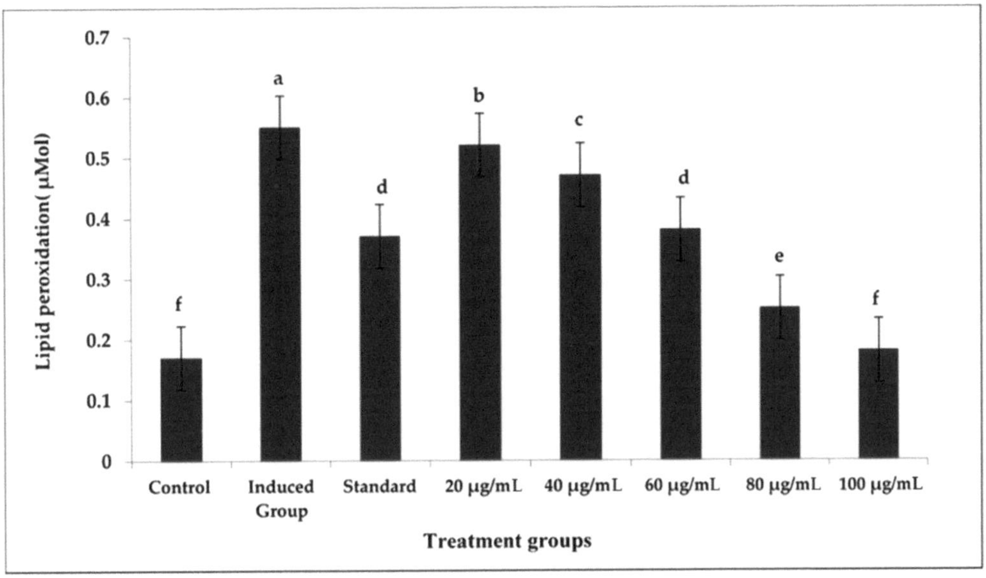

Figure 3. Lipid peroxidation activity of mycobiont extract of *Parmotrema austrosinense*. Bars bearing the same letter(s) do not significantly differ according to Tukey's range test ($p > 0.05$).

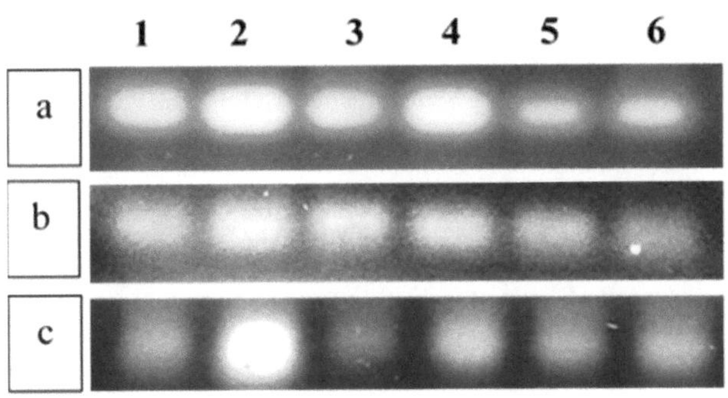

Figure 4. Gene expression analysis of treated fishes: (**a**) Beta-actin gene expression; (**b**) iNOS gene expression; (**c**) TNF-α gene expression. Lane 1—control, Lane 2—induced group, Lane 3—standard group, Lane 4—20 µg/mL, Lane 5—60 µg/mL, Lane 6—100 µg/mL.

4. Discussion

Many earlier reports have described the biological activities of lichen extracts with various organic solvents—particularly from *Parmotrema* species [35,36]. According to previously published studies, most lichen species contain a variety of potentially bioactive secondary metabolites that demonstrate excellent anti-inflammatory and antimicrobial properties [22,37]. These secondary metabolites include phenols, usnic acid, terpenes, steroids, and other depside derivatives, which exhibit several important biological properties, such as antitumor, apoptotic, and cytotoxic activities [36]. Particularly, usnic acid is a highly popular secondary metabolite of lichen, which is known to exert high antimicrobial activity and excellent antiproliferative effects on human leukemia (K562) and endometrial carcinoma [38,39]. Therefore, due to the presence of highly efficient bioactive secondary metabolites, lichen extracts have been employed indigenously for the treatment of several different types of ailments [40]. Particularly, in herbal remedies, several genera of lichens have been used to treat a variety of diseases. A wide variety of species of lichen extracts have been used to control fever, diarrhea, infections, skin diseases, epilepsy, convulsions, and as purgatives [41]. Despite this, the lack of bulk quantity of lichen thalli and their compounds is the reason that various lichen species are still unresearched. Among these unexplored lichens, the species *P. austrosinense* is also one of the less-studied lichens. However, in recent years, mycobiont extracts have gradually been gaining popularity for research purposes, due to their diverse biological prospects [42].

In the present study, we investigated the anti-inflammatory activity of locally available *P. austrosinense*, which is a cosmopolitan lichen distributed in the humid climate of the Tamil Nadu Western Ghats. More specifically, acetone was used to extract the bioactive secondary metabolites from the mycobiont. To study the biopotential of *P. austrosinense* mycobiont extracts, a lipid peroxidation assay was performed using the acetone extract, which was further used to evaluate its anti-inflammatory efficacy using the zebrafish model. The complete study was carried out with the zebrafish model, because of the accumulated knowledge and tools available for the selected fish model, which could be utilized for further exploitation [43].

According to the results, the toxicity assay showed that the mycobiont extract had no toxic effect against zebrafishes up to 800 μg/mL. This result on toxicity is consistent with the results reported by Studzińska-Sroka and Dubino [20]. Another study reported by Poornima et al. [44] also demonstrated that toxicity against the zebrafish model was recorded at the concentration of 800 μg/mL. Thus, the mycobiont extracts were used for anti-inflammatory studies with a zebrafish model of induced inflammation. The study on the anti-inflammatory activity suggested that the mycobiont extract might contain an anti-inflammatory compound, which was found to be highly active against the inflammatory agent TNBS. A significant positive association between lichen compounds and in vivo anti-inflammatory action was demonstrated by Poornima et al. [44] and Nguyen et al. [45]. Furthermore, anti-inflammatory studies with other lichen species also evidently support the present findings [36,46]. The mycobiont acetone extract showed dose-dependent anti-inflammatory action against TNBS-induced inflammation. The lower weight in the zebrafishes with TNBS-induced inflammation was the main indicator of the presence of the anti-inflammatory compound in the extract.

Furthermore, lipid peroxidation test assays were carried out using the acetone extracts of the mycobiont. The results indicated a dose-dependent decrease in LPO levels in the sample-treated TNBS-induced zebrafishes. The TNBS-induced zebrafishes exhibited a decreased level of LPO at the concentration of 100 μg/mL. It was also noted that with increasing extract dose, a gradual decrease in the LPO level was observed. However, the results of the LPO assay in the present study are contradictory when compared to the results of an earlier published report on the acetone extract of *Parmotrema* species [47–49].

The anti-inflammatory activity of the acetone extract of the *P. austrosinense* mycobiont was confirmed by the recovery of neutrophils in the histopathological assays of TNBS-induced zebrafishes. It is possible that the reported anti-inflammatory activity of the extract could be attributed to the presence of lecanoric acid, as reported in an earlier study by Studzińska et al. [21]. According to observations, the present findings are consistent with the results obtained by Poornima et al. [44]. In contrast with our present study, the latter used oxazolone to induce inflammation in the zebrafishes.

Additionally, encouraged by these results, gene expression studies were also carried out after the TNBS-induced inflammation and lipid peroxidation tests. In comparison with the standard, a dose-dependent decrease in TNF-α gene expression was detected. These results also suggest that the aforementioned compound possibly produces a dose-dependent anti-inflammatory response. This was analyzed using TNF gene downregulation, which also resulted in a dose-dependent response, as with the increase in the dose, a greater downregulation of the TNF-α gene was detected with the PCR-amplified products. Notably, the results of our gene expression study are also supported by another previously published report by Lee et al. [50] with LPS-induced macrophages, where they found that after treatment with the test sample, a successful reduction in TNF-α was recorded.

5. Conclusions

In this study, a mycobiont extract of a lichen species was tested for its biopotential properties. Among the various lichen species, *P. austrosinense* is cosmopolitan in its distribution and is known to be widely found in hill stations. However, the anti-inflammatory activity of *P. austrosinense* has been little explored; thus, it was decided to investigate it's in vitro and in vivo anti-inflammatory activities using a zebrafish model in the present research. The in vivo anti-inflammatory studies showed a dose-dependent response. Accordingly, the in vitro LPO assay concluded that the extract showed a decreased LPO level at high doses. The results of PCR gene expression indicated the presence of downregulation of the inflammatory gene marker TNF-α. Therefore, the supportive in vitro and in vivo test results confirmed the anti-inflammatory activity of the mycobiont of *P. austrosinense.* These results are encouraging; however, further extensive investigations are needed to exploit the biopotential of the studied lichen extract for future medical applications.

Author Contributions: K.R.: data curation, formal analysis, and writing—original draft; P.K.: validation, review, and editing; P.P.: conceptualization, supervision, validation, visualization, review, and editing; M.R.S.: formal analysis writing—original draft; M.K.: formal analysis and writing—original draft; T.H.O.: formal analysis and resources; B.S.: formal analysis and writing—review, and editing. All authors have read and agreed to the published version of the manuscript.

Funding: The authors extend their appreciation to the Deputyship for Research and Innovation, "Ministry of Education" in Saudi Arabia for funding this research (IFKSUOR3–103–1).

Institutional Review Board Statement: Not applicable.

Informed Consent Statement: Not applicable.

Data Availability Statement: Data is contained within the article.

Acknowledgments: The authors extend their appreciation to the Deputyship for Research and Innovation, "Ministry of Education" in Saudi Arabia for funding this research (IFKSUOR3–103–1).

Conflicts of Interest: The authors declare no conflict of interest.

References

1. Phi, K.-H.; So, J.E.; Kim, J.H.; Koo, M.H.; Kim, J.-H.; Kim, D.; Lee, J.H.; Lee, S.; Youn, U.J. Chemical constituents from the Antarctic lichen Usnea aurantiaco-atra and their chemotaxonomic significance. *Biochem. Syst. Ecol.* **2023**, *106*, 104581. [CrossRef]
2. Do, T.-H.; Duong, T.-H.; An, T.N.M.; Vo, T.-P.-G.; Do, V.-M.; Nguyen, N.-H.; Sichaem, J. Two new α-glucosidase inhibitory depsidones from the lichen Parmotrema cristiferum (Taylor) Hale. *Chem. Biodivers.* **2023**, *20*, e202201213. [CrossRef]
3. Susithra, E.; Meena, S.; Chamundeeswari, D.; Chekkara, R.; Varalakshmi, E. Molecular docking studies of lichen metabolites as malarial protein inhibitors: Plasmepsin II protease and dihydrofolate reductase. *J. Chem. Pharm. Res.* **2014**, *6*, 501–512.
4. Ureña-Vacas, I.; González-Burgos, E.; Divakar, P.K.; Gómez-Serranillos, M.P. Lichen Depsides and Tridepsides: Progress in Pharmacological Approaches. *J. Fungi* **2023**, *9*, 116. [CrossRef]
5. Bhat, N.B.; Das, S.; Sridevi, B.V.; Nayaka, S.; Birangal, S.R.; Shenoy, G.G.; Joseph, A. Molecular docking and dynamics supported investigation of antiviral activity of Lichen metabolites of Roccella montagnei: An in silico and in vitro study. *J. Biomol. Struct. Dyn.* **2023**, 1–14. [CrossRef]
6. Chauhan, R.; Abraham, J. In vitro antimicrobial potential of the lichen *Parmotrema* sp. extracts against various pathogens. *Iran. J. Basic Med. Sci.* **2013**, *16*, 882.
7. Yañez, O.; Osorio, M.I.; Osorio, E.; Tiznado, W.; Ruíz, L.; García, C.; Nagles, O.; Simirgiotis, M.J.; Castañeta, G.; Areche, C. Antioxidant activity and enzymatic of lichen substances: A study based on cyclic voltammetry and theoretical. *Chem. Biol. Interact.* **2023**, *372*, 110357. [CrossRef]
8. Shukla, I.; Azmi, L.; Rao, C.V.; Jawaid, T.; Kamal, M.; Alkhamees, O.A.; Alaseem, A.M.; Alsanad, S.M. Inclusive roles of protocetraric acid, a secondary metabolite from the common green shield lichen Flavoparmelia caperata in alcohol-induced hepatic injury. *Lat. Am. J. Pharm* **2022**, *41*, 613–623.
9. Morris Kupchan, S.; Kopperman, H.L. L-Usnic acid: Tumor inhibitor isolated from lichens. *Experientia* **1975**, *31*, 625. [CrossRef]

10. Anand Mariadoss, A.V.; Krishnan Dhanabalan, A.; Munusamy, H.; Gunasekaran, K.; David, E. In Silico studies towards enhancing the anticancer activity of phytochemical phloretin against cancer drug targets. *Curr. Drug Ther.* **2018**, *13*, 174–188. [CrossRef]
11. Gupta, A.; Sahu, N.; Singh, A.P.; Singh, V.K.; Singh, S.C.; Upadhye, V.J.; Mathew, A.T.; Kumar, R.; Sinha, R.P. Exploration of Novel Lichen Compounds as Inhibitors of SARS-CoV-2 Mpro: Ligand-Based Design, Molecular Dynamics, and ADMET Analyses. *Appl. Biochem. Biotechnol.* **2022**, *194*, 6386–6406. [CrossRef]
12. Tripathi, A.H.; Negi, N.; Gahtori, R.; Kumari, A.; Joshi, P.; Tewari, L.M.; Joshi, Y.; Bajpai, R.; Upreti, D.K.; Upadhyay, S.K. A review of anti-cancer and related properties of lichen-extracts and metabolites. *Anti-Cancer Agents Med. Chem.* **2022**, *22*, 115–142.
13. Awasthi, D.D. *Key to the Microlichens of India, Nepal and Sri Lanka*; J. Cramer: Stuttgart, Germany, 1991.
14. Gurib-Fakim, A. Medicinal plants: Traditions of yesterday and drugs of tomorrow. *Mol. Asp. Med.* **2006**, *27*, 1–93. [CrossRef]
15. Armin, H.; Georg, B.; Elfie, S.-w. Investigations on secondary metabolites of lichens and liehen mycobionts in culture: Production and biological activity. *Osterreichische* **2008**, *37*, 323–324.
16. Tanas, S.; Odabasoglu, F.; Halici, Z.; Cakir, A.; Aygun, H.; Aslan, A.; Suleyman, H. Evaluation of anti-inflammatory and antioxidant activities of Peltigera rufescens lichen species in acute and chronic inflammation models. *J. Nat. Med.* **2010**, *64*, 42–49. [CrossRef]
17. Shrestha, G.; El-Naggar, A.M.; St. Clair, L.L.; O'Neill, K.L. Anticancer activities of selected species of North American lichen extracts. *Phytother. Res.* **2015**, *29*, 100–107. [CrossRef]
18. Müller, K. Pharmaceutically relevant metabolites from lichens. *Appl. Microbiol. Biotechnol.* **2001**, *56*, 9–16. [CrossRef]
19. Mitrović, T.; Stamenković, S.; Cvetković, V.; Nikolić, M.; Tošić, S.; Stojičić, D. Lichens as source of versatile bioactive compounds. *Biol. Nyssana* **2011**, *2*, 1–6.
20. Studzińska-Sroka, E.; Dubino, A. Lichens as a source of chemical compounds with anti-inflammatory activity. *Herba Pol.* **2018**, *64*, 56–64. [CrossRef]
21. Studzińska, E.; Witkowska-Banaszczak, E.; Bylka, W. Związki biologicznie aktywne porostów. *Herba Pol.* **2008**, *54*, 80–88.
22. Bugni, T.S.; Andjelic, C.D.; Pole, A.R.; Rai, P.; Ireland, C.M.; Barrows, L.R. Biologically active components of a Papua New Guinea analgesic and anti-inflammatory lichen preparation. *Fitoterapia* **2009**, *80*, 270–273. [CrossRef] [PubMed]
23. Umezawa, H.; Shibamoto, N.; Naganawa, H.; Ayukawa, S.; Matsuzaki, M.; Takeuchi, T.; Kono, K.; Sakamoto, T. Isolation of lecanoric acid, an inhibitor of histidine decarboxylase from a fungus. *J. Antibiot.* **1974**, *27*, 587–596. [CrossRef] [PubMed]
24. Rajendran, K.; Poornima, S.; Ponnusamy, P. Antimicrobial and Antiproliferative Activities of Depside Compound Isolated from the Mycobiont Culture of Parmotrema austrosinense (Zahlbr.) Hale. *J. Pure Appl. Microbiol.* **2020**, *14*, 2525–2542. [CrossRef]
25. Kalidoss, R.; Mariraj, M.; Shenbagam, M.; Seles, J.M.; Prasath, K.A.; Rajaprabu, N.; Ponmurugan, P. Anti-microbial and Anti-oxidant Properties of Solvent Extract of Lichen Species Collected from Kodaikanal Hills, Western Ghats of Tamil Nadu. In *Phytomedicine*; CRC Press: Boca Raton, FL, USA, 2020; pp. 53–61.
26. Wolseley, P. A Compendium of the Macrolichens from India, Nepal and Sri Lanka. *Lichenologist* **2008**, *40*, 267–268. [CrossRef]
27. Cordeiro, L.M.; Iacomini, M.; Stocker-Wörgötter, E. Culture studies and secondary compounds of six Ramalina species. *Mycol. Res.* **2004**, *108*, 489–497. [CrossRef]
28. Verma, N.; Behera, B.C. In Vitro culture of lichen partners: Need and Implications. In *Recent Advances in Lichenology*; Springer: Berlin/Heidelberg, Germany, 2015; pp. 147–159.
29. OECD Guideline for Testing Acute Toxicity in Fishes. Available online: http://www.oecd.org/chemicalsafety/risk-assessment/1948241.pdf (accessed on 1 January 2023).
30. Xie, Y.; Meijer, A.H.; Schaaf, M.J. Modeling inflammation in zebrafish for the development of anti-inflammatory drugs. *Front. Cell Dev. Biol.* **2021**, *8*, 620984. [CrossRef]
31. Ducharme, N.A.; Reif, D.M.; Gustafsson, J.-A.; Bondesson, M. Comparison of toxicity values across zebrafish early life stages and mammalian studies: Implications for chemical testing. *Reprod. Toxicol.* **2015**, *55*, 3–10. [CrossRef]
32. Kinkel, M.D.; Eames, S.C.; Philipson, L.H.; Prince, V.E. Intraperitoneal injection into adult zebrafish. *JoVE (J. Vis. Exp.)* **2010**, *42*, e2126.
33. Slaoui, M.; Fiette, L. Histopathology procedures: From tissue sampling to histopathological evaluation. In *Drug Safety Evaluation*; Springer: Berlin/Heidelberg, Germany, 2011; pp. 69–82.
34. Lee, P.Y.; Costumbrado, J.; Hsu, C.-Y.; Kim, Y.H. Agarose gel electrophoresis for the separation of DNA fragments. *JoVE (J. Vis. Exp.)* **2012**, *62*, e3923.
35. Jain, A.; Bhandarkar, S.; Rai, G.; Yadav, A.; Lodhi, S. Evaluation of Parmotrema reticulatum taylor for antibacterial and antiinflammatory activities. *Indian J. Pharm. Sci.* **2016**, *78*, 94–102.
36. Molnár, K.; Farkas, E. Current results on biological activities of lichen secondary metabolites: A review. *Z. Für Nat. C* **2010**, *65*, 157–173. [CrossRef] [PubMed]
37. Goga, M.; Elečko, J.; Marcinčinová, M.; Ručová, D.; Bačkorová, M.; Bačkor, M. Lichen metabolites: An overview of some secondary metabolites and their biological potential. In *Co-Evolution of Secondary Metabolites*; Springer: Berlin/Heidelberg, Germany, 2020; pp. 175–209.
38. Elečko, J.; Vilková, M.; Frenák, R.; Routray, D.; Ručová, D.; Bačkor, M.; Goga, M. A Comparative Study of Isolated Secondary Metabolites from Lichens and Their Antioxidative Properties. *Plants* **2022**, *11*, 1077. [CrossRef] [PubMed]
39. Seçme, M.; Dodurga, Y. Usnic Acid Inhibits Cell Proliferation and Downregulates lncRNA UCA1 Expression in Ishikawa Endometrial Cancer Cells. *Nat. Prod. Biotechnol.* **2021**, *1*, 28–37.

40. Ingelfinger, R.; Henke, M.; Roser, L.; Ulshöfer, T.; Calchera, A.; Singh, G.; Parnham, M.J.; Geisslinger, G.; Fürst, R.; Schmitt, I. Unraveling the pharmacological potential of lichen extracts in the context of cancer and inflammation with a broad screening approach. *Front. Pharmacol.* **2020**, *11*, 1322. [CrossRef]
41. Shukla, V.; Joshi, G.P.; Rawat, M. Lichens as a potential natural source of bioactive compounds: A review. *Phytochem. Rev.* **2010**, *9*, 303–314. [CrossRef]
42. Nguyen, H.-H.; Aree, T.; Nguyen, H.T.; Tran, T.-M.-D.; Nguyen, T.-P.; Vo, T.-P.G.; Nguyen, N.-H.; Duong, T.-H. Diorygmones AB, two new guaiane-sesquiterpenes from the cultured lichen mycobiont of *Diorygma* sp. *Nat. Prod. Res.* **2023**, 1–6. [CrossRef]
43. Patton, E.E.; Tobin, D.M. *Spotlight on Zebrafish: The Next Wave of Translational Research*; The Company of Biologists Ltd.: Cambridge, UK, 2019; Volume 12, p. dmm039370.
44. Poornima, S.; Nagarjun, N.; Ponmurugan, P.; Gnanamangai, B.M.; Narasimman, S. Toxicity and anti-inflammatory study of Parmotrema austrosinense extract against oxozalone induced intestinal inflammation in zebrafish (Danio rerio) model. *Biocatal. Agric. Biotechnol.* **2019**, *21*, 101278. [CrossRef]
45. Nguyen, H.T.; Ketha, A.; Kukavica, B.; Tatipamula, V. Antiinflammatory potential of lichens and its substances. In *Inflammatory Bowel Disease*; MedDocs Publishers: Reno, NY, USA, 2021; pp. 1–9.
46. Güvenç, A.; Akkol, E.K.; Süntar, İ.; Keleş, H.; Yıldız, S.; Çalış, İ. Biological activities of Pseudevernia furfuracea (L.) Zopf extracts and isolation of the active compounds. *J. Ethnopharmacol.* **2012**, *144*, 726–734. [CrossRef]
47. Kosanić, M.M.; Ranković, B.R.; Stanojković, T.P. Antioxidant, antimicrobial and anticancer activities of three Parmelia species. *J. Sci. Food Agric.* **2012**, *92*, 1909–1916. [CrossRef]
48. Jayaprakasha, G.K.; Rao, L.J. Phenolic constituents from the lichen Parmotrema stuppeum (Nyl.) Hale and their antioxidant activity. *Z. Für Nat. C* **2000**, *55*, 1018–1022. [CrossRef]
49. Selvaraj, G.; Tinabaye, A.; Ananthi, R. In vitro antioxidant activities of Salazinic acid and its derivative hexaacetylsalazinic acid. *Int. J. Res. Eng. Technol.* **2015**, *4*, 345–355.
50. Lee, A.S.; Jung, Y.J.; Kim, D.; Nguyen-Thanh, T.; Kang, K.P.; Lee, S.; Park, S.K.; Kim, W. SIRT2 ameliorates lipopolysaccharide-induced inflammation in macrophages. *Biochem. Biophys. Res. Commun.* **2014**, *450*, 1363–1369. [CrossRef] [PubMed]

Journal of *Marine Science and Engineering*

MDPI

Article

Development and Application of a Mechanistic Nutrient-Based Model for Precision Fish Farming

Filipe M. R. C. Soares [1], Ana M. D. Nobre [1], Andreia I. G. Raposo [1,2], Rodrigo C. P. Mendes [1,3], Sofia A. D. Engrola [3], Paulo J. A. P. Rema [4,5], Luís E. C. Conceição [1,*] and Tomé S. Silva [1]

1 Sparos Lda., Área Empresarial de Marim, Lote C, 8700-221 Olhão, Portugal
2 Instituto de Ciências Biomédicas Abel Salazar (ICBAS-UP), Universidade do Porto, Rua de Jorge Viterbo Ferreira 228, 4050-313 Porto, Portugal
3 Centre of Marine Sciences, CCMAR, University of Algarve, 8005-139 Faro, Portugal
4 CIIMAR—Interdisciplinary Centre of Marine and Environmental Research, University of Porto, Novo Edifício do Terminal de Cruzeiros de Leixões, Avenida General Norton de Matos, s/n, 4450-208 Matosinhos, Portugal
5 Departamento de Zootécnia, Universidade de Trás-os-Montes e Alto Douro, Quinta de Prados, 5001-801 Vila Real, Portugal
* Correspondence: luisconceicao@sparos.pt; Tel.: +351-289435145

Abstract: This manuscript describes and evaluates the FEEDNETICS model, a detailed mechanistic nutrient-based model that has been developed to be used as a data interpretation and decision-support tool by fish farmers, aquafeed producers, aquaculture consultants and researchers. The modelling framework comprises two main components: (i) *fish model*, that simulates at the individual level the fish growth, composition, and nutrient utilization, following basic physical principles and prior information on the organization and control of biochemical/metabolic processes; and (ii) *farm model*, that upscales all information to the population level. The model was calibrated and validated for five commercially relevant farmed fish species, i.e., gilthead seabream (*Sparus aurata*), European seabass (*Dicentrarchus labrax*), Atlantic salmon (*Salmo salar*), rainbow trout (*Oncorhynchus mykiss*), and Nile tilapia (*Oreochromis niloticus*), using data sets covering a wide range of rearing and feeding conditions. The results of the validation of the model for fish growth are consistent between species, presenting a mean absolute percentage error (MAPE) between 11.7 and 13.8%. Several uses cases are presented, illustrating how this tool can be used to complement experimental trial design and interpretation, and to evaluate nutritional and environmental effects at the farm level. FEEDNETICS provides a means of transforming data into useful information, thus contributing to more efficient fish farming.

Keywords: aquaculture; fish nutrition; precision fish farming; mechanistic nutrient-based model; numerical model; decision-support tool

Citation: Soares, F.M.R.C.; Nobre, A.M.D.; Raposo, A.I.G.; Mendes, R.C.P.; Engrola, S.A.D.; Rema, P.J.A.P.; Conceição, L.E.C.; Silva, T.S. Development and Application of a Mechanistic Nutrient-Based Model for Precision Fish Farming. *J. Mar. Sci. Eng.* **2023**, *11*, 472. https://doi.org/10.3390/jmse11030472

Academic Editors: Ana Catarina Matias and Carlos Andrade

Received: 16 January 2023
Revised: 20 February 2023
Accepted: 21 February 2023
Published: 22 February 2023

1. Introduction

As fish farming is becoming more digitized, data analytics tools are required to transform data into useful information. Mathematical models that are able to describe and predict the dynamics of fish farming systems are candidate tools in this regard, with great potential to contribute as data interpretation and decision support tools [1,2].

Over the last decades, several mathematical models that describe the state of fish have been developed to support fish farming operations [3–9]. Some models describe the growth of fish based on simple empirical equations that assume that feeding is a non-limiting factor, and is therefore not included as a model input [8,10–15]. Others, based on differential equations, describe the growth and, in some cases, the composition of fish (among other indicators), given time-dependent information about environmental parameters (e.g., water temperature) and feeding operations (e.g., ration size, feed composition) [3,6,16–18].

For a more detailed description of different models that describe fish growth and body composition, see the reviews from Dumas et al. [7] and Chary et al. [9].

Usually, the choice of one type of model over another strongly depends on the availability of data and the purpose for which the model is intended. Within the framework of precision fish farming [1], where highly detailed information systems are required to support a knowledge-based decision-making process, the use of more detailed mathematical models as complementary tools to simpler models may open the door to new insight. For example, detailed mathematical models that account for the effects of ration size and feed composition (e.g., bioenergetic or nutrient-based models) can be used to quantify the impact of different feed formulations or feeding strategies on fish performance, allowing to pre-screen the most promising solutions before practical application [19]. Besides, this type of models can also be used to monitor and predict on a higher time resolution basis (e.g., hourly or daily basis) the state of production using data from past observations and data defining the expected environmental and feeding conditions of the farming unit. Having access to this additional information allow fish farmers to make more informed decisions based on the specific conditions of their farms, and on the current and future state of production; e.g., estimate feed requirements, manage feed stocks, detect disease outbreaks, and plan sales and harvesting dates [16–18,20–22].

Nutrient-based models—defined here as models that explicitly describe the nutritional state of fish based on nutrient intake and utilization—enable a quantitative prediction of the effects of feed composition, in terms of macronutrients, on fish performance [19,23]. This is a particularly relevant aspect, as the range of commercially available feeds for each growth stage can differ in composition. Besides, since feeds with the same energy content may present different nutrient composition, consequently inducing different fish performance [24–26], the use of nutrient-based models may be preferable to bioenergetic models. Within the domain of nutrient-based models, a relevant aspect to be considered is the level of detail included in the modelling approach. Simpler nutrient-based models (also known as "nutrient mass-balance models") that only consider the protein and energy content of the feed are not able, for example, to predict the impact of imbalanced dietary amino acid profiles on fish performance [18]. To work with greater precision, it is therefore necessary to use mechanistic nutrient-based models (also known as "metabolic-flux models") that account for the effects of the feed amino acid profile and that describe the main physiological and metabolic processes of fish (e.g., protein synthesis and degradation, amino acid oxidation and conversion, glycolysis, gluconeogenesis, glycogenesis, and glycogenolysis).

A few detailed mechanistic nutrient-based models of fish have been previously developed by other authors [3–6,19,23,27,28]. For example, Bar et al. [23] developed a highly detailed model of nutrient-pathways, growth, and body composition of fish, calibrated for Atlantic salmon (*Salmo salar*), that considers the effects of limiting amino acids. In turn, Hua et al. [19] adapted a non-ruminant nutrient-based model for rainbow trout (*Oncorhynchus mykiss*), which explicitly describes the utilization of energy-yielding nutrients and metabolites for protein and lipid deposition in the fish body, also considering the effects of limiting amino acids. These previous works set solid foundations that can be further explored to advance on the development of more detailed mechanistic nutrient-based models of fish growth and composition.

This manuscript presents the FEEDNETICS model, a detailed mechanistic nutrient-based model that has been developed to be used as a data interpretation and decision-support tool in the context of precision fish farming. This tool has a wide range of potential applications and it is aimed at fish farmers, aquafeed producers, aquaculture consultants and researchers. The main objectives of this work were:

1. To develop a nutrient-based model that considers the main physiological and metabolic processes of fish;
2. To calibrate and validate the model for five relevant farmed fish species, i.e., gilthead seabream (*Sparus aurata*), European seabass (*Dicentrarchus labrax*), Atlantic salmon (*Salmo salar*), rainbow trout (*Oncorhynchus mykiss*), and Nile tilapia (*Oreochromis niloticus*);

3. To demonstrate the use of the model as a data interpretation and decision-support tool through several use cases.

2. Materials and Methods

2.1. Model Description

FEEDNETICS is a dynamic mechanistic nutrient-based model that simulates fish growth, composition and environmental impact, within the context of fish farming operations, while following basic physical principles (e.g., energy and mass conservation) and prior information on the organization and control of biochemical/metabolic processes. It deterministically simulates the time-dependent physiology and metabolism of a single fish (the "average fish"; *fish model*), being driven by temperature data, feed properties and quantity, which is then upscaled to a population of fish (*farm model*). Figure 1 shows an overview of the input data needed to run the model and the output data it provides.

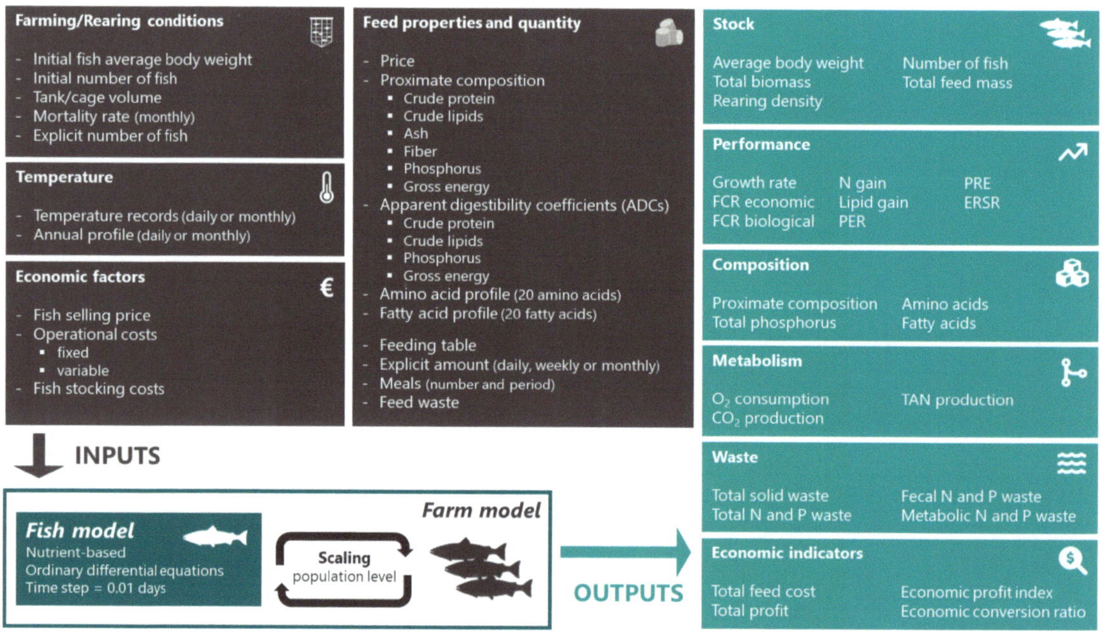

Figure 1. Diagram illustrating the main model inputs and outputs.

In particular, the model considers the most relevant physiological and metabolic processes, including feed intake, digestion, nutrient absorption and transport, central metabolism (i.e., glycolysis, gluconeogenesis, glycogenesis, glycogenolysis, beta-oxidation of fatty acids, Krebs cycle) and nitrogen metabolism (i.e., protein synthesis, protein degradation, synthesis of non-essential amino acids, amino acid oxidation, amino acid conversion) in order to maximize its capability to effectively predict fish body weight and composition along time for general scenarios. A detailed description of this model is provided in Appendix A.

The model has been implemented as a set of difference equations in the Powersim Studio 10 Expert software (Powersim Software AS, Nyborg, Norway), which are solved numerically through forward Euler integration with a fixed timestep of 0.01 days.

2.2. Model Calibration and Validation

The model was calibrated and validated for five fish species: gilthead seabream, European seabass, Atlantic salmon, rainbow trout, and Nile tilapia. The data sets used to calibrate and validate the model, as well as the calibration and validation procedures, are described below.

2.2.1. Data Processing and Data Sets Description

All data sets used in this work include data about several in vivo growth trials, covering a wide range of rearing and feeding conditions (see Table 1). Those were collected from the scientific literature and from data generated by Sparos Lda., and its partners, in R&D projects, some of which were specifically designed to support model calibration and validation.

Table 1. Overview of the data used to calibrate and validate the model for each species.

Attributes	Unit	Gilthead Seabream	European Seabass	Atlantic Salmon	Rainbow Trout	Nile Tilapia
Nr. of data sources	-	19	37	61	33	44
Nr. of observational units	-	118	126	398	110	186
Nr. of diets	-	30	66	350	58	175
Body weight range	g	1–478	5–482	1–6645	2–2080	1–559
Temperature range	°C	11–28	18–26	4–20	4–19	18–30
Diet composition range						
Crude protein	% as fed	37–58	37–56	29–54	26–58	23–46
Crude lipids	% as fed	9–23	10–31	10–47	6–31	3–15
Gross energy	MJ/kg	19–23	18–25	18–29	17–26	13–21
DP/DE	g/MJ	21–26	19–30	12–26	11–28	14–26

For purposes of uniformization and data handling, all data sets were processed into a standard format and stored in a database structured according to 'tidy data' principles [29]. Three types of data are associated to each data set (Figure 2):

- *Raw data*: data in the exact same format/structure as it was collected from the data source;
- *Processed data*: processed data stored in a standard format, where each observational unit is a table (representing a tank or cage), each attribute is a column (including responses and covariates), and each observation is a row (numerical data that describe the state of the observational unit over time, on a daily resolution basis);
- *Metadata*: data that describe the main characteristics of each data set.

During data processing, different methods were applied to convert the *raw data* into *processed data*, including the use of simple conversion factors (e.g., to convert whole-body composition and diet composition from dry weight to wet weight, and to calculate the energy content of diets), linear and non-linear interpolation methods (e.g., to distribute feed intake and temperature data over time, on a daily resolution basis, whenever the data source provided aggregated information), and additive models (e.g., to estimate the dietary amino acid profile and apparent digestibility coefficients based on ingredient composition). In addition, in cases where it was not possible to apply these methods, default values per species were used to ensure that all the minimum required attributes were defined.

After data processing, all *processed data* sets were subjected to quality analysis (e.g., comparison of diet proximate composition and fish whole-body composition data against realistic bounds; analysis of zootechnical indicators, namely nutrient retention efficiencies) in order to identify potential flaws in the data and remove outliers. The different data sets were then partitioned into calibration and validation data sets (roughly 80% for calibration and 20% for validation), based on *metadata* attributes. This process was performed ensuring that the calibration and validation data sets included data covering similar ranges of experimental conditions.

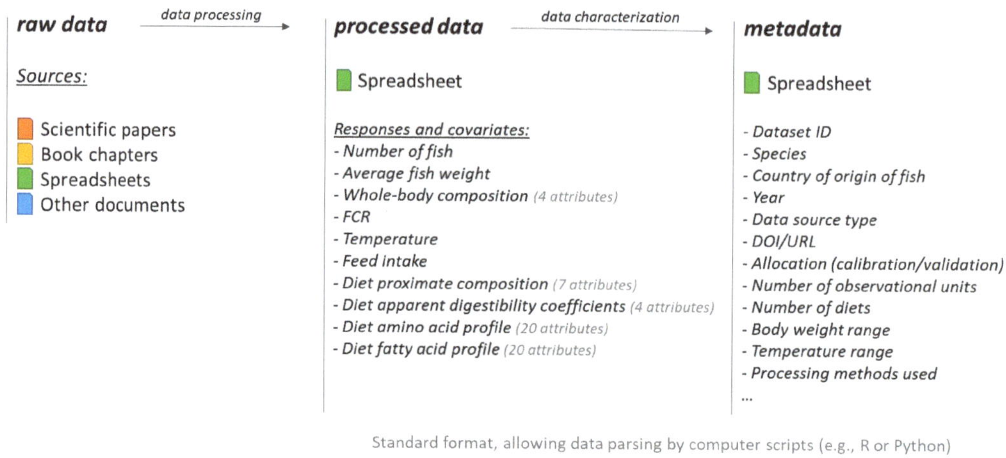

Figure 2. Diagram describing the data set structure, comprising three types of data: *raw data*, *processed data*, and *metadata*.

2.2.2. Model Calibration

For each species, the model was calibrated using the data sets allocated for calibration purposes, in a two-step approach.

In a first step, a subset of model parameters was statically calibrated. This process included the use of linear regression techniques (e.g., least squares, Huber loss minimization, quantile regression) to calibrate body composition and feed intake models, and the assignment of published values to constant or quasi-constant biochemical attributes (e.g., fatty acid molecular weights, energy conversion factors, whole-body amino acid profile).

In a second step, the remaining model parameters were dynamically calibrated in the Powersim Studio 10 Expert software (Powersim Software AS, Nyborg, Norway), using an evolutionary optimization algorithm based on the CMA-ES (covariance matrix adaptation evolutionary strategy) approach [30]. This process comprised the specification of general settings related to the evolutionary algorithm (i.e., number of generations, parents and offspring), the selection of a set of decisions (i.e., model parameters to be calibrated), the adjustment of the calibration space (i.e., minimum and maximum values allowed for each parameter), and the definition of the target set of objectives (weighted loss functions; i.e., $MAPE_{cal\,bw}$ weighing 10.00, CAE_{bw} weighing 0.01, and WE_{CL} weighing 0.10). The following Equations (1)–(3) describe the loss functions:

$$MAPE_{cal\,bw}(\%) = \frac{100}{n}\sum_{i=1}^{n}\left|\frac{P_{bwi} - O_{bwi}}{O_{bwi}}\right|, \tag{1}$$

$$CAE_{bw}(g) = \sum_{i=1}^{n}|P_{bwi} - O_{bwi}|, \tag{2}$$

$$WE_{CL} = \frac{1}{m}\sum_{j=1}^{m}\left(CL_{ref} - CL_{predicted}\right)^2 \times 0.1, \tag{3}$$

where $MAPE_{cal\,bw}$ is the body weight mean absolute percentage error, CAE_{bw} is the body weight cumulative absolute error, n is the number of predicted-observed value pairs, P_{bwi} is the predicted body weight value (g), O_{bwi} is the observed body weight value (g), WE_{CL} is the crude lipids weighted error, $CL_{predicted}$ is the predicted whole-body crude lipids content (%), CL_{ref} is a reference value for whole-body crude lipids content (%) derived from quantile regression of body composition data, and m is the number of time steps.

Model calibration was run until the evolutionary optimization algorithm reached the minimum convergence rate or limit of generations specified, in order to find a plausible parameterization that minimizes the prediction error for the calibration data set.

2.2.3. Model Validation

For each species, the model was validated using the independent data sets allocated for validation purposes (i.e., *processed data* files not used for model calibration). The validation results were evaluated qualitatively, through visual inspection of the model behavior over time in comparison with point observations, and quantitatively, by estimating the body weight mean absolute percentage error for the validation data set ($MAPE_{val\,bw}$) (4), as follows:

$$MAPE_{val\,bw}(\%) = \frac{100}{n}\sum_{i=1}^{n}\left|\frac{P_{bwi} - O_{bwi}}{O_{bwi}}\right|, \quad (4)$$

where n is the number of predicted-observed value pairs, P_{bwi} is the predicted body weight value (g), and O_{bwi} is the observed body weight value (g).

2.3. Model Application

Various applications are presented to illustrate the use of the model to complement experimental trial design and interpretation, and to support the aquaculture industry in evaluating the impact of nutritional and environmental effects at the farm level.

2.3.1. Complement Trial Design and Interpretation

The experiment conducted by dos Santos et al. [31] aimed to evaluate the performance of different Nile tilapia strains (i.e., GIFT-1, GST-24, GIFT-2, GST-14) reared under similar conditions. The authors used the Gompertz model (sigmoid function) as a tool to support the analysis of the results, namely, to characterize the growth pattern and growth rate of the different strains evaluated. The nutrient-based model presented here can be used to further complement the analysis of the results. In particular, and because the model includes feeding as an input, it provides a method to ascertain if the differences in weight gain are due to differences in feed intake. To this end, we ran the model considering as inputs data made available by dos Santos et al. [31] in the published version of the article (i.e., initial weight, temperature, FCR), as well as other additional information provided by the main author (personal communication; i.e., feeding rates and diet proximate composition). The results of the model were compared with the experimental results to make further inferences about the performance of the aforementioned Nile tilapia strains.

Additionally, we considered the experiment presented by Farinha et al. [32], which focused on the evaluation of European seabass juveniles that were fed diets with varying methionine levels. To formulate the diets and define the methionine levels to be considered, the authors relied on the findings of a previous study carried out with similar diets and with fish of about the same size. The nutrient-based model presented here was applied to this study to show how it could have contributed as a complementary tool to support the design of the experiment, namely, in defining the different methionine levels to be tested. This illustrates the usefulness of the model for cases where there are no previous scientific findings available to support the design of the experiment. To this end, we ran the model considering as inputs the detailed records of the experiment presented by Farinha et al. [32]. The results of the model were compared with experimental data to illustrate the model's ability to predict the growth of European seabass when fed diets with varying methionine levels.

2.3.2. Evaluate Nutritional and Environmental Effects at the Farm Level

To demonstrate further uses of the model by the industry, three distinct use cases are made available in Appendix B:

1. Use case 1: Evaluate the impact of different commercial feeds on trout production performance;

2. Use case 2: Evaluate the impact of different temperature profiles on post-smolt production performance;
3. Use case 3: Predict the long-term effects of marginal changes in diet digestibility on bream production performance.

These model applications were carried out considering typical fish farming production scenarios, generally using information at the level that is available to the typical user (e.g., diet composition and feeding rates were defined based on information available in official technical sheets provided by feed producers). In each application, the different scenarios were evaluated by comparing several key-performance indicators provided by the model (e.g., time to reach harvest weight, growth rate, feed conversion ratio—FCR, economic conversion ratio—ECR).

3. Results

3.1. Model Calibration and Validation

3.1.1. Model Performance for Calibration Data Sets

Table 2 presents the final estimates of the three loss functions (error metrics) used to dynamically calibrate the model for each species (i.e., $MAPE_{cal\,bw}$, CAE_{bw} and WE_{CL}; see Equations (1)–(3) for formulas).

Table 2. Error metrics for the calibration data set for each species.

Loss Functions/Error Metrics	Gilthead Seabream	European Seabass	Atlantic Salmon	Rainbow Trout	Nile Tilapia
$MAPE_{cal\,bw}$ (%)	6.82	5.15	6.33	8.28	18.18
CAE_{bw} (g)	91.29	75.70	588.83	162.12	166.04
WE_{CL} (−)	0.24	0.16	0.17	0.27	0.20

Overall, the performance of the model in predicting the body weight of fish for the calibration data sets is similar for most fish species in terms of $MAPE_{cal\,bw}$. However, it must be pointed out that the model presents worse performance for Nile tilapia ($MAPE_{cal\,bw}$ = 18.18%) compared to other species ($MAPE_{cal\,bw}$ between 5.15% and 8.28%).

The worse performance for Nile tilapia may be related to several reasons linked to the data sets of this species, two of which are worth mentioning: (i) the different publications on the experiments carried out with Nile tilapia appear to show a much greater variability in terms of rearing conditions (e.g., rearing density, water quality) than the other species, which may explain part of the deviation between model predictions and observations; and (ii) the high variability of Nile tilapia strains used in different studies can also contribute to the higher calibration error in Nile tilapia, since, if the differences between strains are manifested in different metabolic rates or nutrient utilization efficiencies, those differences may not be fully replicated by the current model parametrization as the model calibration was done at the species level and not at the strain level.

The CAE_{bw} and WE_{CL} directly depend on the size, structure, and content of the calibration data set; therefore, they are presented here for reference only and not used to make any considerations about the model performance.

Given the outcomes of the calibration procedure, the resulting parameter sets were accepted and used for the model validation stage.

3.1.2. Model Performance for Validation Data Sets

Figure 3 shows the model validation results for each species in terms of body weight prediction. In general, the model accuracy is consistent across species, presenting a mean absolute percentage error ($MAPE_{val\,bw}$) ranging between 11.7% and 13.8%. The relative error distribution shows a symmetrical pattern, meaning that the scale and sign of model deviations are homogeneous over the body weight range (i.e., there is no clear bias for under- or over-estimation). The higher density of points for lower body weight classes (more evident in the case of rainbow trout due to higher n), is explained by the fact that most of the data sets used to validate (and calibrate) the model for each species are related to experiments carried out with small fish.

Visual inspection of the model behavior over time indicates that the predictions are representative of the observed growth pattern of fish (see Section 3.2. Model Application for examples). However, as expected, for some validation data sets, the magnitude of the deviation between model predictions and observations tends to increase throughout the simulation, since simulation errors are cumulative and generally multiplicative, and the model is not re-initialized when there are intermediate sampling points.

It is also worth mentioning that, although $MAPE_{val\,bw}$ can be seen as a proxy for the model generalization error, one cannot always expect an accuracy according to this error metric when extrapolating fish growth with this model. Model accuracy is not only dependent on the model itself, but also on the quality of the input data used to drive it. Thus, in order to obtain accurate predictions, we cannot fail to mention that the input data used to drive the model must be as representative as possible of the scenarios to be evaluated.

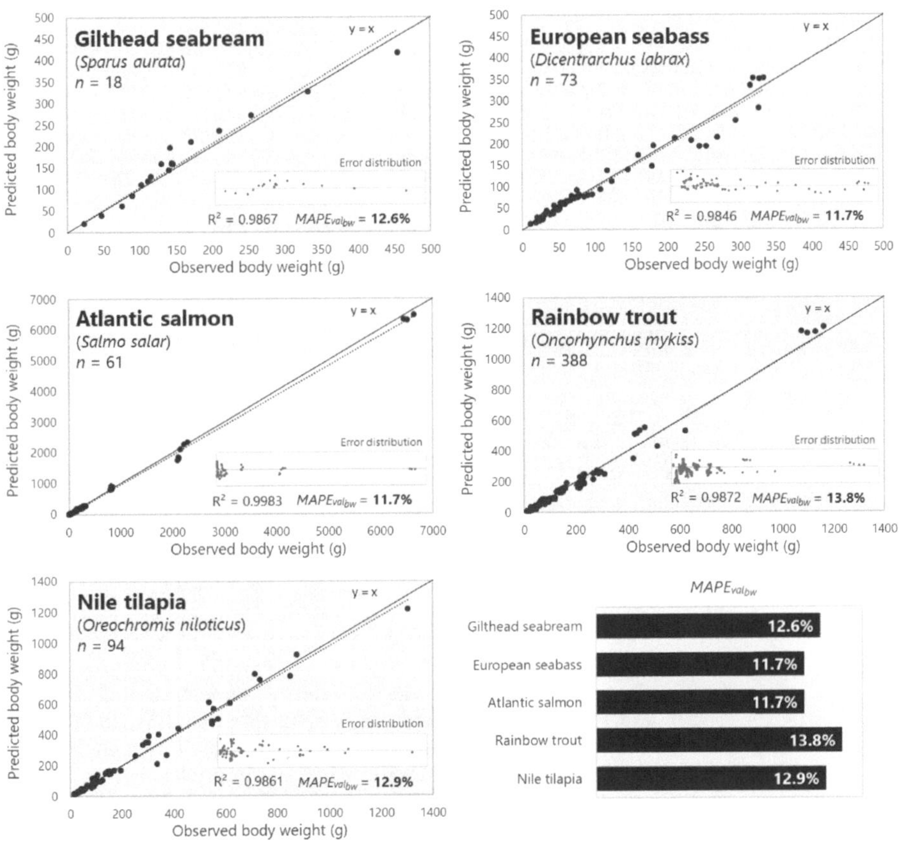

Figure 3. Scatter plots showing predicted versus observed body weight for the validation data set of each species, where n is the number of predicted-observed value pairs and $MAPE_{val\,bw}$ is the mean absolute percentage error. The solid line denotes the line of equality ($y = x$) and the dotted line is the linear trendline. Inside each plot, in a smaller frame, is shown the relative error (y-axis) distributed over the body weight range (x-axis). On the lower-right corner, a bar plot shows a comparison between the $MAPE_{val\,bw}$ values obtained for the different species.

3.2. Model Application

3.2.1. Complement Trial Design and Interpretation

Figure 4 shows the time series of predicted and observed average body weight values for the dos Santos et al. [31] data set, which includes data about different Nile tilapia strains (i.e., GIFT-1, GST-24, GIFT-2, GST-14) reared under similar conditions. For all Nile tilapia strains, the model predictions for body weight are consistent with the observations, with a MAPE ranging between 11.4% and 16.9%. The predicted relative growth rate (RGR, %/day) is also in line with the observed values.

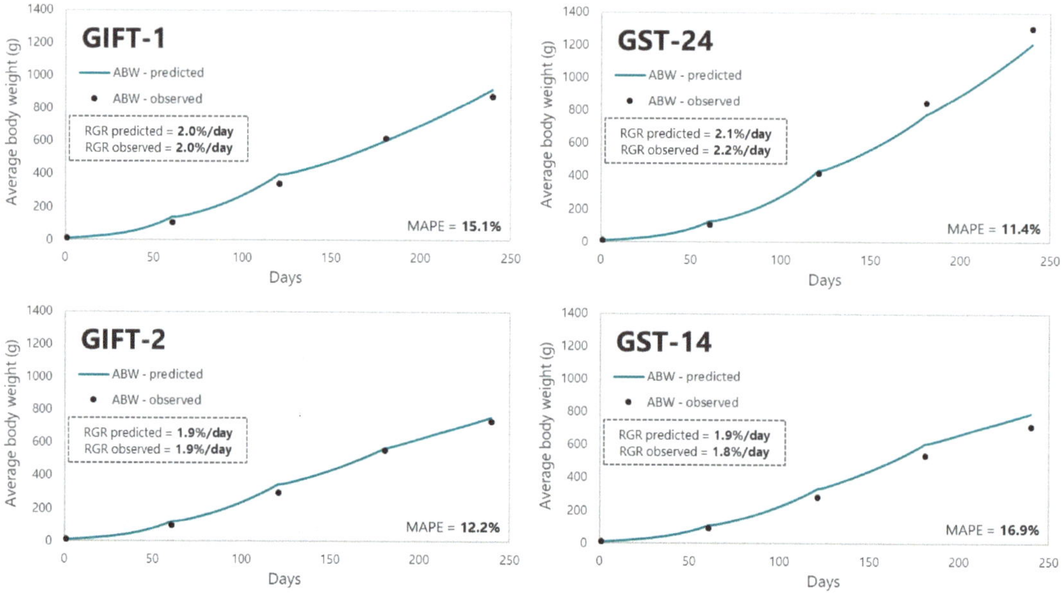

Figure 4. Time series of predicted (solid line) and observed (points) average body weight (ABW) for the dos Santos et al. [31] data set, which includes experimental data about different Nile tilapia strains (i.e., GIFT-1, GST-24, GIFT-2, GST-14) reared under similar conditions. MAPE stands for mean absolute percentage error (%) and RGR for relative growth rate (%/day), the latter calculated as: $RGR = \left(e^{(ln(ABW_{i+n})-ln(ABW_i))/(day_{i+n}-day_i)} - 1\right) \times 100$.

The ability of the model to predict the growth of these four strains suggests that differences in the growth performance of these strains are mainly related to feed intake, i.e., higher growth rates were achieved due to the capacity of some strains (i.e., GST-24 and GIFT-1) to ingest more feed. This is something that is not apparent when analyzing the experimental data or the results from simpler growth models, such as the Gompertz model, and that can complement the research presented by dos Santos et al. [31]. If differences in growth performance were manifested due to different metabolic efficiencies, the nutrient-based model should clearly reflect this, by predicting the growth of some strains expressively better than others, over the entire period. Information about the observed voluntary feed intake (VFI, % body weight/day; GIFT-1 = 5.92%/day, GST-24 = 6.45%/day, GIFT-2 = 5.33%/day, GST-14 = 5.64%/day) also indicate a direct positive relationship between feed consumption and growth rate. However, these values alone do not support the hypothesis that the main factor driving growth performance is feed intake capacity, since this indicator does not aggregate information on the nutrient utilization capacity. This application illustrates how this nutrient-based model can be a useful tool to support the interpretation of experimental results, contributing to new insights about the main factors that explain fish performance.

Figure 5 shows the time series of predicted and observed average body weight values for the Farinha et al. [32] data set, which includes experimental data about European seabass that were fed diets with different methionine levels (i.e., M0.65, M0.85, M1.25, M1.50). For all dietary treatments, the model predictions are consistent with the observations, with a MAPE ranging between 2.6% and 11.5%. This means that the model is able to predict the effects of different methionine levels on the growth performance of European seabass accurately.

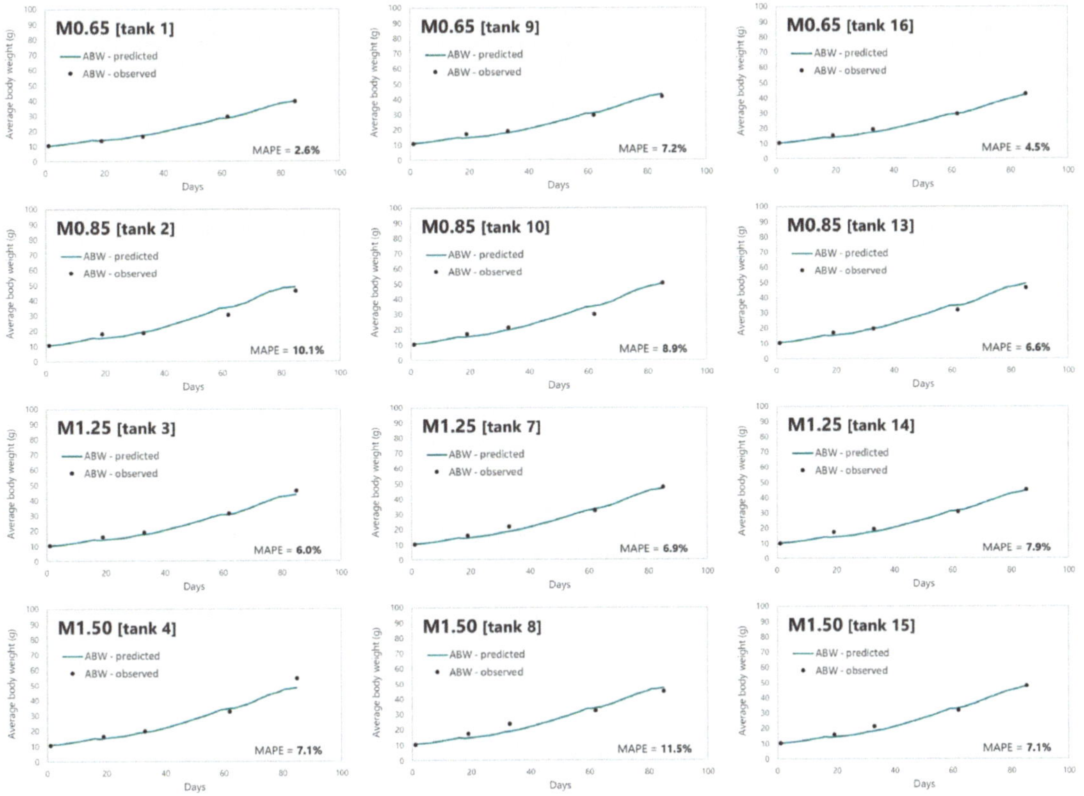

Figure 5. Time series of predicted (solid line) and observed (points) average body weight (ABW) for the Farinha et al. [32] data set, which includes experimental data about European seabass that were fed diets with different methionine levels (i.e., M0.65, M0.85, M1.25, M1.50). MAPE stands for mean absolute percentage error (%).

As described by Farinha et al. [32], the analysis of experimental data through one-way ANOVA indicated significant differences between the dietary treatments M0.65 and M1.50 for final average body weight, pointing to methionine deficiency in fish fed with the M0.65 diet. The authors further carried out a linear broken-line regression analysis to find the breakpoint that defines the methionine level below which growth performance is affected (i.e., methionine requirement). Usually in this type of assessment (broken-line regression), it is recommended (not mandatory though) to have at least two observed points in each segment so as to ensure that the slope of each segment is estimated based on points directly related to observed data and that the breakpoint is then defined with better precision. However, this was something that did not happen, as M0.65 was the only diet that displayed a clear methionine deficiency, meaning that the broken-line was adjusted considering only one directly observed point below the breakpoint. The use of the nutrient-based model presented in this work could have been useful during the design of this experiment as it could have allowed a better definition of the methionine levels to be tested (prescreening of diets in silico), in order to ensure at least two points (methionine levels) below the breakpoint. This illustrates how this nutrient-based model can be a useful tool to support the design of experiments, helping to anticipate fish performance, and thus, make more informed decisions to ensure full effectiveness of the experiment to be performed. Furthermore, in nutrient requirements studies similar to the one presented by Farinha et al. [32], the model can also be used to extrapolate the performance of fish subjected to different rearing conditions (e.g., different temperatures) or diet compositions (e.g., different protein or energy content) to verify whether the nutrient requirements determined apply in those cases or not.

3.2.2. Evaluate Nutritional and Environmental Effects at the Farm Level

Additional model applications directed to industrial end users (e.g., fish farmers, aquafeed producers, and aquaculture consultants) are presented in Appendix B and include the following use cases: 1. evaluate the impact of different commercial feeds on trout production performance; 2. evaluate the impact of different temperature profiles on post-smolt production performance; and 3. predict the long-term effects of marginal changes in diet digestibility on bream production performance. These applications illustrate how the nutrient-based model presented here can be used as a decision-support tool, helping to define the best feeding and farming strategies for each specific case, through the analysis of the impact of nutritional and environmental effects at the farm level.

4. Discussion

This manuscript describes and evaluates the FEEDNETICS model, a mechanistic nutrient-based model intended to be used as a data interpretation and decision-support tool by different end users, i.e., fish farmers, aquafeed producers, aquaculture consultants, and researchers. The framework behind FEEDNETICS comprises two main components: (i) *fish model*, that simulates at the individual level the fish growth, composition, and nutrient utilization, and (ii) *farm model*, that upscales all information to the population level.

The *fish model* is based upon a mechanistic and deterministic modelling approach, following basic physical principles and prior information on the organization and control of biochemical/metabolic processes. It assumes a multi-compartmental structure describing the flow of nutrients and metabolites between the gut, blood, and body (single compartment representing the main body tissues, e.g., brain, liver, muscle, adipose tissue, bone). Similar mechanistic nutrient-based models have been previously developed by other authors for larval [6,28], juvenile, and adult fish stages [3–5,19,23,27]. However, not all of these models explicitly consider different compartments to describe the flow of nutrients and metabolites. Although not a requisite to predict the growth and body composition of fish, assuming a multi-compartmental structure enables a more realistic description of nutrient flow and utilization rates. In addition, its modular nature facilitates the continuous improvement of the model through the simplification or complexification of specific (sub-)processes. For example, in the current version of the model, the *gut* compartment does not mechanistically simulate the digestive process, since the apparent digestibility coefficients (ADCs) of nutrients are provided as input. However, as nutrient digestibility of feeds may vary with fish size, water temperature, and ration size [33–37], among other factors, it may be relevant to refine the model in order to account for such effects. As the gut is considered as a separate compartment in this model, future improvements aimed at incorporating a mechanistic description of the digestive process can be achieved without the need to completely reformulate the entire structure of the model. The same is applicable to other physiological and metabolic processes (e.g., appetite and ingestion).

The *farm model* was implemented based on the assumption that the entire fish population can be described based on a linear scaling of the outputs provided by the *fish model*. Therefore, the existing inter-individual variability is not accounted for in the current model implementation. Although it is a strong assumption, this approach aims to simplify the required input data from the user. Furthermore, it must be pointed out that accurately modelling fish size distribution within a population is challenging, as it is not only affected by genetic variability among individuals, but may also be influenced by production practices (e.g., ration size and stocking density) [38–40]. Nevertheless, if relevant for future applications, some outputs provided by the *farm model* can be modified in order to scale from individual to population level based on fish size distribution functions, where the parameters of such functions would be considered as input.

In general terms, and despite the simplifications that were assumed to ensure greater usability, the model has a complex and detailed structure. Considering such a high level of detail has the advantage of making the model usable for a wider range of purposes. However, it also means that the input data needed to run the model are considerably more detailed than in the case of simpler models (e.g., bioenergetic models) [20]. While this may not be an issue for some end users (e.g., aquafeed producers, aquaculture consultants and researchers), it may compromise usability for others who may not have access to detailed information, namely in terms of the nutrient composition of feeds (e.g., fish farmers). This usability issue is highly relevant and should be considered when implementing this tool as a user-friendly application. To overcome it, it is suggested to define default values and implement auxiliary data processing methods to deal with missing input data. Default values can be defined under different assumptions, but it must be ensured that they meet the requirements of the species and that they are representative, as far as possible, of common practices applied in the industry. Auxiliary data processing methods can be implemented on top of different data structures, for example, to convert monthly resolution data to daily resolution data, or to estimate amino acids and apparent digestibility coefficients of feeds based on information about the composition of its ingredients.

The calibration and validation of the model was carried out at the species level, for five commercially relevant farmed fish species, i.e., gilthead seabream, European seabass, Atlantic salmon, rainbow trout, and Nile tilapia. In these procedures, we made use of data sets related to in vivo experimental trials covering a wide range of rearing and feeding conditions (see Section 2.2.1. Data Processing and Data Sets Description). Most of these data sets were collected from the scientific literature and include valuable knowledge effortfully generated over the last few decades by the academic community. Calibrating the model with these data sets means that part of this prior knowledge is being integrated through mathematical functions and made easily available to support new advances in fish farming and nutrition. The extension of the model to other species can be done following the same calibration and validation procedures, once sufficient data is collected and processed to do so. Furthermore, the model can also be calibrated in order to accommodate physiological and metabolic differences of different fish lineages/strains.

Overall, the results of the validation of the model for fish growth are consistent between species, presenting a mean absolute percentage error (MAPE) between 11.7 and 13.8%. An objective characterization of the model precision is something difficult to generalize, as the acceptable error range is often subjective and strongly depends on the application for which the model is intended to be used. Therefore, it is advisable for users to carry out more specific assessments in this regard, for example, by comparing the precision of this model for a particular application against other available methods that can be applied for the same purpose. While doing so, it is important to ensure that the input data used to run the model is representative and as accurate as possible, and that the complexity of the problem being solved is adequate for all methods being compared.

Various model applications are presented in Section 3.2. Model Application and in Appendix B. Those illustrate how the FEEDNETICS model can be used to complement trial design and interpretation and to support in the evaluation of nutritional and environmental effects at the farm level. However, the applications illustrated here represent only a small picture of the potential uses that can be made out of this model. Further potential applications include, for example, using the model to optimize finishing feeding strategies that aim to enrich fish with a high content of omega-3 fatty acids, or using it as a monitoring and forecasting tool to provide information about the current and future state of production on a daily basis.

Future model developments can be made at several levels, some of which have already been mentioned above. The multi-compartmental structure of some physiological and metabolic processes included in the *fish model* allows simplifying or complexifying the modeling approach to improve the model's ability to describe fish performance. In addition, optimization algorithms can be implemented on top of the model to move from what-if scenario analysis to automated criteria-based optimization, where the algorithm would find the best solutions based on a set of criteria (e.g., key-performance indicators) pre-defined by the user.

5. Conclusions

In an era of increasing digitization, where it is important to transform data into useful information, the development and deployment of mathematical models as data interpretation and decision-support tools will contribute in the movement towards the implementation of precision fish farming practices. The ultimate goal is to increase production efficiency and productivity through greater knowledge and control of the systems. FEEDNETICS provides a means of transforming data into useful information, thus contributing to further advances in the fish farming industry and more efficient production.

Author Contributions: Conceptualization, T.S.S. and L.E.C.C.; methodology, T.S.S., F.M.R.C.S., A.M.D.N. and L.E.C.C.; software, T.S.S. and F.M.R.C.S.; validation, F.M.R.C.S., T.S.S. and A.M.D.N.; formal analysis, T.S.S. and F.M.R.C.S.; investigation, F.M.R.C.S., T.S.S., R.C.P.M., A.I.G.R., A.M.D.N., L.E.C.C., S.A.D.E. and P.J.A.P.R.; resources, T.S.S., F.M.R.C.S., L.E.C.C., S.A.D.E. and P.J.A.P.R.; data curation, F.M.R.C.S., T.S.S., R.C.P.M., A.I.G.R. and A.M.D.N.; writing—original draft preparation, F.M.R.C.S. and T.S.S.; writing—review and editing, F.M.R.C.S., T.S.S., A.M.D.N., L.E.C.C., R.C.P.M., A.I.G.R., S.A.D.E. and P.J.A.P.R.; visualization, F.M.R.C.S., T.S.S. and A.M.D.N.; supervision, T.S.S. and L.E.C.C.; project administration, L.E.C.C.; funding acquisition, L.E.C.C., A.M.D.N., F.M.R.C.S., T.S.S., S.A.D.E. and P.J.A.P.R. All authors have read and agreed to the published version of the manuscript.

Funding: This work results from activities on the following projects: AquaIMPACT, funded by the European Union's Horizon 2020 research and innovation program under grant agreement no. 818367; FEEDNETICS 4.0, funded by EUROSTARS-2 program, and by Portugal and the European Union through FEDER/ERDF, CRESC Algarve 2020 and NORTE 2020, in the framework of Portugal 2020 under reference E!12516-FEEDNETICS 4.0_40813; NoviFEED, funded by Iceland, Liechtenstein and Norway, through EEA grants, in the scope of the program Blue Growth, operated by the Directorate General for Maritime Policy (DGPM), Portugal, under reference PT-INNOVATION-0099; and from FCT—Foundation for Science and Technology—through projects UIDB/04326/2020, UIDP/04326/2020, and LA/P/0101/2020.

Institutional Review Board Statement: Not applicable.

Informed Consent Statement: Not applicable.

Data Availability Statement: The processed data sets used for model calibration, validation, and application are not publicly available due to proprietary and commercial use restrictions. All processed data sets used in this work are the property of Sparos Lda., and/or third-parties who have established data transfer agreements (DTAs) with Sparos Lda.

Acknowledgments: We would like to thank Vander Bruno dos Santos (Instituto de Pesca/APTA/SAA, Brazil) by kindly sharing detailed data on the experiment carried out by him and other authors on the performance of different Nile tilapia strains.

Conflicts of Interest: F.M.R.C.S., A.M.D.N., A.I.G.R., R.C.P.M., L.E.C.C. and T.S.S. work for Sparos Lda. The other authors declare that they have no conflicts of interest. The views expressed in this work are the sole responsibility of the authors and do not necessarily reflect the views of any third party. The funders had no role in the design of the study, in the collection, analyses, or interpretation of data, in the writing of the manuscript, or in the decision to publish the results.

Appendix A.

Appendix A.1. High-Level Overview of the FEEDNETICS Model

Conceptually, the model can be subdivided in two large components: a *fish model* and a *farm model*. Most of the complexity of the overall FEEDNETICS model is related to the *fish model*, which is the focus here. The roles of the *farm model* are mostly population management (e.g., deterministically remove fish in order to replicate a given time-dependent mortality rate), feeding regime management (e.g., scale given feed according to how many fish exist) and the calculation of secondary outputs (e.g., performance measures, environmental impact measures, economic indicators). All other calculations are performed within the *fish model*, which considers a set of mass pools representing different metabolites in different body compartments (in the current version: gut, blood, and body) and different states, along with a set of mass flows representing different biological processes (digestion, absorption, conversion, synthesis, degradation, oxidation) along with their time-dependent rates. Mathematically, the *fish model* can be seen as a set of ordinary difference equations, which are, in practice, numerically solved in the Powersim Studio 10 Expert software (Powersim Software AS, Nyborg, Norway) using the "forward Euler" integration approximation with a fixed timestep of 0.01 days (i.e., in a general sense, the model follows the dynamics of an externally-driven deterministic system of first-order ordinary difference equations, in practice).

A diagram detailing the overall high-level structure of the FEEDNETICS model (*fish model* and *farm model*) can be seen in Figure A1.

Unless stated otherwise, the use of the "FEEDNETICS model" throughout the rest of this appendix refers specifically to the *fish model* part of the general model.

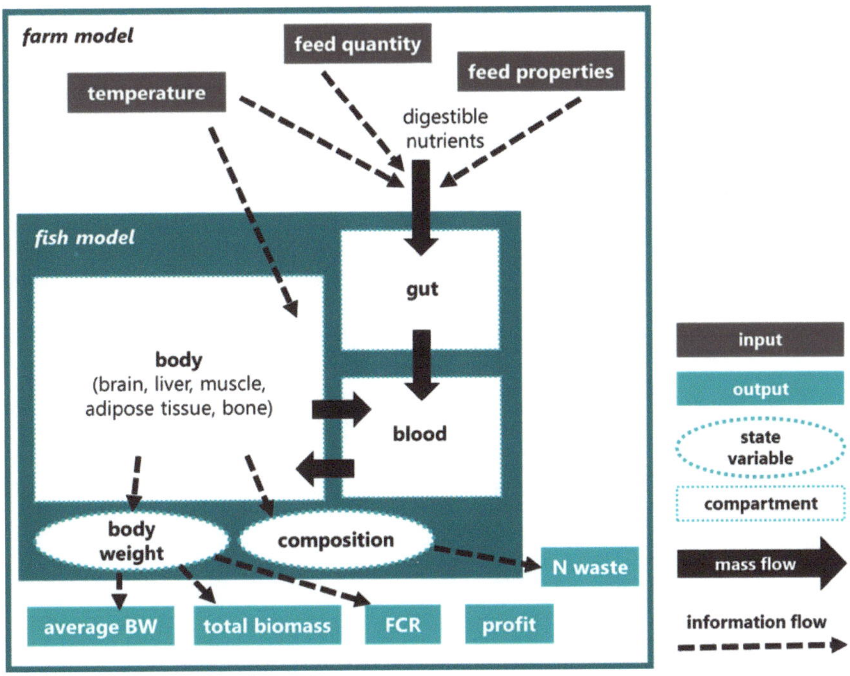

Figure A1. Diagram detailing a simplified high-level structure of FEEDNETICS, with emphasis on the *fish model* component. Inputs to the *fish model* (previously pre-processed by the *farm model*) drive its time-dependent behavior, while its state variables are used in the calculation of all outputs (only a small subset of which are depicted in the diagram). Each compartment contains a set of mass pools (not depicted) which represent the amount of each metabolite in each compartment.

Appendix A.2. Model Inputs

Given that the behavior of the model is strongly driven by its inputs, it is important to consider them as a starting point. The FEEDNETICS model can be seen as driven by basically three time-dependent inputs, all with a time resolution of 0.01 days:

1. Temperature (°C), single value. In practice, the user provides "daily temperature average" and "daily temperature amplitude" and a typical daily temperature curve is generated by the *farm model* using a sinusoidal shape, assuming that the lowest temperature occurs around midnight.
2. Feed given (g/day), single value. In practice, the user provides either this information on a daily resolution level or a feeding table (matrix defining the feeding rates, expressed as % body weight/day, per fish boy weight class and temperature class) from which this information is estimated on a daily resolution level. The daily given feed is then distributed along the day using information on the frequency and distribution of meals (i.e., number of meals, time of first meal, and time between meals).
3. Feed properties, many values (often constant along time). These are:
    - Macronutrient composition (i.e., crude protein, crude lipids, ash, fiber, gross energy, phosphorus);
    - Apparent digestibility coefficients (ADCs; i.e., crude protein, crude lipids, gross energy, phosphorus);
    - Amino acid profile (i.e., the standard 20 proteinogenic amino acids);
    - Fatty acid profile (i.e., 20 different fatty acids).

Appendix A.3. Feed Intake Control and Gut Compartment

Regarding feed intake control, the FEEDNETICS model follows the modelling approach of Lupatsch [41], with some adaptations:

$$FI_{max} = a \times BW^b \times e^{cT} \times I(T > T_{low}) \times I\left(T < T_{high}\right), \tag{A1}$$

where FI_{max} represents maximum feed intake in grams/day, a, b, c, T_{low}, and T_{high} are species-specific parameters, BW represents fish body weight in grams, T represents the current temperature, and I represents the indicator function (returns 1 if the comparison is true and 0 if the comparison is false).

This implies that the maximum feed intake is assumed to depend linearly on a power of the body weight and exponentially on temperature, being reduced to zero if temperature is too low or too high. This step (among others) attempts to induce the loss of weight observed in fish under extreme temperatures, which is not seen for high temperatures, if strictly following the feed intake model suggested by Lupatsch [41].

To prevent discontinuities in the model (which can make model optimization more challenging), we adopt an analytical approximation of the above expression (which converges to the above expression as the parameter β tends to infinity):

$$FI_{max} = a \times BW^b \times e^{cT} \times \frac{1}{1 + \left(\frac{T_{low}}{T}\right)^{\beta}} \times \frac{1}{1 + \left(\frac{T}{T_{high}}\right)^{\beta}}. \tag{A2}$$

As such, actual feed intake is then calculated as:

$$FI = \min(FI_{max}, feedgiven), \tag{A3}$$

Taking into account the actual feed intake, along with the (possibly time-dependent) feed composition and apparent digestibility coefficients (ADCs), the amount of digestible nutrients being introduced in the *gut* compartment is calculated. The presence of indigestible nutrients in the gastrointestinal tract is not explicitly modeled in FEEDNETICS, though these values (along with wasted/non-ingested feed) are tracked for the purpose of environmental impact estimation.

At this point, the mass of ingested feed is converted to molar amounts of nutrients:

$$DI_{nutrient} = \frac{FI \times ADC_{nutrient} \times feed_{nutrient}}{Mw_{nutrient}}, \quad \text{(A4)}$$

where $DI_{nutrient}$ represents the digestible intake of a nutrient per unit of time, $ADC_{nutrient}$ is the given apparent digestibility coefficient of the nutrient, $feed_{nutrient}$ is the percentage of the nutrient in the feed, and $Mw_{nutrient}$ is the molecular weight of the nutrient.

To model digestion and absorption rates ($digestion_{nutrient}$ and $absorption_{nutrient}$, respectively) of ingested feed, we considered a simplified model with a single enzyme that converts digestible nutrients ($digestible_{nutrient}$) to digested nutrients ($digested_{nutrient}$), then a single receptor/transporter that converts digested nutrients to absorbed nutrients, using second-order kinetics, as follows:

$$digestion_{nutrient} = k_{digestion} \times enzyme \times digestible_{nutrient}, \quad \text{(A5)}$$

$$absorption_{nutrient} = k_{absorption} \times receptor \times digested_{nutrient}, \quad \text{(A6)}$$

where the $k_{digestion}$ and $k_{absorption}$ rate constants, and *enzyme* and *receptor* indicate their abundance.

To model enzyme and receptor abundance, we considered simple first-order processes, where enzyme/receptor production is proportional to substrate abundance, while enzyme/receptor degradation is proportional to enzyme/receptor abundance.

$$\frac{d(enzyme)}{dt} = \left(k_{enz\_prod} \times \sum_{nutrient} digestible_{nutrient}\right) - k_{enz\_deg} \times enzyme, \quad \text{(A7)}$$

$$\frac{d(receptor)}{dt} = \left(k_{rec\_prod} \times \sum_{nutrient} digested_{nutrient}\right) - k_{rec\_deg} \times receptor, \quad \text{(A8)}$$

where k_{enz_prod} and k_{enz_deg} are rates of enzyme production and degradation, respectively, and k_{rec_prod} and k_{rec_deg} are rates of receptor production and degradation, respectively.

The *gut* compartment does not effectively affect nutrient digestibility (since ADCs are provided as input), but simply delays the availability of ingested nutrients for metabolic processes (anabolism or catabolism).

A diagram representing the internal structure of the *gut* compartment, with all the considered mass pools and mass flows, can be seen in Figure A2, which shows that the structure is generally linear and one-way.

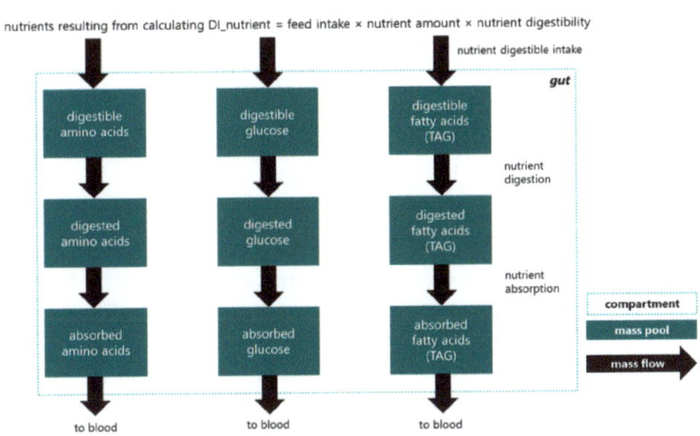

Figure A2. Diagram detailing the internal structure of the *gut* compartment, showing the different pools considered, along with mass flows between pools. For simplicity, the amino acid and fatty acid pools are displayed in vectorized form (i.e., they appear as a single unit, but actually represent 20 different pools each).

Additionally, a set of calculations is performed to provide the fish with information on short-term feeding history. The result is two variables between 0 and 1 ("fed" and "fasting") that are used in other downstream calculations.

Appendix A.4. Core Metabolic Model (Blood and Body Compartments)

The FEEDNETICS model is built around a core metabolic model that represents the main anabolic and catabolic processes in fish. To ensure a simple model, all main body tissues (e.g., liver, brain, muscle) are considered under a single *body* compartment, which interfaces with the *blood* compartment. Although movement of metabolites between the *body* and *blood* compartments is two-way, mass flow tends to occur mostly in the blood-to-body direction due to the regular entry of nutrients from the gut to the bloodstream. In this model, the role of the *blood* compartment is two-fold: to deliver absorbed nutrients to the *body* compartment, delaying them, and to regulate metabolite availability for use in metabolic processes (*body* compartment).

A diagram showing the general structure of the *blood* and *body* compartments, along with the considered mass flows, can be seen in Figure A3.

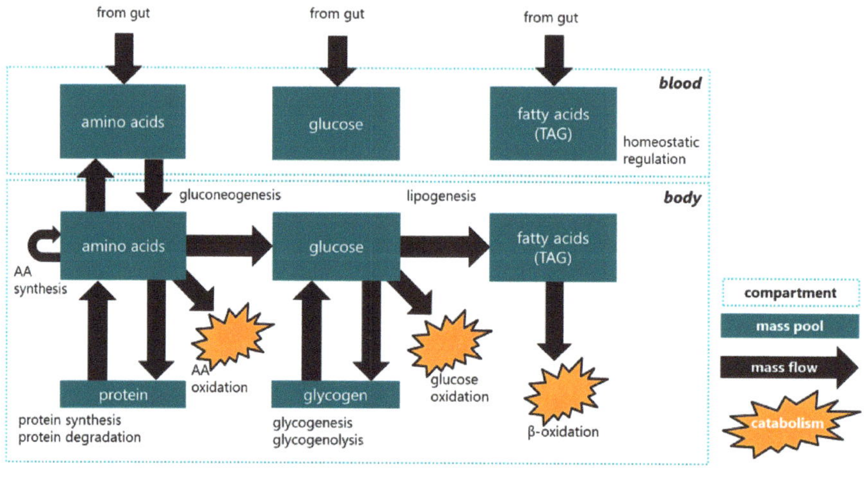

Figure A3. Diagram detailing the internal structure of the *blood* and *body* compartments, showing the different pools considered, along with mass flows between pools. For simplicity, the amino acid and fatty acid pools are displayed in vectorized form (i.e., they appear as a single unit, but actually represent 20 different pools each).

Appendix A.5. Body Weight and Composition Model

A central point of the FEEDNETICS model relates to the control of body composition. As a starting point, fish body is assumed to be mostly composed of water, crude protein (which includes protein, amino acids, and other nitrogen-containing compounds), crude lipids (fatty acids, triacylglycerides, phospholipids, steroids), and minerals/ash. The amount of glucose and glycogen is generally below 1% of whole-body weight [42] and considered negligible in FEEDNETICS, and all dietary and metabolism-derived glucose is re-directed to lipogenesis.

First, total protein ($protein_{total}$) is directly calculated (in amino acids equivalents) by adding the contributions of all amino acids (AA):

$$protein_{total} = \sum_{i=1}^{20} proteinAA_i \times AA\_Mw_i, \tag{A9}$$

where $proteinAA_i$ represents the mass of the ith amino acid present in the *body* protein pool and AA_Mw_i represents the molecular weight of the ith amino acid.

For the components for which an explicit mass budget cannot be established (e.g., water, ash), their quantities are assumed to depend mostly on structural biomass and, as such, directly or indirectly estimated as an allometric function of $protein_{total}$.

For all other components, their quantities are calculated by summing the contributions of the different corresponding mass pools. For example, the total amount of lipids ($lipid_{total}$) is estimated as:

$$lipid_{total} = \sum_{i=1}^{20} (TAG\_body\_FA_i + TAG\_blood\_FA_i) \times FA\_Mw_i, \tag{A10}$$

where $TAG\_body\_FA_i$ represents the mass of the ith fatty acid present in the *body* lipids pool, $TAG\_blood\_FA_i$ represents the mass of the ith fatty acid present in the *blood* lipids pool, and FA_Mw_i represents the molecular weight of the ith fatty acid.

Total body mass is estimated by summing the mass of the different components, while energy density is estimated based on the amounts of protein, lipids, and total carbohydrates, using fixed coefficients.

Furthermore, an important variable (CL_q) is also calculated. This variable controls the amount of lipids that is used to produce energy (as opposed to protein), by comparing the amount of lipids present in the body with (species and body weight dependent) reference values obtained by quantile regression:

- If the amount of lipids is below the lowest reference value, $CL_q = 0$ (i.e., use protein to produce energy and not lipids);
- If the amount of lipids is above the highest reference value, $CL_q = 1$ (i.e., as far as possible, use lipids to produce energy and spare protein);
- Otherwise, it gives an intermediate value, using a reference value that depends on the "fed state" of the fish:

$$CL_q = \frac{1}{1 + \left(\frac{lipid_{ref}}{lipid_{total}}\right)^{\beta}}, \tag{A11}$$

where β controls the shape of the curve and $lipid_{ref}$ is calculated based on the *fasting* variable, as an interpolation between the lowest and highest reference values (i.e., when *fasting* is 0, the reference value is the maximum value; when *fasting* is 1, the reference value is the minimum value).

Appendix A.6. Energetic Model

An important aspect of the accurate modelling of fish growth is to ensure that estimates are consistent with basic physical principles, such as energy conservation. As such, we consider the following assumptions:

- Energy production and energy expenditure can be represented in ATP (adenosine triphosphate) equivalents, given that ATP is (along with other nucleoside phosphates) the most widespread energy-yielding metabolite in cells;
- During any given step (timescale ≈ 14.4 min per timestep), we assume that ATP levels are essentially in steady-state (i.e., the rate of ATP production must match the rate of ATP degradation over the course of any given timestep);

- ATP expenditure is assumed to result from a combination of:
    - Anabolism energy costs—energy expenditure due to anabolic reactions, which is implicitly defined by their rates (protein synthesis, glycogenesis, lipogenesis, non-essential AA synthesis);
    - Catabolism energy costs—energy expenditure due to energy-consuming catabolic reactions, which is implicitly defined by their rates (protein degradation);
    - Basal energy costs—fixed feed-independent costs that depend on fish body weight and temperature (and accounts for fixed costs not included in the previous two points);
    - Feeding energy costs—variable costs that depend on the fish's "fed state" (and accounts for feed-dependent costs not included in the previous three points);
- A fixed upper limit on ATP expenditure rate is assumed (600 μmol.g^{-1}.h^{-1}), to ensure that the actual values remain within physiologically-reasonable bounds [43];
- ATP production is constrained to match ATP expenditure and results from a combination of:
    - Metabolite conversion (e.g., gluconeogenesis from AA and interconversion of AA);
    - Oxidative catabolism (glucose oxidation, beta-oxidation of fatty acids, amino acid oxidation);
- The rates of AA oxidation and FA beta-oxidation are determined by the difference between "required ATP production to match ATP expenditure" and "ATP resulting from metabolite conversion and glucose oxidation", with the relative weight of the two catabolic processes being defined according to the CL_q variable.

A system of valves is used to ensure that ATP expenditure and ATP production always match, even in unusual situations (e.g., not enough substrate to ensure required ATP production; not enough ATP expenditure to balance out energy-yielding reactions). If we ignore this system, the resulting balances can be described as:

$$ATP_{exp} = ATPcost_{anab} + \left(1 + fed_{scaling} \times feed_{cost_{scale}}\right) \times ATPcost_{basal}(BW, T), \quad \text{(A12)}$$

$$ATP_{req} = ATP_{exp} - ATPprod_{catab} - ATPprod_{glucox}, \quad \text{(A13)}$$

where ATP_{exp} represents ATP expenditure (in mol/hr), $ATPcost_{anab}$ represents ATP costs due to explicit metabolic reactions, $fed\_scaling$ is a value between 0 and 1 that represents the fish's fed state, $feed\_cost\_scale$ is a parameter that controls feeding costs, $ATPcost_{basal}$ represents (otherwise unaccounted) basal energy costs, BW is the current body weight of fish (in grams), T is the current temperature (in °C), ATP_{req} represents the required ATP from amino acid oxidation and fatty acid beta-oxidation, $ATPprod_{catab}$ represents the ATP produced by energy-yielding metabolite conversion reactions, and $ATPprod_{glucox}$ represents the ATP production due to glucose oxidation.

The remaining ATP requirements (ATP_{req}) are met using a combination of amino acid oxidation and fatty acid beta-oxidation that is controlled by the CL_q variable. The way to interpret this variable (which is bounded between 0 and 1) is as the ratio between "mass of lipid oxidized" and "total mass of protein + lipid oxidized".

$$m_{oxAA} = \frac{1 - CL_q}{CL_q} m_{oxFA}, \quad \text{(A14)}$$

$$ATP_{req} = m_{oxAA} \times ATPstoich_{AA} + m_{oxFA} \times ATPstoich_{FA}, \quad \text{(A15)}$$

$$m_{oxAA} = \frac{ATP_{req}}{\frac{CL_q}{1 - CL_q} ATPstoich_{FA} + ATPstoich_{AA}}, \quad \text{(A16)}$$

where m_{oxAA} represents the rate of amino acid oxidation, m_{oxFA} represents the rate of fatty acid beta-oxidation, and $ATPstoich_{AA}$ and $ATPstoich_{FA}$ represent the profile-dependent

ATP yield of the catabolic reactions (in mol ATP/g fuel) for amino acids and fatty acids, respectively.

The value of m_{oxAA} therefore provides an estimate of the required mass of amino acids to be oxidized. On the other hand, it is important to ensure that the consumption of amino acids complies with the minimum protein requirements (see below Appendix A.7. Nitrogen metabolism—amino acids and proteins). As such, the actual value used is calculated as such:

$$m_{oxAA} = \max(\frac{ATP_{req}}{\frac{CL_q}{1-CL_q} ATPstoich_{FA} + ATPstoich_{AA}}, \text{min\_AA\_loss}), \quad \text{(A17)}$$

where min_AA_loss is calculated as detailed below (in the Appendix A.7. Nitrogen metabolism—amino acids and proteins).

The actual rate of required fatty acid beta-oxidation is then calculated as:

$$m_{oxFA} = \frac{ATP_{req} - ATP_{AA\_ox}}{ATPstoich_{FA}}, \quad \text{(A18)}$$

where ATP_{AA_ox} represents the ATP generated from amino acid oxidation.

Appendix A.7. Nitrogen Metabolism—Amino Acids and Proteins

Perhaps the most important aspects of FEEDNETICS have to do with how it deals with nitrogen metabolism (namely, protein synthesis and degradation, as well as amino acid interconversion and oxidation). The current implementation of protein synthesis and degradation processes are heavily inspired by the ones described in the models developed by Conceição et al. [6] and Bar et al. [23,27,44], and the mainstream literature on amino acid metabolism in fish [45–47] and other vertebrates [48,49]. It assumes a fixed (species-specific) amino acid profile (i.e., the rates of accumulation and loss of amino acids from the protein pool relative to each other are constant).

Appendix A.7.1. Protein Synthesis

Regarding protein synthesis, first, a maximum rate of protein synthesis (max_$prot_{synth}$) is calculated using a simplified version of Bar et al. [23,27,44] protein synthesis model which assumes a fixed amino acid profile for the synthesized protein:

$$\text{max\_}prot_{synth} = \min\left(kRNA \times vs_T \times Cs \times protein_{total}, \text{lim\_}prot_{synth}\right), \quad \text{(A19)}$$

where lim_$prot_{synth}$ represents the maximum rate of protein synthesis in terms of available substrate (i.e., free amino acids), $kRNA$ represents translation rate, vs_T represents the temperature effect on protein synthesis, and Cs represents the transcription rate (i.e., quantity of RNA present per gram of protein).

The calculation of the temperature effect on protein synthesis (vs_T) is straightforward, since it assumes a linear model:

$$vs_T = temperature_{effect} \times T, \quad \text{(A20)}$$

where $temperature_{effect}$ is a scalar parameter and T is the current temperature (in °C).

The calculation of $kRNA$ aims to emulate the effects of ribosome occupancy on the limitation of protein synthesis. First, ribosomes are assumed to exist in two states: occupied ($ribo_{occupied}$) or unoccupied ($ribo_{unoccupied}$). Then, the transitions between the two states are calculated as such:

$$ribo_{activation} = k_{ribo} \times ribo_{occupied} \times valve_{activation}, \quad \text{(A21)}$$

$$ribo_{deactivation} = k_{ribo} \times ribo_{unoccupied} \times valve_{deactivation}, \quad \text{(A22)}$$

$$\frac{d}{dt}\left(ribo_{occupied}\right) = ribo_{deactivation} - ribo_{activation}, \tag{A23}$$

$$\frac{d}{dt}\left(ribo_{unoccupied}\right) = ribo_{activation} - ribo_{deactivation}, \tag{A24}$$

where k_{ribo} is a time constant, while $valve_{activation}$ and $valve_{deactivation}$ are values between 0 and 1 that control the balance between the two states. $valve_{activation}$ is close to 1 if the fat levels are above the reference values and if free amino acids are also above their reference values (and closer to 0 otherwise), while $valve_{deactivation}$ is closer to 1 the closer actual protein synthesis rate approaches the maximum protein synthesis rate (serving as a negative feedback regulation signal).

Total ribosome activity ($ribo_{act}$; as a value between 0 and 1) is then estimated and used to interpolate (linearly, on a log-scale) between two constants ($kRNA_{min}$ and $kRNA_{max}$), as such:

$$ribo_{act} = \frac{ribo_{unoccupied}}{ribo_{occupied} + ribo_{unoccupied}}, \tag{A25}$$

$$kRNA = e^{(1-ribo_{act})\times\ln(kRNA_{min})+ribo_{act}\times\ln(kRNA_{max})}. \tag{A26}$$

After determining the maximum rate of protein synthesis, two valve values (i.e., values between 0 and 1) are calculated: $prot_{synt_{regulator}}$ and $AA_{synt_{valv}}$. The first one suppresses protein synthesis in situations of prolonged fasting and whenever lipid levels are abnormally low (thus inducing an increase in de novo lipid synthesis):

$$prot_{synt_{regulator}} = 0.05 + 0.95 \times \min(1, \max(0, CL_q + fed - starving)), \tag{A27}$$

while the second one suppresses protein synthesis according to how low the free amino acid levels are (according to their set reference levels):

$$AA_{synt_{valv}} = \min\left(\frac{1}{1 + \frac{1}{\left(\frac{AA_{free}}{refAA_{free}}\right)^{AA_{synt_{beta}}}}}\right), \tag{A28}$$

where min() is a function that accepts a vector as input and returns a scalar (the minimum) as output, AA_{free} is a vector containing the free amount of each amino acid, $refAA_{free}$ is a vector containing the reference values for the free amount of each amino acid, and $AA_{synt_{beta}}$ is an exponent that modifies the valve behavior.

The actual rate of protein synthesis is then calculated as a product of these factors, being afterwards converted into amino acids equivalents using a fixed species-dependent amino acid profile:

$$Aprot_{synth} = \max\_prot_{synth} \times prot_{synt_{regulator}} \times AA_{synt_{valv}}. \tag{A29}$$

Appendix A.7.2. Protein Degradation

Regarding protein degradation, it is also assumed to follow a fixed amino acid profile (i.e., when protein is converted into amino acids, it follows the same amino acid profile as used for amino acid synthesis to ensure that the total amino acid profile of the fish approximates the set profile). First, the upper bound for protein degradation ($max_{prot_{deg}}$) is calculated and assumed to be proportional to the total amount of protein:

$$max_{prot_{deg}} = k_{deg} \times deg_{temp_{factor}} \times protein_{total}, \tag{A30}$$

where k_{deg} represents a scaling factor and $deg_{temp_{factor}}$ represents the quadratic effect of temperature on protein degradation:

$$deg_{temp_{factor}} = Vd_b + Vd_m \times \left(temperature - T_{optimal}\right)^2, \tag{A31}$$

where Vd_b, Vd_m, and $T_{optimal}$ control the shape of the parabola.

The lower bound for protein degradation ($min_{prot_{deg}}$) is calculated assuming that it is proportional to the protein requirements:

$$min\_prot_{deg} = prot\_deg\_min\_factor \times \text{min\_AA\_loss}, \tag{A32}$$

where min_AA_loss is defined below by equation (A35).

Finally, a valve value is calculated and used to interpolate between the two bounds, based on the available amounts of free amino acids:

$$AA_{deg_{valv}} = \min\left(\frac{1}{1+\left(\frac{AA_{free}}{refAA_{free}}\right)^{AA\_deg\_beta2}}\right)^{AA\_deg\_beta1}, \tag{A33}$$

where min() is a function that accepts a vector as input and returns a scalar (the minimum) as output, while AA_deg_beta1 and AA_deg_beta2 control the shape of the response. The actual rate of protein degradation ($prot_{deg}$) is then calculated, being afterwards converted into amino acid equivalents using a fixed species-dependent amino acid profile:

$$prot_{deg} = min_{prot_{deg}} + \left(max_{prot_{deg}} - min_{prot_{deg}}\right) \times AA_{deg_{valv}}. \tag{A34}$$

Appendix A.7.3. Amino Acid Oxidation

Regarding amino acid oxidation, it is assumed that there is a basal amino acid oxidation rate that follows the same logic as the "protein fasting maintenance" calculations of Lupatsch [41]:

$$\text{min\_AA\_loss} = \text{req\_prot\_a} \times e^{req\_prot\_b \times temperature} \times \left(\frac{bw}{1000}\right)^{req\_prot\_c}, \tag{A35}$$

Where req_prot_a, req_prot_b, and req_prot_c are parameters describing the effects of temperature and body weight on protein fasting maintenance.

The total amount of amino acids oxidized (m_{oxAA}) is then calculated, as detailed in the Appendix A.6. Energetic model above, and distributed by the different amino acids such that those that exist in higher amounts (in relation to their reference values) are preferably used. First, two auxiliary vector quantities are calculated:

$$AA\_free\_max\_norm = \frac{AA_{free}}{max\_refAA_{free}}, \tag{A36}$$

$$AA\_free\_min\_norm = \frac{AA_{free}}{min\_refAA_{free}}, \tag{A37}$$

and then they are combined as such:

$$AA_{ox_{valv}} = \min\left(\frac{AA\_free\_max\_norm}{\sum AA\_free\_max\_norm, AA\_free\_min\_norm}, \frac{AA\_free\_min\_norm}{\sum AA\_free\_min\_norm}\right), \tag{A38}$$

and normalized to ensure that resulting vector is a partition:

$$AA_{ox_{weights}} = \frac{AA_{ox_{valv}}}{\sum AA_{ox_{valv}}}. \tag{A39}$$

The resulting rate of amino acid oxidation is then calculated as:

$$AA\_ox_{rate} = AA_{ox_{weights}} \times m_{oxAA}. \tag{A40}$$

Appendix A.7.4. Gluconeogenesis

Regarding gluconeogenesis (i.e., synthesis of glucose from amino acids), first, a maximum overall rate of glucose synthesis from amino acids ($V_{\max gluconeo}$) is calculated, assuming an exponential effect of temperature and a linear effect of body weight, and having into account possible substrate limitations:

$$\begin{aligned} V_{\max gluconeo} = \; & \min\Big(AA_{gluco_{weights}} \times stoich_{glucose_{AA \to glucose}} \times a_{gluconeo} \times bw \\ & \times e^{b \times temperature}, stoich_{glucose_{AA \to glucose}} \times \tfrac{AA_{free}}{timestep}, \Big) \end{aligned} \tag{A41}$$

where $AA_{gluco_{weights}}$ is a partitioning vector (similar to $AA_{ox_{weights}}$) that ensures that gluconeogenesis is suppressed, unless there are abundant amino acids, and $a_{gluconeo}$ and b are parameters.

The actual rate of gluconeogenesis ($V_{gluconeo}$) is then calculated by having into account an inhibiting effect of glucose on its synthesis (negative feedback) and ensuring that abundant amino acids are given preference as substrates:

$$V_{gluconeo} = V_{\max gluconeo} \times \frac{1}{1 + \frac{glucose}{ref_{glucose}}} \times AA_{gluco_{weights}}. \tag{A42}$$

Appendix A.7.5. Synthesis of Non-Essential Amino Acids

Regarding the synthesis of non-essential amino acids, reactions are simulated with stoichiometries recalculated to assume glucose as a source of carbon and ensuring that the nitrogen balance is not violated (i.e., the amount of nitrogen incorporated into synthesized amino acids is upper bounded by the amount of nitrogen lost during that timestep, from, for example, amino acid oxidation). For the stoichiometries, it uses mostly the work of Olsen [50] as an information source. Here, reactions are separated into "ATP-producing" and "ATP-consuming", but are calculated in a similar way, using matrix operations. The formulas ensure that the reactions are stimulated by glucose (substrate) and suppressed by the products of the reactions. In a similar way, reaction rates for the conversion of methionine in cysteine and of phenylalanine in tyrosine are also calculated.

Appendix A.8. Carbon Metabolism—Glucose, Glycogen, and Fatty Acids

Besides nitrogen metabolism, it is important to also accurately model central (carbon) metabolism, which encompasses the metabolism of carbohydrates (glucose and glycogen) and lipids (as fatty-acid equivalents).

Appendix A.8.1. Glucose Oxidation

Regarding glucose oxidation, first, a maximum value ($V_{\max glucox}$) is calculated, having into account substrate availability:

$$V_{\max glucox} = \min\left(a_{glucox} \times bw \times e^{b \times temperature}, \frac{glucose}{timestep}\right). \tag{A43}$$

Then, two valves are applied to ensure that glucose oxidation is suppressed whenever glucose levels become particularly low and whenever energy availability is too high.

Appendix A.8.2. Glycogenesis and Glycogenolysis

Regarding glycogenesis (i.e., synthesis of glycogen from glucose) and glycogenolysis (i.e., degradation of glycogen into glucose), they are both calculated based on a common maximum rate ($V_{\max glycogen}$) proportional to the total amount of body protein:

$$V_{\max glycogen} = constant \times protein_{total}. \tag{A44}$$

This value is then used to calculate the reaction rate, assuming Michaelis-Menten-like kinetics, and taking, in the case of glycogenesis, glucose as a stimulant and glycogen as an inhibitor, and, in the case of glycogenolysis, glycogen as a stimulant and glucose as an inhibitor.

Appendix A.8.3. Lipogenesis

Regarding lipogenesis (i.e., synthesis of lipids from glucose), first, a maximum lipogenesis rate ($V_{\max lipogen}$) is calculated:

$$V_{\max lipogen} = \min\left(a_{lipogen} \times bw \times e^{b \times temperature}, \frac{glucose}{timestep}\right). \tag{A45}$$

Then, additional factors are added to ensure that lipogenesis is suppressed when glucose becomes unavailable and when lipid levels become abnormally high.

Appendix A.8.4. Beta-Oxidation

Regarding the rate of beta-oxidation (i.e., oxidation of fatty acids), it is defined based on the m_{oxFA} value, calculated as detailed in the above Appendix A.6. Energetic model, multiplied by the absolute availability of the different fatty acids (i.e., it assumes a pure dilution model: all fatty acids enter the body in the proportion they exist in the diet and they are lost in the proportion they exist in the body):

$$TAG_{betox} = TAG_{ox_{weights}} \times m_{oxFA}. \tag{A46}$$

Appendix B.

In this appendix, we present some additional use cases that illustrate the potential application of the FEEDNETICS model to support the industry to: (i) evaluate the impact of different commercial feeds; (ii) evaluate the impact of different temperature profiles; and (iii) predict the long-term effects of marginal changes in diet digestibility.

Use case 1: Evaluate the impact of different commercial feeds on trout production performance

Use case 1 illustrates how FEEDNETICS can be used to evaluate the impact of two high energy feeds on rainbow trout production performance, and to quantify savings on feed obtained by the best performing scenario. This use case was set up for a generic RAS farm and two commercial feeds used by the rainbow trout RAS industry were considered. The key results and outcomes are presented in Figure A4 and are only applicable to the input data specified. Changes in rearing temperature, feed properties, feeding rates, and target harvest weight will alter results and main outcomes. In this case, the main outcomes identify Aquafeed 1 as leading to an overall better performance, including a significant decrease in the total N and P wastes, as well significant better economic conversion (Figure A4). This information is highly relevant for optimizing RAS production as it implies a balance between fish growth, feed efficiency, water quality, and profitability. Evaluating feeding efficiency indicators is very important, not only for feed conversion economics, but also for planning and managing the biofilter.

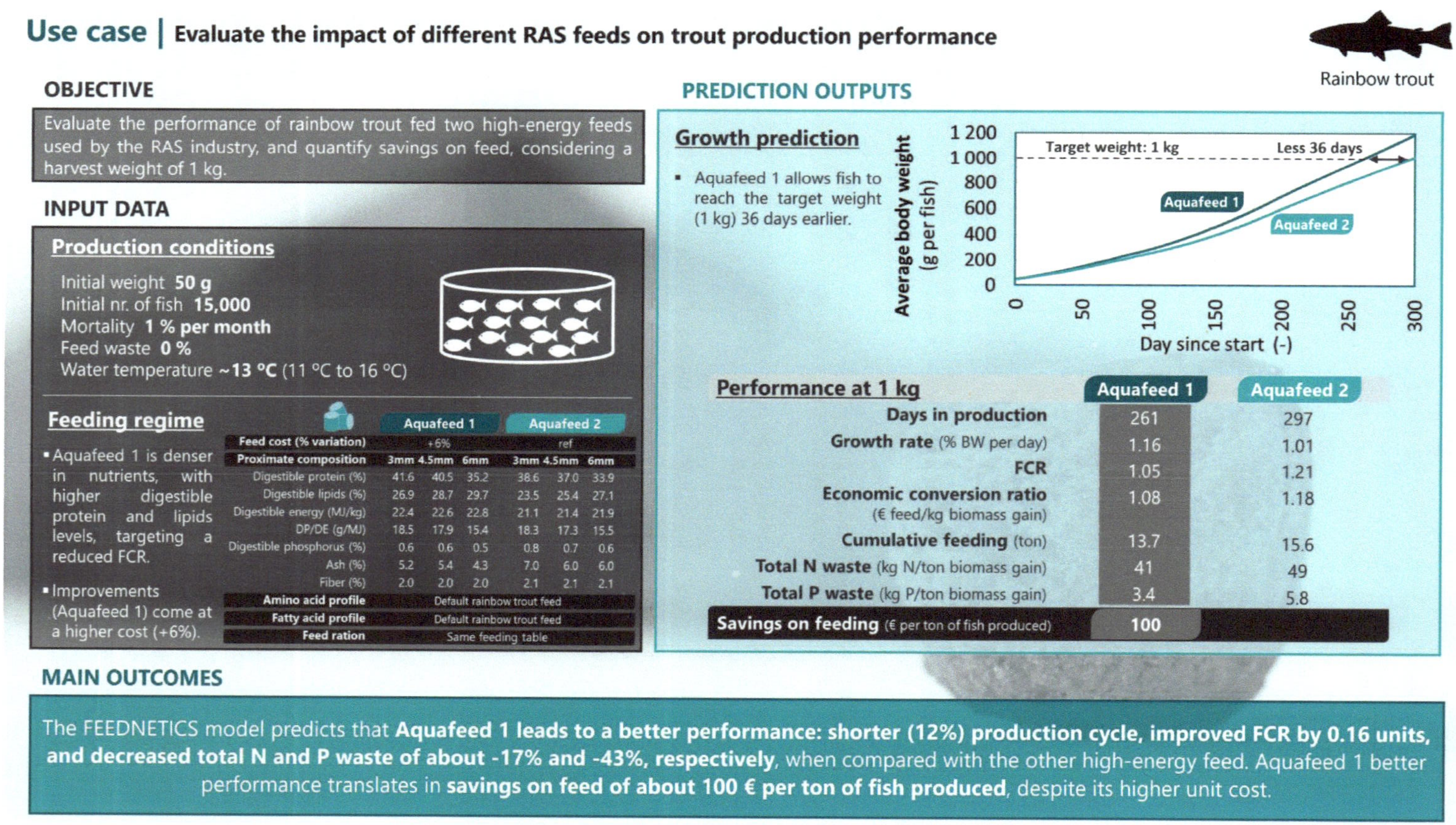

Figure A4. Use case 1: Evaluate the impact of different RAS feeds on trout production performance. All results are FEEDNETICS model predictions.

Use case 2: Evaluate the impact of different temperature profiles on post-smolt production performance

Use case 2 illustrates how FEEDNETICS can be used to evaluate post-smolt salmon performance produced in sites with different temperature profiles, including controlled temperature conditions, as can be found in recirculating aquaculture systems (RAS). This is a relevant topic for the salmon industry, as post-smolt production (up to 1 kg) has become more common to shorten time in sea cages [51]. In particular, this use case was set up to reproduce the Atlantic salmon (*Salmo salar*) growth data obtained from a semi-commercial scale research work [52] upscaled to a 1 million post-smolt operation, considering two different temperature scenarios: (i) yearly temperature profile at Ålesund, Norway (data from seatemperature.org), and (ii) a constant average temperature of 13.4 °C to represent the conditions in a RAS system and match the post-smolt rearing conditions described by Crouse et al. [52] for Cohort B2. The key results and outcomes are presented in Figure A5.

In addition to evaluating different temperature profiles, this use case can be further developed and customized considering:

- Different feeds and feeding rates used (or planned) for RAS and cage operations;
- Production costs, besides feed costs, and the fish price, for RAS and cage operations;
- Different mortality rates for RAS and cage operations.

These different data inputs will impact not only the predicted fish performance but also the model output indicators for production profitability. They can be inserted into FEEDNETICS according to each user's specific farming conditions in order to perform a cost-benefit analysis on producing post-smolts in RAS up to larger sizes before moving fish to sea cages.

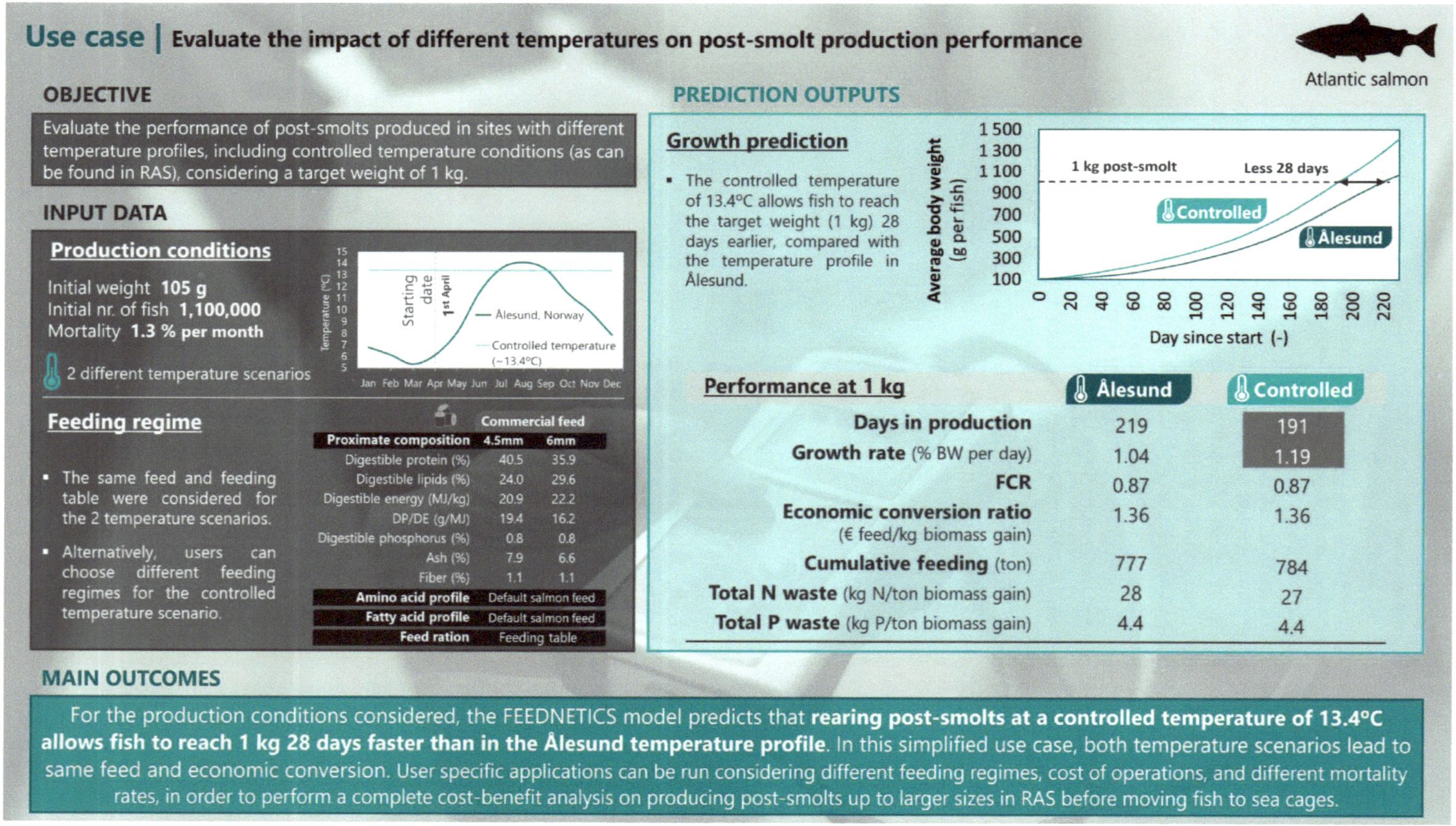

Figure A5. Use case 2: Evaluate the impact of different temperature profiles on post-smolt production performance. All results are model predictions.

Use case 3: Predict the long-term effects of marginal changes in diet digestibility on bream production performance

Use case 3 illustrates how FEEDNETICS can be used to complement trials that test, for example, the effects of digestibility enhancers such as lipid emulsifiers, phytogenetic additives, gut health promoters, and feed enzymes. Due to the particularity of such trials, where most rearing conditions are controlled to be within optimal levels, and the fact that the effects of improved digestibility are not non-linear over the long-term, translating the better performance induced by digestibility enhancers to the commercial scale cannot be simply done by linear upscaling. In this regard, the use of nutrient-based models, such as FEEDNETICS, may be useful in allowing to extrapolate the long-term impact of digestibility enhancers on commercial-scale settings.

This use case was set up considering a feed additive inclusion at low level (0.2%), with minor revisions to feed formulations in order to maintain feed price. The main differences considered in the diet with the additive, compared to the baseline diet, were a marginal increase of approximately 1.5% in the apparent digestibility of crude protein, crude lipids, and gross energy. The production conditions followed typical seabream cage production settings, reared in a warm water temperature profile from Madeira Island in Portugal. The main results and outcomes are presented in Figure A6.

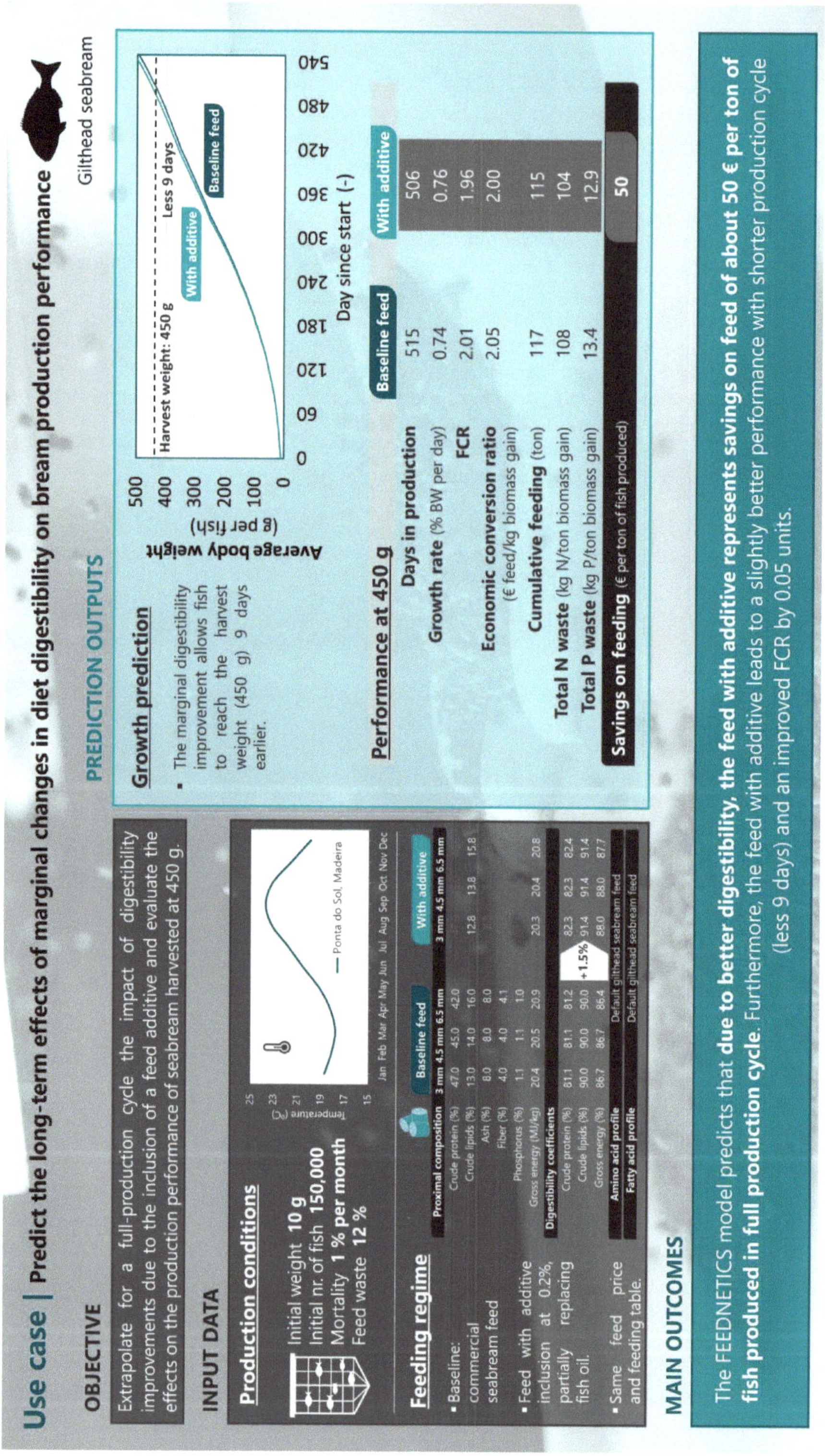

Figure A6. Use case 3: Predict the long-term effects of marginal changes in diet digestibility on bream production performance. All results are model predictions.

Use case 3: Predict the long-term effects of marginal changes in diet digestibility on bream production performance

Use case 3 illustrates how FEEDNETICS can be used to complement trials that test, for example, the effects of digestibility enhancers such as lipid emulsifiers, phytogenetic additives, gut health promoters, and feed enzymes. Due to the particularity of such trials, where most rearing conditions are controlled to be within optimal levels, and the fact that the effects of improved digestibility are not non-linear over the long-term, translating the better performance induced by digestibility enhancers to the commercial scale cannot be simply done by linear upscaling. In this regard, the use of nutrient-based models, such as FEEDNETICS, may be useful in allowing to extrapolate the long-term impact of digestibility enhancers on commercial-scale settings.

This use case was set up considering a feed additive inclusion at low level (0.2%), with minor revisions to feed formulations in order to maintain feed price. The main differences considered in the diet with the additive, compared to the baseline diet, were a marginal increase of approximately 1.5% in the apparent digestibility of crude protein, crude lipids, and gross energy. The production conditions followed typical seabream cage production settings, reared in a warm water temperature profile from Madeira Island in Portugal. The main results and outcomes are presented in Figure A6.

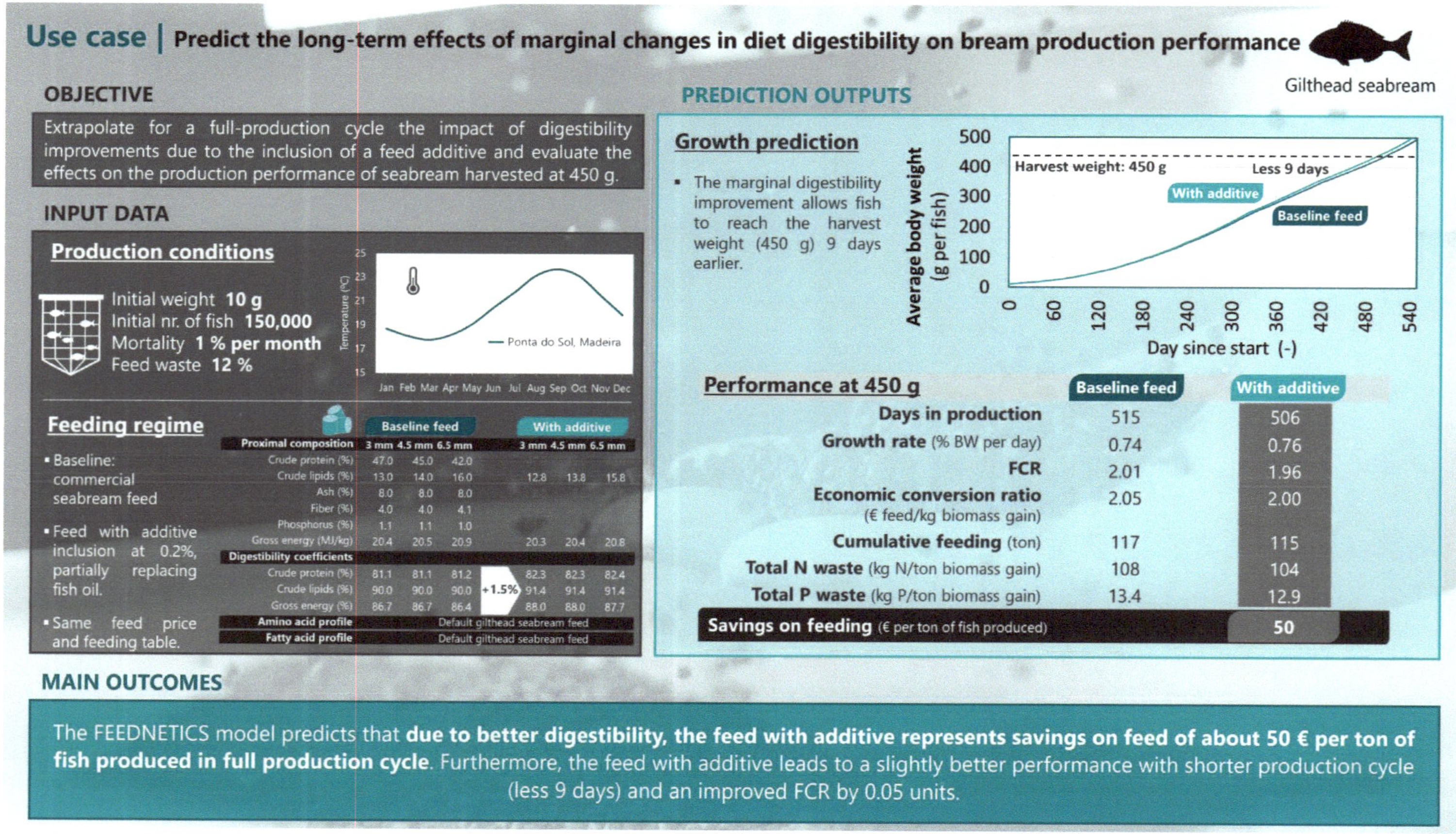

Figure A6. Use case 3: Predict the long-term effects of marginal changes in diet digestibility on bream production performance. All results are model predictions.

References

1. Føre, M.; Frank, K.; Norton, T.; Svendsen, E.; Alfredsen, J.A.; Dempster, T.; Eguiraun, H.; Watson, W.; Stahl, A.; Sunde, L.M.; et al. Precision fish farming: A new framework to improve production in aquaculture. *Biosyst. Eng.* **2018**, *173*, 176–193. [CrossRef]
2. Zhou, C.; Xu, D.; Lin, K.; Sun, C.; Yang, X. Intelligent feeding control methods in aquaculture with an emphasis on fish: A review. *Rev. Aquac.* **2018**, *10*, 975–993. [CrossRef]
3. Machiels, M.A.M.; Henken, A.M. A dynamic simulation model for growth of the African catfish, *Clarias gariepinus* (Burchell 1822): I. Effect of feeding level on growth and energy metabolism. *Aquaculture* **1986**, *56*, 29–52. [CrossRef]
4. Machiels, M.A.M.; Henken, A.M. A dynamic simulation model for growth of the African catfish, *Clarias gariepinus* (Burchell 1822): II. Effect of feed composition on growth and energy metabolism. *Aquaculture* **1987**, *60*, 33–53. [CrossRef]
5. Machiels, M.A.M.; Van Dam, A.A. A dynamic simulation model for growth of the African catfish, *Clarias gariepinus* (Burchell 1822): III. The effect of body composition on growth and feed intake. *Aquaculture* **1987**, *60*, 55–71. [CrossRef]
6. Conceição, L.E.C.; Verreth, J.A.J.; Verstegen, M.W.A.; Huisman, E.A. A preliminary model for dynamic simulation of growth in fish larvae: Application to the African catfish (*Clarias gariepinus*) and turbot (*Scophthalmus maximus*). *Aquaculture* **1998**, *163*, 215–235. [CrossRef]
7. Dumas, A.; France, J.; Bureau, D. Modelling growth and body composition in fish nutrition: Where have we been and where are we going? *Aquac. Res.* **2010**, *41*, 161–181. [CrossRef]
8. Lugert, V.; Thaller, G.; Tetens, J.; Schulz, C.; Krieter, J. A review on fish growth calculation: Multiple functions in fish production and their specific application. *Rev. Aquac.* **2016**, *8*, 30–42. [CrossRef]
9. Chary, K.; Brigolin, D.; Callier, M.D. Farm-scale models in fish aquaculture-An overview of methods and applications. *Rev. Aquac.* **2022**, *14*, 2122–2157. [CrossRef]
10. Springborn, R.R.; Jensen, A.L.; Chang, W.Y.B.; Engle, C. Optimum harvest time in aquaculture: An application of economic principles to a Nile tilapia, *Oreochromis niloticus* (L.), growth model. *Aquac. Res.* **1992**, *23*, 639–647. [CrossRef]
11. Mallet, J.P.; Charles, S.; Persat, H.; Auger, P. Growth modelling in accordance with daily water temperature in European grayling (*Thymallus thymallus* L.). *Can. J. Fish. Aquat. Sci.* **1999**, *56*, 994–1000. [CrossRef]
12. De Graaf, G.; Prein, M. Fitting growth with the von Bertalanffy growth function: A comparison of three approaches of multivariate analysis of fish growth in aquaculture experiments. *Aquac. Res.* **2005**, *36*, 100–109. [CrossRef]
13. Dumas, A.; France, J.; Bureau, D.P. Evidence of three growth stanzas in rainbow trout (*Oncorhynchus mykiss*) across life stages and adaptation of the thermal-unit growth coefficient. *Aquaculture* **2007**, *267*, 139–146. [CrossRef]
14. Baer, A.; Schulz, C.; Traulsen, I.; Krieter, J. Analysing the growth of turbot (*Psetta maxima*) in a commercial recirculation system with the use of three different growth models. *Aquac. Int.* **2011**, *19*, 497–511. [CrossRef]
15. Mayer, P.; Estruch, V.D.; Jover, M. A two-stage growth model for gilthead sea bream (*Sparus aurata*) based on the thermal growth coefficient. *Aquaculture* **2012**, *358*, 6–13. [CrossRef]
16. Brigolin, D.; Pastres, R.; Tomassetti, P.; Porrello, S. Modelling the biomass yield and the impact of seabream mariculture in the Adriatic and Tyrrhenian Seas (Italy). *Aquac. Int.* **2010**, *18*, 149–163. [CrossRef]
17. Føre, M.; Alver, M.; Alfredsen, J.A.; Marafioti, G.; Senneset, G.; Birkevold, J.; Willumsen, F.V.; Lange, G.; Espmark, Å.; Terjesen, B.F. Modelling growth performance and feeding behaviour of Atlantic salmon (*Salmo salar* L.) in commercial-size aquaculture net pens: Model details and validation through full-scale experiments. *Aquaculture* **2016**, *464*, 268–278. [CrossRef]
18. Nobre, A.M.; Valente, L.M.; Conceição, L.; Severino, R.; Lupatsch, I. A bioenergetic and protein flux model to simulate fish growth in commercial farms: Application to the gilthead seabream. *Aquac. Eng.* **2019**, *84*, 12–22. [CrossRef]
19. Hua, K.; Birkett, S.; De Lange, C.F.M.; Bureau, D.P. Adaptation of a non-ruminant nutrient-based growth model to rainbow trout (*Oncorhynchus mykiss Walbaum*). *J. Agric. Sci.* **2010**, *148*, 17–29. [CrossRef]
20. Cho, C.Y.; Bureau, D.P. Development of bioenergetic models and the Fish-PrFEQ software to estimate production, feeding ration and waste output in aquaculture. *Aquat. Living Resour.* **1998**, *11*, 199–210. [CrossRef]
21. Chowdhury, M.K.; Siddiqui, S.; Hua, K.; Bureau, D.P. Bioenergetics-based factorial model to determine feed requirement and waste output of tilapia produced under commercial conditions. *Aquaculture* **2013**, *410*, 138–147. [CrossRef]
22. Stavrakidis-Zachou, O.; Papandroulakis, N.; Lika, K. A DEB model for European sea bass (*Dicentrarchus labrax*): Parameterisation and application in aquaculture. *J. Sea Res.* **2019**, *143*, 262–271. [CrossRef]
23. Bar, N.S.; Sigholt, T.; Shearer, K.D.; Krogdahl, Å. A dynamic model of nutrient pathways, growth, and body composition in fish. *Can. J. Fish. Aquat. Sci.* **2007**, *64*, 1669–1682. [CrossRef]
24. Weihe, R.; Dessen, J.E.; Arge, R.; Thomassen, M.S.; Hatlen, B.; Rørvik, K.A. Improving production efficiency of farmed Atlantic salmon (*Salmo salar* L.) by isoenergetic diets with increased dietary protein-to-lipid ratio. *Aquac. Res.* **2018**, *49*, 1441–1453. [CrossRef]
25. Peres, H.; Oliva-Teles, A. Influence of temperature on protein utilization in juvenile European seabass (*Dicentrarchus labrax*). *Aquaculture* **1999**, *170*, 337–348. [CrossRef]
26. Wang, L.; Zhang, W.; Gladstone, S.; Ng, W.K.; Zhang, J.; Shao, Q. Effects of isoenergetic diets with varying protein and lipid levels on the growth, feed utilization, metabolic enzymes activities, antioxidative status and serum biochemical parameters of black sea bream (*Acanthopagrus schlegelii*). *Aquaculture* **2019**, *513*, 734397. [CrossRef]
27. Bar, N.S.; Radde, N. Long-term prediction of fish growth under varying ambient temperature using a multiscale dynamic model. *BMC Syst. Biol.* **2009**, *3*, 107. [CrossRef]

28. Rønnestad, I.; Conceição, L.E.C. Artemia protein is processed very fast in *Solea senegalensis* larvae: A dynamic simulation model. *Aquaculture* **2012**, *350*, 154–161. [CrossRef]
29. Wickham, H. Tidy Data. *J. Stat. Softw.* **2014**, *59*, 1–23. [CrossRef]
30. Hansen, N. The CMA evolution strategy: A tutorial. *arXiv* **2016**, arXiv:1604.00772. [CrossRef]
31. Dos Santos, V.B.; Silva, V.V.; de Almeida, M.V.; Mareco, E.A.; Salomão, R.A. Performance of Nile tilapia *Oreochromis niloticus* strains in Brazil: A comparison with Philippine strain. *J. Appl. Anim. Res.* **2019**, *47*, 72–78. [CrossRef]
32. Farinha, A.P.; Schrama, D.; Silva, T.; Conceição, L.E.; Colen, R.; Engrola, S.; Rodrigues, P.; Cerqueira, M. Evaluating the impact of methionine-enriched diets in the liver of European seabass through label-free shotgun proteomics. *J. Proteom.* **2021**, *232*, 104047. [CrossRef]
33. Windell, J.T.; Foltz, J.W.; Sarokon, J.A. Effect of fish size, temperature, and amount fed on nutrient digestibility of a pelleted diet by rainbow trout, *Salmo gairdneri*. *Trans. Am. Fish. Soc.* **1978**, *107*, 613–616. [CrossRef]
34. Watanabe, T.; Takeuchi, T.; Satoh, S.; Kiron, V. Digestible crude protein contents in various feedstuffs determined with four freshwater fish species. *Fish. Sci.* **1996**, *62*, 278–282. [CrossRef]
35. Kim, J.D.; Breque, J.; Kaushik, S.J. Apparent digestibilities of feed components from fish meal or plant protein based diets in common carp as affected by water temperature. *Aquat. Living Resour.* **1998**, *11*, 269–272. [CrossRef]
36. Olsen, R.E.; Ringø, E. The influence of temperature on the apparent nutrient and fatty acid digestibility of Arctic charr, *Salvelinus alpinus* L. *Aquac. Res.* **1998**, *29*, 695–701. [CrossRef]
37. Fernández, F.; Miquel, A.G.; Guinea, J.; Martínez, R. Digestion and digestibility in gilthead sea bream (*Sparus aurata*): The effect of diet composition and ration size. *Aquaculture* **1998**, *166*, 67–84. [CrossRef]
38. Fontaine, P.; Gardeur, J.N.; Kestemont, P.; Georges, A. Influence of feeding level on growth, intraspecific weight variability and sexual growth dimorphism of Eurasian perch *Perca fluviatilis* L. reared in a recirculation system. *Aquaculture* **1997**, *157*, 1–9. [CrossRef]
39. Irwin, S.; O'halloran, J.; FitzGerald, R.D. Stocking density, growth and growth variation in juvenile turbot, *Scophthalmus maximus* (Rafinesque). *Aquaculture* **1999**, *178*, 77–88. [CrossRef]
40. Aijun, M.; Chao, C.; Jilin, L.; Siqing, C.; Zhimeng, Z.; Yingeng, W. Turbot *Scophthalmus maximus*: Stocking density on growth, pigmentation and feed conversion. *Chin. J. Oceanol. Limnol.* **2006**, *24*, 307–312. [CrossRef]
41. Lupatsch, I. Factorial approach to determining energy and protein requirements of gilthead seabream (*Sparus aurata*) for optimal efficiency of production. Ph.D. Thesis, Universitäts-und Landesbibliothek, Bonn, Germany, 2004. Available online: https://hdl.handle.net/20.500.11811/2005 (accessed on 6 January 2023).
42. Breck, J.E. Body composition in fishes: Body size matters. *Aquaculture* **2014**, *433*, 40–49. [CrossRef]
43. Norin, T.; Clark, T.D. Measurement and relevance of maximum metabolic rate in fishes. *J. Fish Biol.* **2016**, *88*, 122–151. [CrossRef] [PubMed]
44. Bar, N.S. Dynamic model of fish growth. Ph.D. Thesis, Norges teknisk-naturvitenskapelige universitet (NTNU), Trondheim, Norway, 2007.
45. Wilson, R.P. Amino acid requirements of finfish. *Amino Acids Farm Anim. Nutr.* **1994**, 377–399.
46. Mambrini, M.; Kaushik, S.J. Indispensable amino acid requirements of fish: Correspondence between quantitative data and amino acid profiles of tissue proteins. *J. Appl. Ichthyol.* **1995**, *11*, 240–247. [CrossRef]
47. Conceição, L.E.C.; Grasdalen, H.; Rønnestad, I. Amino acid requirements of fish larvae and post-larvae: New tools and recent findings. *Aquaculture* **2003**, *227*, 221–232. [CrossRef]
48. Reeds, P.J. Regulation of protein turnover. In *Animal Growth Regulation*; Campion, D.R., Hausman, G.J., Martin, R.J., Eds.; Springer: Boston, MA, USA, 1989; pp. 183–210. [CrossRef]
49. Simon, O. Metabolism of proteins and amino acids. In *Protein Metabolism in Farm Animals*; Bock, H.D., Eggum, B.O., Low, A.G., Simon, O., Zebrowska, T., Eds.; Oxford University Press: Oxford, UK, 1989; pp. 273–366.
50. Olsen, O.A.S. Structured modelling of fish physiology. Ph.D. Thesis, Norges Tekniske Høgskole (NTH), Trondheim, Norway, 1989.
51. Mowi, A.S.A. *Salmon Farming Industry Handbook*; Mowi A.S.A: Bergen, Norway, 2020.
52. Crouse, C.; Davidson, J.; May, T.; Summerfelt, S.; Good, C. Production of market-size European strain Atlantic salmon (*Salmo salar*) in land-based freshwater closed containment aquaculture systems. *Aquac. Eng.* **2021**, *92*, 102138. [CrossRef]

Journal of
*Marine Science
and Engineering*

Article

Salinity Tolerance and the Effect of Salinity and Algal Feed on the Demographics of Cultured Harpacticoid Copepods *Tisbe holothuriae* and *Tigriopus* sp. from the Messolonghi Lagoon (W. Greece)

George N. Hotos *, Evi Kourelea and Ioannis Fotodimas

Plankton Culture Laboratory, Department of Animal Production, Fisheries and Aquaculture, University of Patras, 30200 Messolonghi, Greece

* Correspondence: ghotos@upatras.gr; Tel.: +30-26310-58252

Abstract: The locally isolated harpacticoid copepods *Tigriopus* sp. and *Tisbe holothuriae* were subjected to salinity tolerance experimentation at salinities under and above of 40 ppt, and presented high halotolerances in *Tigriopus* LC50 (24 h) of 1 ± 4.43 ppt and 132 ± 5.35 ppt, respectively, and in *Tisbe* of 15 ± 2.41 ppt and 93 ± 3.23 ppt, respectively. *Tetraselmis suecica,* among other microalgal feeds (*Asteromonas gracilis, Rhodomonas salina*, *Dunaliella salina* and *Isochrysis galbana*), resulted in the higher production of nauplii in *Tigriopus* and *R. salina* and *D. salina* in *Tisbe* (also close to *T. suecica* in *Tigriopus*). The demographics (number of nauplii, egg sacs, completion of hatching) of both copepods, using combinations of salinities in the range of 22–60 ppt and *D. salina* and *R. salina* as feeds, exhibited almost the same preference for microalgae but were negatively affected by the salinity of 60 ppt. The present experiments showed that these local copepods that have extreme salinity tolerance and a wide preference for easily cultured microalgae can be used in ecological studies and for mass production as live feed in marine fish hatcheries.

Keywords: copepods; harpacticoid; *Tigriopus*; *Tisbe holothuriae*; salinity tolerance; microalgal feed

Citation: Hotos, G.N.; Kourelea, E.; Fotodimas, I. Salinity Tolerance and the Effect of Salinity and Algal Feed on the Demographics of Cultured Harpacticoid Copepods *Tisbe holothuriae* and *Tigriopus* sp. from the Messolonghi Lagoon (W. Greece). *J. Mar. Sci. Eng.* **2022**, *10*, 1663. https://doi.org/10.3390/jmse10111663

Academic Editor: Ana Catarina Matias

Received: 14 October 2022
Accepted: 1 November 2022
Published: 4 November 2022

1. Introduction

The rotifers (*Brachionus* spp.) and the nauplii of the crustacean *Artemia* sp. are the most widespread live foods for mass production of larvae in fish farms. However, the larvae of certain marine fish species such as those of the Lutjanidae and Serranidae families are not nutritionally covered by these foods and require the use of other live food that covers all of their metabolic needs [1]. In recent years, interest has focused on the nauplii of various species of copepods, mainly of the classes Calanoida and Harpacticoida, and although their use in commercial feed for fish larvae is currently very limited, they present many advantages, leaving many prospects for their wider use in the future [2]. With the use of copepods as live food, there is ground for the introduction of new species of fish into fish farms that until now was not possible to grow, since commercially available foods were not suitable even after enrichment to cover them nutritionally [3]. Many species of copepods are highly tolerant to fluctuations in salinity, temperature and other physicochemical environmental factors. However, their productivity can vary significantly at different values of these factors. When mass producing a species for use as live feed in aquaculture, it is important to ensure optimal conditions to achieve maximum productivity [4]. Copepods of the order Harpacticoida in their nauplii stages have a small size, high reproductive potential, rapid population growth and are nutritionally flexible and tolerant to a wide range of environmental factors, such as temperature and salinity [5,6], with the result that they present themselves advantageously as food for cultured fish larvae. Compared to other orders of copepods such as calanoids, they appear more advantageous for mass production since sustainable cultures of them can be achieved at high densities [7]. At the same time, harpacticoid copepods help keep fish larval rearing tanks clean by consuming the algae and food debris that adheres to their walls [8]. The conversion of organic molecules resulting from the catabolism of copepod's food into essential fatty acids (EFAs) makes them a more complete source of nutrients for fish larvae than more common foods used today such as *Artemia* or rotifers [9]. According to Camus and Zeng, [7] the content of chemical compounds in tissues is mainly determined by the species of microalgae that constitute their food. The ability of harpacticoid copepods to biosynthesize long-chain polyunsaturated fatty acids from saturated fatty acids gives them the adaptive capacity to cope with adverse situations related either to the reduced availability of nutrients in their environment, or to gradual environmental changes due to climate change [10]. Harpacticoid copepods, usually living as benthic organisms in coastal habitats, feed mainly on algae and organic particles [11]. Depending on the species and the availability of food, the harpacticoid species are found as epibenthic, endobenthic or mesobenthic organisms. Calliari et al. [12] studied the effect of salinity on the reproduction of two species of copepods of the genus *Acartia* (*A. tonsa* and *A. clausi*). In both species, egg hatching success was low at the lowest salinities tested. Furthermore, the variable size of the eggs between the different salinity values suggests that the egg shell is permeable to water and thus the embryos at the different salinity values faced different osmotic pressures, whereas the higher embryonic mortality at low salinities for *A. clausi* suggests that embryos are more sensitive than adults to salinity reduction. The narrow range of tolerance to salinity changes during early nauplii stages is a common pattern in several copepod species [13,14]. In a study [15] on the effect of different salinity values on the copepod *Pseudodiaptomus pelagicus*, fecundity was indirectly affected by the change in salinity, as fewer eggs were counted in the egg sacs of females reared at lower and higher salinities versus optimal salinity. In another study [16], in individuals of the copepod *Tisbe holothuriae*, the deviation from the optimal salinity value (38 ppt) resulted in a decrease in productivity (a lower number of nauplii from each egg sac), as well as an increase in mortality.

This research work seeks to further contribute to the required knowledge for the use of copepods as food in larval stages of cultured marine fish, against common live foods used today. The interest in the mass cultivation of copepods with the aim of using them as food for the larval stages of fish farming, but also for ornamental fish in aquaria, is growing both in Greece and in other countries in recent years. The mass production of copepods with high nutritional value could lead to the reduction of production costs in fish farming units. In addition, the cultivation of specific copepod species as food could contribute to the production of aquaculture fish of higher nutritional value. In order to achieve this, however, optimal physicochemical growth conditions are required, which are determined not only by the species of copepods, but in some cases also by the particular characteristics of the populations from which they originate. The purpose of the present study is to investigate the possibility of cultivating two genera of copepods that are widely found in Greece and more generally in the Mediterranean, *Tisbe* and *Tigriopus*, as well as to investigate the effect of four levels of salinity and two types of microalgae (as food) on their development and production. In addition, investigating their tolerance to extreme salinities will add insight into the survival of these organisms under potential extreme climate change impacts on coastal aquatic ecosystems.

2. Materials and Methods

Tigriopus sp. and *Tisbe holothuriae* (Figures 1 and 2 and Videos S1 and S2, respectively) originated from a screening survey of the saline waters (35–40 ppt) of the Messolonghi lagoon (38°20′05.16″ N, 21°25′28.51″ E) on the Ionian coastline of Greece [17]. The experimentation was comprised of 3 experimental series accomplished separately. The first experiment concerns the salinity tolerance of the copepods. The second experiment concerns the influence of several algal feeds on progeny production. The third experiment concerns the combined influence of salinity and algal feed on the demographics of the copepods. For simplicity, the three experiments are named hereafter as: "Experiment on salinity tolerance", "Experiment on feed influence" and "Experiment on salinity-feed demographics", respectively.

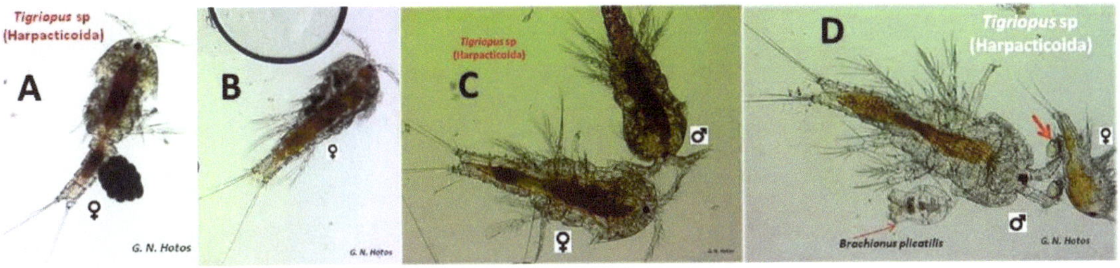

Figure 1. (**A**) *Tigriopus* sp. female with egg sac, (**B**) female without egg sac, (**C**) Female and male, (**D**) Grasping of a young female by a fully grown male; arrow indicates the characteristic lumps on the first pair of antennas of the male.

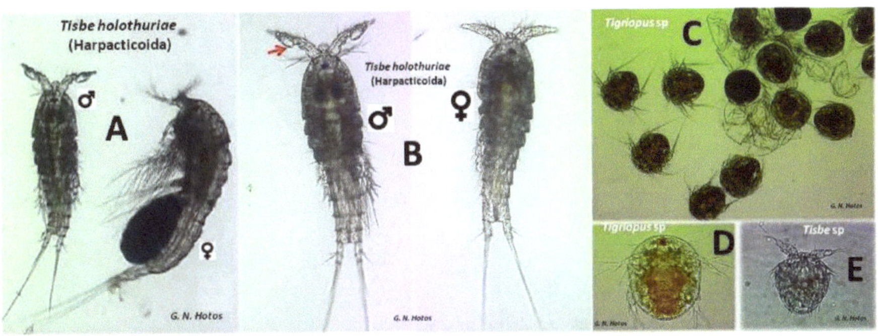

Figure 2. (**A**) Male and female of *Tisbe holothuriae* with the female carrying a big egg sac, (**B**) Male and female in juxtaposition; arrow indicates the characteristic lumps on the first pair of antennas of the male, (**C**) Hatching of *Tigriopus* sp. egg mass, (**D**) Newly hatched nauplius of *Tigriopus* sp., (**E**) Newly hatched nauplius of *T. holothuriae*.

For each experiment for both copepod species, the required number of individuals or ovipositor females were obtained from stock cultures acclimatized and maintained in the laboratory under optimal conditions. More specifically, each population of each copepod species was maintained in 1 L glass cylindrical containers with 850 mL of 37 ppt salinity water at room temperature and lighting (~19 °C and ~500 lux, respectively), with a daily supply of a small but sufficiently concentrated by centrifugation (paste) mixture of phytoplankton (*Tetraselmis suecica*, *Rhodomonas salina*, *Isochrysis galbana*, *Dunaliella salina* and *Asteromonas gracilis*), cultivated in the laboratory and used in their exponential growth phase. From the container of the stock culture, amounts of 10 mL were taken with a glass pipette, from which, after microscopic inspection, the copepods required in the experiments were taken. In particular, the adult copepods (without egg sac) used for the salinity tolerance experiment and the ovulated females (with egg sac) used in the food type effect experiment were obtained.

In the salinity tolerance experiment (Figure 3A), small plastic containers of 20 mL were used in which water with salinities of 0, 2, 4, 6, 8, 10, 20, 30, 60, 70, 80, 90, 100, 105, 110, 115, 120, 125, 130, 135, 140, 145 and 150 ppt was placed. Twenty-five (25) copepods were placed directly in each number with the defined salinity in order to apply the probit method [18], i.e., to find the 50% mortality rate of the population after the first 24 h (LC50). At the end of the first 24 h, the number of live and dead animals was recorded. Presumed dead animals for determining whether they were indeed dead were collected with a Pasteur pipette into a dish containing salt water of 35–37 ppt and after a reasonable interval were re-checked for liveness. The probit method calculates the 50% mortality rate of experimental animals in a series of escalating concentrations of some substance (in our case salinity). In total, 50% lethality is calculated as LC50 (Lethal Concentration 50%) ± 95% confidence limits (CL) and corresponds to that concentration of the test substance which causes death in 50% of the number of experimental animals exposed to it within a certain period of time (usually 24 h as in our case). The 95% confidence limits of the 24-h LC50, (lethal salinity for 50% of subjects in 24 h) are given as: LC50 1.96[SE(LC50)].

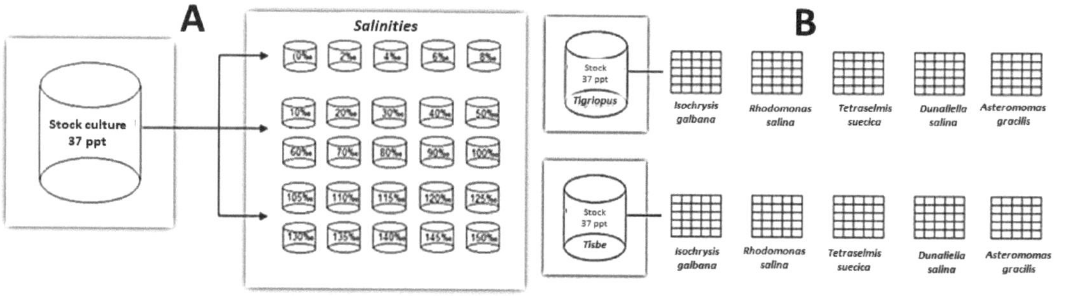

Figure 3. Schematization of the salinity tolerance experiment (**A**) and the feeding experiment (**B**) for both copepod species.

The standard error (SE) of the LC50 was calculated using the formula:

$$\mathrm{SE(LC50)} = \frac{1}{b\sqrt{pn\overline{w}}} \tag{1}$$

where b = the slope of the probit-salinity regression line, p = the number of salinities used for each case, n = the number of subjects used in each salinity group = 25, $\overline{w}$ = the mean "weight of probit observations" [18].

The feeding experiment (Figure 3B) used multi-chambered plastic plates (20 chambers of 3 mL per plate) in which one ovulated individual per chamber was placed in 37 ppt salinity water, ~20 °C and 500 lux diffuse illumination. Each chamber according to the protocol was given the selected species of microalgae from cultures in the exponential phase in a dose of ~0.1 mL of culture per 3 days. The algae used were *Tetraselmis suecica*, *Asteromonas gracilis*, *Isochrysis galbana*, *Rhodomonas salina* and *Dunaliella salina*. In terms of cell biovolume, *Rhodomonas* was ~400 μm^3, *Isochrysis* ~ 140 μm^3, *Asteromonas* ~ 1200 μm^3, *Dunaliella* ~ 350 μm^3 and *Tetraselmis* ~ 450 μm^3. At their exponential phase of their culture, they attained: *Rhodomonas* ~ 5 × 10^6 cells/mL, *Isochrysis* ~ 12 × 10^6 cells/mL, *Asteromonas* ~ 0.8 × 10^6 cells/mL, *Dunaliella* ~ 9 × 10^6 cells/mL and *Tetraselmis* ~ 7.5 × 10^6 cells/mL. They were cultured in 2 L Erlenmeyer flasks under 8000 lux, 16 hL:8 hD, 19 °C, salinity 40 ppt, supply of filtered air and Walne's nutrient medium. The progress of each culture was monitored by counting cells with a Fuchs–Rosenthal haematocytometer and a quantity of culture was collected from the middle of the exponential phase. This quantity was centrifuged (3000 rpm, 3 min) and the algal mass was treated appropriately to create a 5–20 × 10^5 cells/mL stock culture depending on each algae's cell biovolume and culture density. A 0.1 mL drop of the concentrated algal stock culture of 5–20 × 10^5 cells/mL of the specific microalgae was poured into each well of the appropriate plate. The quantity of the algae in the wells was estimated to be 0.5–2 × 10^5 cells/mL per 0.1 mL drop, divided by 3 mL (volume of the well) = 0.16–0.66 × 10^5 cells. The lower values correspond to bigger biovolumes, and the inverse applies to smaller biovolumes. Overall, approximately the same algal mass was introduced in every well irrespective of the algal type. The experiment lasted 28 days and at the end, the total number of individuals in each chamber was counted.

For the experiment of salinity-food demographics (Figure 4), one 1st egg sac female was put in each of the thirty 15 mL chamber of the plates used and filled with 35 ppt salt water. After the complete hatching of the eggs, the mother copepods were pipetted out and removed leaving only the F1 nauplii. The nauplii fed with a proper mixture of microalgae were monitored daily and the females that appeared with their 1st egg sac were sucked, washed in clear 35 ppt seawater and placed in new plates filled with 20, 32, 44, 60 ppt sterilized seawater using minute amounts of concentrated microalgae of either *R. salina* or *D. salina* as food thereafter. For each combination of salinity-algae, 36 replicates were created. The temperature was kept at ~21 °C and with a ~500 lux diffuse illumination. The values of pH initially were 7.9 ± 1 but thereafter fluctuated between 7.6 and 8.1. The daily inspection of the populations was performed under a dissecting microscope and salinity was checked by a refractory salinometer. When an increase in salinity (due to mild evaporation) exceeded 2 ppt, the proper addition of distilled water was done. During the experiment, the following parameters were recorded: (i) sum of nauplii from all egg sacs produced by each individual female, (ii) number of F1 nauplii produced by the 1st egg sac, (iii) number of egg sacs produced by each female, (iv) time in days for the complete hatching of the eggs of 1st egg sac, without counting rejected egg sacs (by some females) that were not hatched, (v) time (in days) between the creation of the 1st and 2nd egg sac, without counting females that did not create 2nd egg sac and (vi) percentage of females among adult copepods in each combination of salinity-food. The percentage of females was calculated out of the total number of adults hatched from all 36 female egg sacs at each salinity. Individuals that died at a stage before that of the adults (nauplii-copepodites) were not included for the calculation of the % proportion of female individuals.

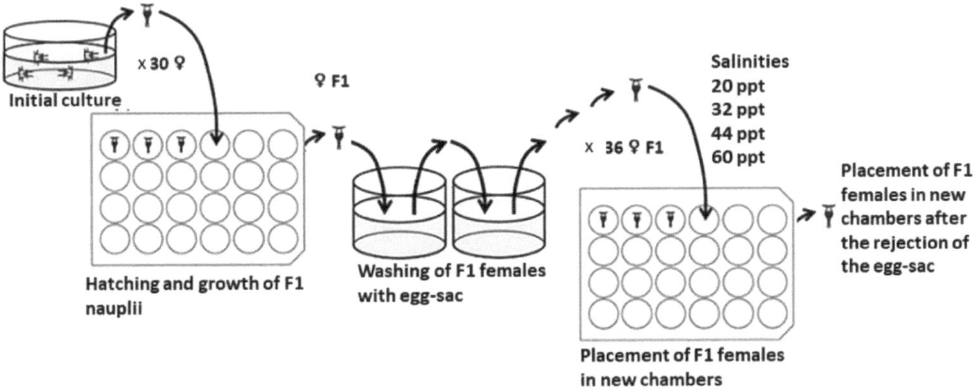

Figure 4. Schematization of the salinity-food demographics experiment for both copepod species.

Statistical analysis was performed using ANOVA and paired-wise comparisons at the 0.05 level of significance by Tukey's test using the free PAST3 software (Øyvind Hammer, University of Oslo, Oslo, Norway).

3. Results

3.1. Experiment on Salinity Tolerance

Tigriopus sp. (Figure 5) proved to be more resistant than *Tisbe* sp. (Figure 6) at both low and high salinities. From the processing of the probit methodology (Table 1, Figures 5 and 6) for low salinities we found: *Tigriopus* LC50 = 1 ppt ± 4.43 ppt (95% CL) and *Tisbe* LC50 = 15 ppt ± 2.41(95% CL). That is, in *Tigriopus*, the salinity that causes 50% mortality is the very low one of 1 ppt, whereas in *Tisbe*, the relevant salinity is 15 ppt. For high–very high salinities we have: *Tigriopus* LC50 = 132 ppt ± 5.35 (95% CL) and *Tisbe*

LC50 = 93 ppt ± 3.23 (95% CL). That is, in *Tigriopus*, the salinity that causes 50% mortality is very high one at 132 ppt, whereas in *Tisbe*, it is 93 ppt.

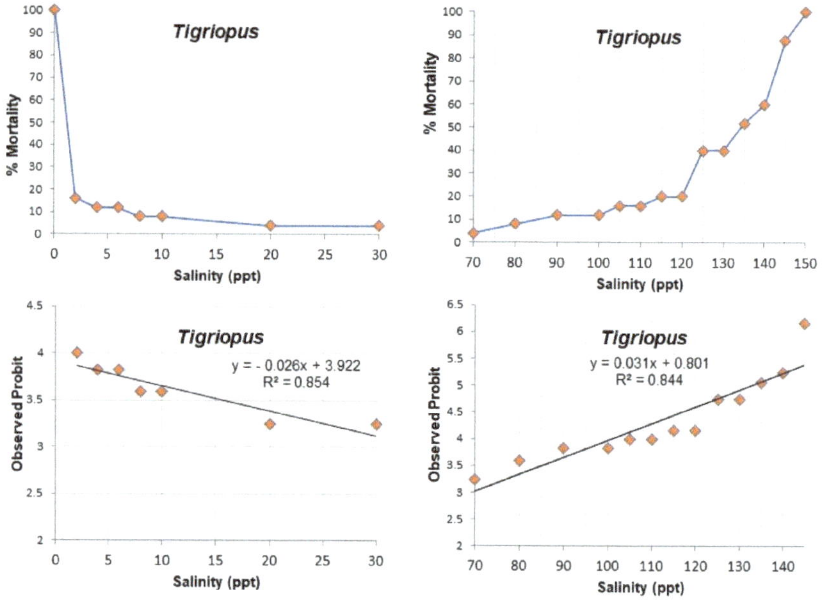

Figure 5. Mortality (%) and probits of mortality vs. salinity of *Tigriopus* sp. in the LC50 experiment. Also depicted are the regression equations and R^2 values.

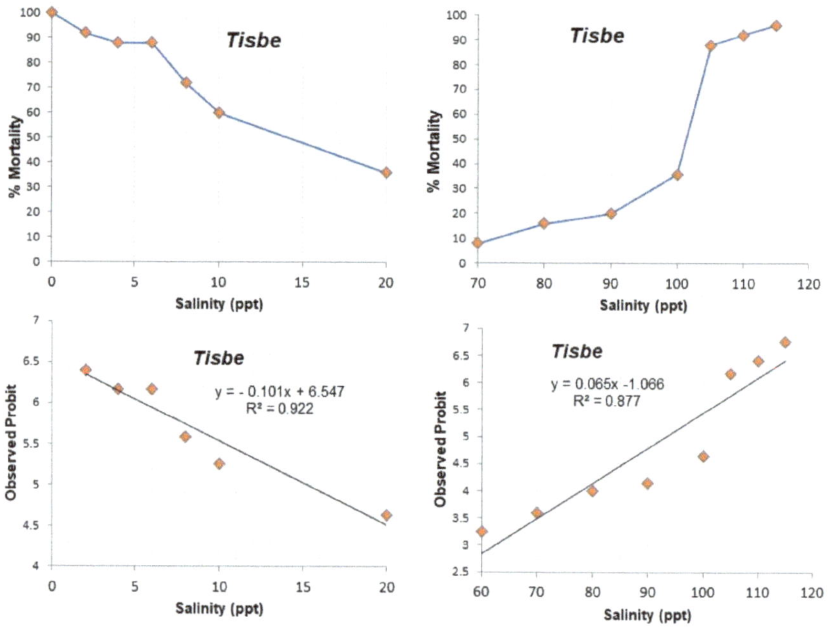

Figure 6. Mortality (%) and probits of mortality vs. salinity of *T. holothuriae* in the LC50 experiment. Also depicted are the regression equations and R^2 values.

Table 1. Records on probit analysis of the influence of salinity on mortality of *T. holothuriae* and *Tigriopus* sp.

Salinities (ppt)	Mortality *Tisbe*	Mortality *Tigriopus*	% Mortality *Tisbe*	% Mortality *Tigriopus*	Observed Probit *Tisbe*	Observed Probit *Tigriopus*	Expected Probit *Tisbe*	Expected Probit *Tigriopus*	Weight *Tisbe*	Weight *Tigriopus*
0	(25/25)	(25/25)	100	100	0	0	0	0	0	0
2	(23/25)	(4/25)	92	16	6.4	4	6.34	3.88	0.336	0.405
4	(22/25)	(3/25)	88	12	6.17	3.82	6.15	3.82	0.37	0.37
6	(22/25)	(3/25)	88	12	6.17	3.82	5.95	3.77	0.439	0.37
8	(18/25)	(2/25)	72	8	5.58	3.59	5.73	3.7	0.532	0.336
10	(15/25)	(2/25)	60	8	5.25	3.59	5.53	3.65	0.581	0.336
20	(9/25)	(1/25)	36	4	4.64	3.24	4.52	3.38	0.581	0.237
30	0	(1/25)	0	4	0	3.24	0	3.12	0	0.154
40	0	0	0	0	0	0	0	0	0	0
50	0	0	0	0	0	0	0	0	0	0
60	(1/25)	0	4	0	3.24	0	2.83	0	0.092	0
70	(2/25)	(1/25)	8	4	3.59	3.24	3.48	3.1	0.269	0.154
80	(4/25)	(2/25)	16	8	4	3.59	4.13	3.33	0.471	0.208
90	(5/25)	(3/25)	20	12	4.15	3.82	4.78	3.64	0.627	0.302
100	(9/25)	(3/25)	36	12	4.64	3.82	5.43	3.95	0.601	0.439
105	(22/25)	(4/25)	88	16	6.17	4	5.76	4.12	0.503	0.471
110	(23/25)	(4/25)	92	16	6.4	4	6.08	4.28	0.401	0.532
115	(24/25)	(5/25)	96	20	6.75	4.15	6.41	4.43	0.302	0.558
120	(25/25)	(5/25)	100	20	0	4.15	0	4.59	0	0.601
125		(10/25)		40		4.74		4.75		0.627
130		(10/25)		40		4.74		4.9		0.634
135		(13/25)		52		5.05		5.06		0.634
140		(15/25)		60		5.25		5.22		0.627
145		(22/25)		88		6.17		5.38		0.601
150		(25/25)		100		0		0		0
								SUM Weight *Tisbe*		6.105
								MEAN Weight *Tisbe*		0.436
								SUM Weight *Tigriopus*		8.596
								MEAN Weight *Tigriopus*		0.429

3.2. Experiment on Feed Influence

Tetraselmis resulted in the highest production of offspring in *Tigriopus* (36 ± 0.9 SE), followed by statistically equal *Dunaliella* (29 ± 1.7) and *Rhodomonas* (26 ± 1.55), which were also statistically equal for *Tisbe* (27 ± 1.45 and 28 ± 1.34, respectively) (Figure 7). *Asteromonas* and *Isochrysis* were the species with the highest and lowest cell biovolume, respectively, which resulted in substantially and significantly fewer offspring in *Tisbe* (21 ± 0.56 and 16 ± 0.44, respectively) and even fewer in *Tigriopus* (16 ± 0.62 and 7 ± 0.48, respectively). However, concerning *Tetraselmis* in *Tisbe*, the production of offspring was significantly much lower (22 ± 1.08) than the relevant one of *Tigriopus* (36 ± 0.91). Based on these findings, *Rhodomonas* and *Dunaliella* were selected as the microalgae to be tested for the next experiment on copepod demographics, as they resulted in equal numbers of progenies in both copepod species.

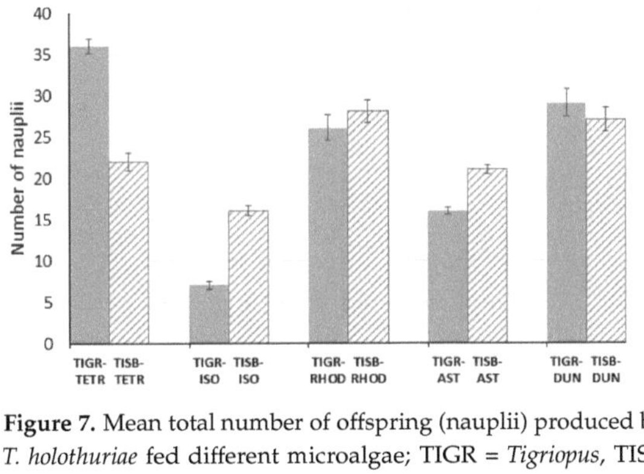

Figure 7. Mean total number of offspring (nauplii) produced by a single female of *Tigriopus* sp. and *T. holothuriae* fed different microalgae; TIGR = *Tigriopus*, TISB = *Tisbe*, TETR = *Tetraselmis*, ISO = *Isochrysis*, RHOD = *Rhodomonas*, AST = *Asteromonas*, DUN = *Dunaliella*. Error bars stand for Standard Error (SE).

3.3. Experiment on Salinity-Feed Demographics

Rhodomonas and *Dunaliella* equally affected ($p > 0.05$) the total number of produced nauplii of *Tigriopus* at each of the four salinities (Figure 8A). The significantly lowest values across salinities for both microalgae were recorded at the highest salinity of 60 ppt (18 nauplii for *Rhodomonas*, 16 for *Dunaliella*), whereas the highest numbers (37 for *Rhodomonas*, 33 for *Dunaliella*) were recorded at the salinity of 32 ppt. In *Tisbe*, although the pattern concerning salinity is alike to that of *Tigriopus* (most progenies at 32 ppt, least at 60 ppt), the effect of the microalgal feed is totally different with *Dunaliella*'s significant higher values ($p > 0.05$) than their counterparts of *Rhodomonas* at all salinities (Figure 8B).

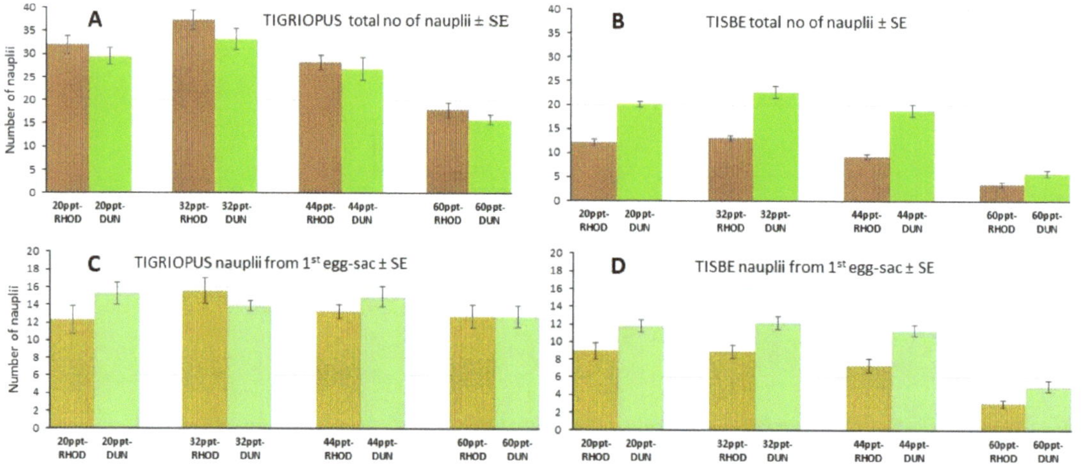

Figure 8. (**A**,**B**) Total number of nauplii produced by a female of *Tigriopus* and *Tisbe*, respectively, across treatments of different salinity types of algal feed. (**C**,**D**) number of nauplii produced from the first egg sac of *Tigriopus* and *Tisbe*, respectively, across treatments of different salinity types of algal feed. Values are means ± SE. RHOD = *Rhodomonas*, DUN = *Dunaliella*.

The number of nauplii produced by the 1st egg sac was equal ($p > 0.05$) across all salinities and type of food (min. 12, max. 15 nauplii) in *Tigriopus* (Figure 8C). In *Tisbe*, (Figure 8D) *Dunaliella* effected the production of significantly more nauplii compared to *Rhodomonas* ($p < 0.05$) and produced the same number (11–12 nauplii) statistically across the salinity range of 20–44 ppt, with the salinity of 60 ppt being far less productive in number of nauplii both across the other salinities and compared to *Rhodomonas* (5 vs. 3, respectively). The number of nauplii was also statistically equal across the salinity range of 20–44 ppt (7–9 nauplii) for *Rhodomonas*.

In both species of copepods (Figure 9A,B), the highest salinity of 60 ppt for both microalgae resulted in far fewer numbers of total egg sacs produced by a single female in its life span, compared to all other lower salinities ($p < 0.05$). Whereas in *Tigriopus* (Figure 9A), in all salinities, the number of egg sacs produced was statistically equal for both microalgae at each particular salinity, in *Tisbe* (Figure 9B), at the salinities of 32 and 44 ppt, *Dunaliella* resulted in a significantly higher number of egg sacs compared to *Rhodomonas*. In general, *Tigriopus* at all salinities for both microalgae produced more egg sacs than each corresponding case for *Tisbe* (max. mean 3.8 egg sacs vs. 2.7 respectively).

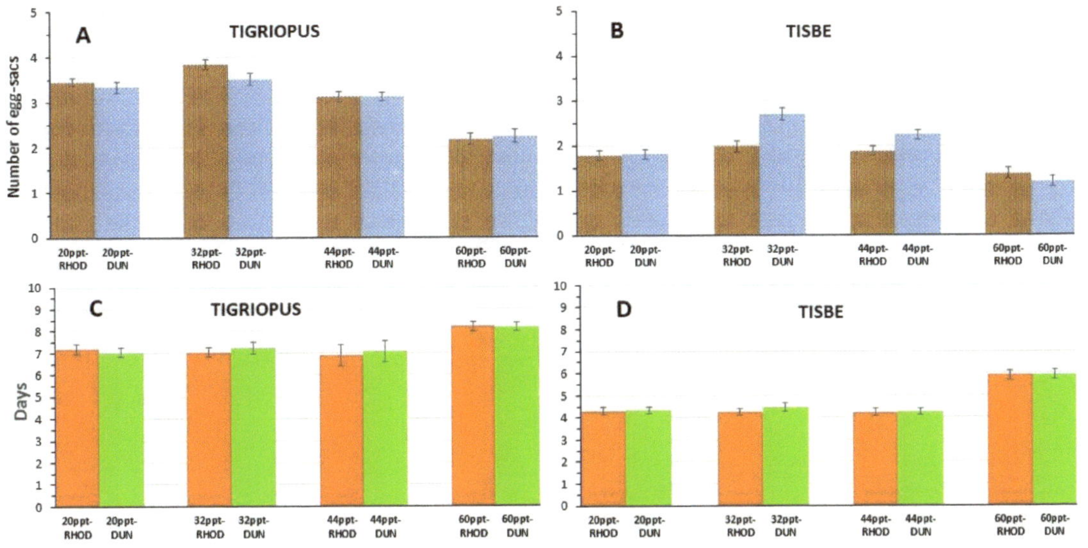

Figure 9. (**A**,**B**) total number of egg sacs produced during the lifetime of a female of *Tigriopus* and *Tisbe*, respectively, across treatments of different salinity types of algal feed. (**C**,**D**) days required for the complete hatching of the eggs of the first egg sac of *Tigriopus* and *Tisbe*, respectively, across treatments of different salinity types of algal feed. Values are means ± SE. RHOD = *Rhodomonas*, DUN = *Dunaliella*.

The days required for the complete hatching of all the eggs of the 1st egg sac were in general clearly more in *Tigriopus* (~7–8 days, Figure 9C) compared to *Tisbe* (~4–6, Figure 9D). In both species and at all salinities, this time period was statistically equal among the two microalgae at each particular salinity and across salinities in the range of 20–44 ppt. Only at the highest salinity of 60 ppt the required days were statistically greater. The same pattern also occurred concerning the days required for the appearance of the 2nd egg sac from the appearance of the first (Figure 10A,B). Across salinities and type of microalgae, the number of days were statistically equal at the salinity range of 20–44 ppt and significantly less than the days required at 60 ppt in every particular treatment in both copepods, the only difference being that overall values were less in *Tisbe* (Figure 10B) as compared to *Tigriopus* (Figure 10A).

The percentage of females was not affected by the type of food in *Tigriopus* (Figure 10C) as at every salinity, the percentage was statistically equal between *Rhodomonas* and *Dunaliella*. On the contrary, in *Tisbe* (Figure 10D), at all salinities except the one at 20 ppt, *Dunaliella* effected significantly more females compared to *Rhodomonas*.

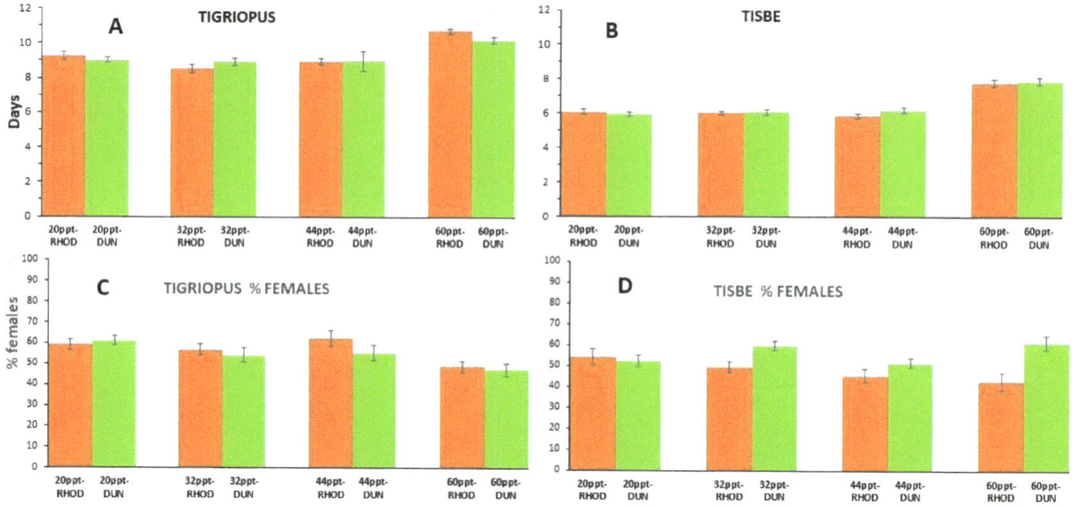

Figure 10. (**A**,**B**) days required for appearance of the 2nd egg sac since the appearance of the 1st egg sac in *Tigriopus* and *Tisbe*, respectively, across treatments of different salinity types of algal feed. (**C**,**D**) percentage of females in *Tigriopus* and *Tisbe*, respectively, across treatments of different salinity types of algal feed. Values are means ± SE. RHOD = *Rhodomonas*, DUN = *Dunaliella*.

4. Discussion

Copepods of saline environments in general and harpacticoids in particular are euryhaline [17,19–24], but to what extent is not clearly known. Existing studies may provide an optimal salinity at which copepods increase in population and additionally provide a range of salinity at which their population maintains satisfactory dynamics, but they do not clearly answer the upper and lower limits of salinity at which they can survive. For *Tisbe holothuriae* [25], 38 ppt is given as the optimum salinity [16,26]. For other *Tisbe* species such as *T. biminiensis*, the limits are reported as very narrow, i.e., 27–34 ppt and even as a prohibitive salinity at that of 20 ppt [27]. For various species of *Tigriopus*, the survival range is 20–40 ppt with the optimal being at 30 ppt [11], and especially for *T. californicus* [19] where 100% survival in the range 30–70 ppt is reported. These values are initially considered consistent with the value of 37 ppt in which the copepods lived during the experiments of this work, but we cannot claim that we have exhausted the topic for the optimal development of cultured populations of the specific copepod species. The issue of the presence of copepods in hypersaline waters is important from an ecological point of view as, on the one hand, hypersaline systems are among the most extreme environments on Earth with seasonally dramatic changes in salinity [28], and on the other hand, copepods play an important ecological role, participating largely in the food web and indirectly in the biological recycling of elements [20]. In the literature, there are no corresponding values for *T. holothuriae* and *Tigriopus* spp. and therefore ours can be considered as the first data ever presented. Based on our findings, we assume that these copepods will not have any difficulty adapting to salinities that exceed seawater salinity, and based on Figures 5 and 6, we can assume that they will show minimal mortality due to osmoregulatory stress at salinities up to about 70 ppt, especially if their acclimatization is gradual. The importance of copepods resistance to hypersaline environments such as that of the Messolonghi salterns has not been investigated but may be important because they may contribute to good salt production by consuming the microalga *Dunaliella* sp., whose presence degrades the salt (personal communication). It is clear both from our experimentation and from similar work (but only for *Tigriopus*, [19,21]) that absolutely fresh water produces almost immediately 100% mortality in both species. However, in brackish water, the situation is radically different. Wheras *Tigriopus* shows a great adaptation to salinities of 4–10 ppt with almost negligible mortality and practically absolute survival above 10 ppt salinity, *Tisbe* only after 20 ppt shows 0% mortality. These numbers for *Tigriopus* especially are in agreement with those found by Hawkins [21]. From the above, we consider that although *Tigriopus* is shown to withstand a surprising range of salinity (~4–120 ppt) compared to *Tisbe* (~20–90 ppt), both species can colonize a variety of environments, even extreme water systems (e.g., estuaries, lagoons, hypersaline basins, etc.) and from the point of view of cultivation, any source of supply of salt water can be appropriately used.

The issue of the most suitable microalgal feed for copepods is rather complicated and fragmented in the literature, e.g., it was found [29,30] that *Tetraselmis* is inferior to *Isochrysis* as feed for the calanoid *Gladioferens imparipes* and several species of *Acartia*, respectively, whereas the opposite occurred in our study for both *Tigriopus* and *Tisbe*. As it was demonstrated [31] that *T. japonicus* does not present any kind of selectivity between *Tetraselmis suecica* and *Isochrysis galbana*, equally consuming them in a mixture, it is probable that underlying digestive mechanisms govern the growth of copepods. Many researchers have carried out studies on the correlation of the reproduction of various species of copepods with the species of phytoplankton used for their diet, both in the natural environment and under laboratory conditions. The results demonstrated a strong correlation between reproductive traits of copepods (fecundity, hatching success, nauplii survival) and algae characteristics such as cell size, morphology, toxicity and biochemical composition [1]. In our study, there were clearly different responses among the two copepod species when fed different microalgae (Figure 7). Two of the microalgae (*A. gracilis* and *D. salina*) used as food in the present work came from the Messolonghi lagoon [17], the same natural environment as the copepods, so it is possible to draw conclusions about the suitability of these species to be introduced into intensive copepod farming systems. As *Dunaliella* together with *Rhodomonas* presented the same nauplii productivity for both copepods next to *Tetraselmis* (which was more effective in *Tigriopus*), *Dunaliella* and *Rhodomonas* combined with various salinities were selected for the next experimentation on demographics of the culture.

It must be noticed that this work was not carried out under controlled laboratory conditions, without the mass cultivation of copepods, but instead, with the reproduction of individual females and subsequent study of the development of their nauplii progenies. However, we feel that as a first trial, in order to answer some issues on the response of these copepod species to microalgal feeds at different salinities, the present study can reflect quite well what could happen in bigger culture tanks.

The mortality of the female individuals of the F1 generation after their placement in the individual plates for the study of their reproductive activity at the different salinity values was zero. However, at the extreme salinity value (60 ppt) for both species, the time for the transition from the nauplii stages to the adult stage was much longer than that observed in the other salinities. According to Hong et al. [32], the osmoregulatory mechanisms of some marine organisms during their life cycle are differentiated. Studies related to the effect of salinity on individuals of the species *Tigriopus japonicus* report that although adult individuals have a great capacity to adapt to changing salinity values, the developmental stage between nauplii and copepods appears to be the most sensitive and presents the highest mortality [33,34], whereas according to Paiva et al. [35], adult crustacean individuals are more tolerant to salinity stress than individuals at the other developmental stages.

In the present work, the average value of the total number of offspring produced at the optimal salinity values for both species (*Tigriopus* and *Tisbe*) was significantly lower than the values reported by other researchers for female individuals of the species of the same genera grown under controlled laboratory conditions. The reasons why a reduced production of nauplii was observed are likely to be:

High population density. In the present work, for the reproduction and development of copepods, 15 mL chambers were used, resulting in higher population densities compared to those reported by other researchers who used vessels with a larger capacity. In fact, due to the benthic nature of *Tisbe* and *Tigriopus*, almost all of the nauplii, especially the first nauplii stages, were concentrated at the bottom of the chambers (diameter of chanber 4cm), and as a result their density increased even more. According to Mauchline [36], the bodies of most copepods are denser than the seawater they live in, regardless of salinity. For this reason, staying in the benthos requires less energy consumption than staying in the water column. The phenomenon became even more intense with the proceeding of the experiment as a rudimentary substrate, which was created at the bottom of the chambers and attracted the copepods. It has been reported that density can affect the growth, survival and fecundity of predatory copepods [37]. In the study of Punnarak et al. [11], in lower density cultures, higher survival rates were observed for copepods collected from natural environments, whereas according to Fava and Crotti [38], copepod cultures at high densities result in increased excrements because of animal stress and, therefore, their survival, development and reproduction are not favored.

Cannibalism. Cannibalism of young individuals is a strategy particularly widespread for many species of animal organisms in nature [39]. In many crustacean species that inhabit both marine and estuarine systems, intraspecific predation of larval individuals is often one of the main reasons for their mortality, helping to maintain their population sizes at a constant level [40,41]. Cannibalism has been reported for calanoid copepods [42,43] as well as harpacticoid copepods of the species *Tigriopus fulvus* [44,45], *T. californicus* [46] and *T. brevicornis* [45]. According to Gallucci and Ólafsson [45], the phenomenon of cannibalism is observed when we have high population densities even when in the environment there is a sufficiency of other types of food. In the present work, it is possible that cannibalistic behaviors were manifested by the adults to the young nauplii, because of the high population density at the bottom of the chambers, with the result that the measurement of the number of nauplii conducted daily gave a lower number of them than those hatched from the eggs of the egg sacs.

Not renewing the development medium. Copepod reproduction is directly affected by abiotic factors in their environment, such as water temperature, turbidity and pH [22]. In the present work, no renewals of the culture medium (water) was made, since such a practice was not possible due to the high density of copepods. During the development of the experiments, the temperature was controlled, but the turbidity and the pH changed, mainly in the salinity values that were not favorable for the development of copepods. At the same time, during the course of the experiment, a rudimentary substrate was created at the bottom of the chambers from the remains of the excess food, from the exoskeletons and from the excreta of the copepods. Excess food remaining in small-volume copepod cultures can create problems such as: accumulation of dead algal cells in the water and on the walls of the container (with an increased possibility of bacterial growth), adhesion of a mass of dead microalgae to the swimming appendages of the copepods resulting in obstruction of free their movement [47], selective feeding by copepods (avoidance of consumption of dead microalgae cells) [2], increase in water turbidity, etc.

Genetic factors. In harpacticoid copepods, even among populations of the same species originating from different regions, significant discrepancies have been reported regarding optimal salinity conditions and abiotic factors in general. For populations of the species *Tisbe holothuriae* in the gulf of Marseille, there was reproductive activity at salinity values from 20 ppt to 48 ppt, whereas for populations of the same species originating from the Saronic Gulf at salinities of 20 ppt and 48 ppt, not only was no reproductive activity observed but instead, individuals showed high mortality. Based on the researchers who studied the above behaviors, the salinity tolerance of different populations was determined by the salinity of the environment from which they came [16,48]. According to Edmands [49] in the genus *Tigriopus*, a wide range of genetic divergence has been found in the populations of its species. The individuals studied in the present work came from the Messolonghi lagoon. Although there was acclimation to laboratory conditions over several generations, the genotype of wild populations determined the reproductive behavior of females and the total number of nauplii hatched from each egg sac at different salinity values, differentiating it from that reported in studies with corresponding salinity values for populations of the same genus originating from different regions.

For both copepod species studied in this work, the mean value of the hatching time of their first egg sac and the mean value of the time between the appearance of the first and the appearance of the second egg sac are significantly longer at the extreme value of salinity (60 ppt) in relation to the values in the other salinities. Growth time is an important feature of zooplankton fitness and is closely related to the time of initiation of production of new individuals and the rate of population growth, with faster growth leading to higher abundances in a shorter time [23]. Thus, fast-growing populations in natural environments have a competitive advantage over slower-growing ones, whereas under controlled conditions of production, fast growth is associated on the one hand with cost reduction and increased production, and on the other hand with the production of more robust individuals due to achieving the ideal growth conditions.

The reproduction of copepods is directly affected by the quantity and quality of the ingested food [22], whereas they show great adaptability to the available types of food present in their environment, changing their physiological processes according to the amount of energy they can obtain through their food. In fact, according to Hasset and Landry [50], the lack of food from the environment of copepods (natural or laboratory) can lead to changes in their feeding behavior and in the activity of their digestive enzymes, increasing their survival over time (even for a period of 3 weeks in starvation conditions [51]). In the present work, in the individuals of the genus *Tigriopus*, for all the parameters studied, no significant differences were presented for the two different types of food used. On the contrary, for the individuals of *T. holothuriae*, for all parameters at the most favorable salinity value (32 ppt), the microalga *D. salina* appeared more favorable as food than in *R. salina*. In a corresponding study [52] on copepods of the species *T. holothuriae* that grew in salinity 28–33 ppt, the diet with microalgae of the genera *Dunaliella* and *Rhodomonas* varied the composition of their tissues in long-chain HUFAs. Copepods fed *Rhodomonas* showed a lower concentration of HUFAs than those fed *Dunaliella*. Miles et al. [53] suggest a mixture of microalgae for the optimal development of copepods of the genus *Tisbe* and for the production of a large number of offspring, whereas according to Cutts [37,54], food must change between the various developmental stages of copepods as the size of their mouthparts change along with their nutrient requirements.

From the first evidence of the results of this work, it appears that both the development and reproduction of copepods are negatively affected by salinity values that deviate from that found in the natural environment of the populations of the species studied (35–40 ppt). The total production of nauplii fluctuated at lower levels, compared to those reported in similar studies with individuals of same or different species of the same genus. The differences in the ability of copepods to adapt to salinity values, different from those found in their natural environment, are related to genetic factors that must be investigated in order to apply (in the cases of copepod mass production systems) the optimal conditions for growth and reproduction.

The knowledge obtained from the study of the effect of different salinity values on the growth and reproduction of specific copepod species, in addition to being used in the production process of fish farming units, could also be used for ecological practices through the creation of predictive models on the effects of changing environmental conditions, not only at the species level but also at the population level. The local adaptation of populations is the evolution of their characteristics, which have been optimized for specific habitats, so that the genotypes of these populations endow them with better fitness than genotypes of the same species from other habitats [55]. In recent decades, genetic data demonstrate that marine populations of coastal habitats are less homogeneous than previously thought, suggesting that local selective forces may be strong enough to offset ongoing gene flow. Information about adaptive divergence is particularly important for predicting the effects of climate change and improving prediction models, which often assume that all populations of a species have the same range of tolerance to environmental conditions (e.g., salinity, temperature, water chemistry). Such an assumption, however, may underestimate the risk of species extinction from a habitat if individual populations have a smaller tolerance range than the species as a whole [24].

Supplementary Materials: The following are available online at https://www.youtube.com/watch?v=zmwXeoN1wTM, Video S1: *Tigriopus* copepods—adults, egg hatching-nauplii (accessed on 13 October 2022); https://www.youtube.com/watch?v=mKWNr5dhJGY, Video S2: Copepods—harpacticoids vs. calanoids—*Tisbe* (accessed on 13 October 2022).

Author Contributions: Conceptualization, G.N.H.; methodology, G.N.H., E.K. and I.F.; investigation, G.N.H., E.K. and I.F.; resources, G.N.H.; data curation, G.N.H., E.K. and I.F.; writing—original draft preparation, G.N.H.; writing—review and editing, G.N.H.; supervision, G.N.H. All authors have read and agreed to the published version of the manuscript.

Funding: This research received no external funding.

Institutional Review Board Statement: The study was conducted according to the guidelines of the Declaration of Helsinki and approved by the Ethics Committee of University of Patras (protocol code 13932—9 July 2022).

Informed Consent Statement: Not applicable.

Acknowledgments: The authors thank Despoina Avramidou and the technical staff at the plankton culture laboratory for collecting and keeping the copepods from the lagoon in stock cultures.

Conflicts of Interest: The authors declare no conflict of interest.

References

1. Hussain, M.B.; Laabir, M.; Daly Yahia, M.N. A Novel Index Based on Planktonic Copepod Reproductive Traits as a Tool for Marine Ecotoxicology Studies. *Sci. Total Environ.* **2020**, *727*, 138621. [CrossRef] [PubMed]
2. Alajmi, F.; Zeng, C. Evaluation of Microalgal Diets for the Intensive Cultivation of the Tropical Calanoid Copepod, *Parvocalanus crassirostris*. *Aquac. Res.* **2015**, *46*, 1025–1038. [CrossRef]
3. Baensch, F.; Tamaru, C.S. Spawning and Development of Larvae and Juveniles of the Rare Blue Mauritius Angelfish, *Centropyge debelius* (1988), in the Hatchery. *J. World Aquac. Soc.* **2009**, *40*, 425–439. [CrossRef]
4. Støttrup, J.G. The Elusive Copepods: Their Production and Suitability in Marine Aquaculture. *Aquac. Res.* **2000**, *31*, 703–711. [CrossRef]
5. Sun, B.; Fleeger, J.W. Sustained Mass Culture of Amphiascoides Atopus a Marine Harpacticoid Copepod in a Recirculating System. *Aquaculture* **1995**, *136*, 313–321. [CrossRef]
6. Støttrup, J.G.; Norsker, N.H. Production and Use of Copepods in Marine Fish Larviculture. *Aquaculture* **1997**, *155*, 231–247. [CrossRef]
7. Camus, T.; Zeng, C. Reproductive Performance, Survival and Development of Nauplii and Copepodites, Sex Ratio and Adult Life Expectancy of the Harpacticoid Copepod, Euterpina Acutifrons, Fed Different Microalgal Diets. *Aquac. Res.* **2012**, *43*, 1159–1169. [CrossRef]
8. Delbare, D.; Dhert, P.; Lavens, P. Zooplankton. In *Manual on the Production and Use of Live Food for Aquaculture*; Lavens, P., Sorgelos, P., Eds.; FAO Fisheries Technical Paper; FAO: Rome, Italy, 1996; Volume 361, pp. 252–282.
9. Nanton, D.A.; Castell, J.D. The Effects of Dietary Fatty Acids on the Fatty Acid Composition of the Harpacticoid Copepod, *Tisbe* sp., for Use as a Live Food for Marine Fish Larvae. *Aquaculture* **1998**, *163*, 251–261. [CrossRef]
10. Werbrouck, E.; Bodé, S.; van Gansbeke, D.; Vanreusel, A.; de Troch, M. Fatty Acid Recovery after Starvation: Insights into the Fatty Acid Conversion Capabilities of a Benthic Copepod (Copepoda, Harpacticoida). *Mar. Biol.* **2017**, *164*, 217–226. [CrossRef]
11. Punnarak, P.; Jarayabhand, P.; Piumsomboon, A. Cultivation of Harpacticoid Copepods (Families Harpacticidae and Laophontidae) under Selected Environmental Conditions. *Agric. Nat. Resour.* **2017**, *51*, 278–285. [CrossRef]
12. Calliari, D.; Andersen, C.; Thor, P.; Gorokhova, E.; Tiselius, P. Salinity Modulates the Energy Balance and Reproductive Success of Co-Occurring Copepods *Acartia tonsa* and *A. clausi* in Different Ways. *Mar. Ecol. Prog. Ser.* **2006**, *312*, 177–188. [CrossRef]
13. Chinnery, F.E.; Williams, J.A. The Influence of Temperature and Salinity on Acartia (Copepoda: Calanoida) Nauplii Survival. *Mar. Biol.* **2004**, *145*, 733–738. [CrossRef]
14. Devreker, D.; Souissi, S.; Seuront, L. Development and Mortality of the First Naupliar Stages of *Eurytemora affinis* (Copepoda, Calanoida) under Different Conditions of Salinity and Temperature. *J. Exp. Mar. Biol. Ecol.* **2004**, *303*, 31–46. [CrossRef]
15. Ohs, C.L.; Rhyne, A.L.; Grabe, S.W.; DiMaggio, M.A.; Stenn, E. Effects of Salinity on Reproduction and Survival of the Calanoid Copepod *Pseudodiaptomus pelagicus*. *Aquaculture* **2010**, *307*, 219–224. [CrossRef]

16. Miliou, H.; Moraitou-Apostolopoulou, M. Combined Effects of Temperature and Salinity on the Population Dynamics of *Tisbe holothuriae* HUMES (Copepoda: Harpacticoida). *Arch. Hydrobiol.* **1991**, *121*, 431–448. [CrossRef]
17. Hotos, G.N. A Preliminary Survey on the Planktonic Biota in a Hypersaline Pond of Messolonghi Saltworks (W. Greece). *Diversity* **2021**, *13*, 270. [CrossRef]
18. Wardlaw, C.A. (Ed.) *Practical Statistics for Experimental Biologists*; John Wiley & Sons: Plymouth, UK, 1989.
19. Herbert, J. *Salinity and Upper Temperature Tolerances of a Rockpool Copepod, Tigriopus Californicus (Baker, 1912)*; University of Oregon: Eugene, OR, USA, 1976.
20. Frangoulis, C.; Christou, E.D.; Hecq, J.H. Comparison of Marine Copepod Outfluxes: Nature, Rate, Fate and Role in the Carbon and Nitrogen Cycles. *Adv. Mar. Biol.* **2004**, *47*, 253–309. [CrossRef]
21. Hawkins, B.L. *The Biology of the Marine Copepod, Tigriopus californicus (Baker)*; Humboldt State University: Arcata, CA, USA, 1962.
22. Turner, J.T. The Importance of Small Planktonic Copepods and Their Roles in Pelagic Marine Food Webs. *Zool. Stud.* **2004**, *42*, 255–266.
23. Allan, J.D. Life History Patterns in Zooplankton. *Am. Nat.* **1976**, *110*, 165–180. [CrossRef]
24. Sanford, E.; Kelly, M.W. Local Adaptation in Marine Invertebrates. *Ann. Rev. Mar. Sci.* **2011**, *3*, 509–535. [CrossRef]
25. Heath, P.; Moore, C. Rearing Dover Sole Larvae on *Tisbe* and *Artemia* Diets. *Aquac. Int.* **1997**, *5*, 29–39. [CrossRef]
26. Miliou, H. The Effect of Temperature, Salinity and Diet On Final Size of Female *Tisbe holothuriae* (Copepoda, Harpacticoida). *Crustaceana* **1996**, *69*, 742–754. [CrossRef]
27. Souza-Santos, L.P.; Pastor, J.M.O.; Ferreira, N.G.; Costa, W.M.; Araújo-Castro, C.M.; Santos, P.J.P. Developing the Harpacticoid Copepod *Tisbe biminiensis* Culture: Testing for Salinity Tolerance, Ration Levels, Presence of Sediment and Density Dependent Analyses. *Aquac. Res.* **2006**, *37*, 1516–1523. [CrossRef]
28. Belmonte, G.; Moscatello, S.; Batogova, E.A.; Pavlovskaya, T.; Shadrin, N.V.; Litvinchuk, L.F. Fauna of Hypersaline Lakes of the Crimea (Ukraine). *Thalass. Salentina* **2012**, *34*, 11–24.
29. El-Tohamy, W.; Qin, J.; Abdel-Aziz, N.; El-Ghobashy, A.; Dorgham, M. Suitable Algal Species and Density for the Culture of Copepod *Gladioferens imparipes* as a Potential Live Food for Fish Larvae. *Aquac. Int.* **2021**, *29*, 105–125. [CrossRef]
30. Pan, Y.-J.; Souissi, S.; Souissi, A.; Wu, C.-H.; Cheng, S.-H.; Hwang, J.-S. Dietary Effects on Egg Production, Egg-Hatching Rate and Female Life Span of the Tropical Calanoid Copepod *Acartia bilobata*. *Aquac. Res.* **2014**, *45*, 1659–1671. [CrossRef]
31. Lee, M.-C.; Choi, H.; Park, J.C.; Yoon, D.-S.; Lee, Y.; Hagiwara, A.; Park, H.G.; Shin, K.-H.; Lee, J.-S. A Comparative Study of Food Selectivity of the Benthic Copepod *Tigriopus japonicus* and the Pelagic Copepod *Paracyclopina nana*: A Genome-Wide Identification of Fatty Acid Conversion Genes and Nitrogen Isotope Investigation. *Aquaculture* **2020**, *521*, 734930. [CrossRef]
32. Hong, H.; Wang, J.; Shi, D. Effects of Salinity on the Chronic Toxicity of 4-Methylbenzylidene Camphor (4-MBC) in the Marine Copepod *Tigriopus japonicus*. *Aquat. Toxicol.* **2021**, *232*, 105742. [CrossRef]
33. Raisuddin, S.; Kwok, K.W.H.; Leung, K.M.Y.; Schlenk, D.; Lee, J.-S. The Copepod Tigriopus: A Promising Marine Model Organism for Ecotoxicology and Environmental Genomics. *Aquat. Toxicol.* **2007**, *83*, 161–173. [CrossRef]
34. Kwok, K.W.H.; Leung, K.M.Y. Toxicity of Antifouling Biocides to the Intertidal Harpacticoid Copepod *Tigriopus japonicus* (Crustacea, Copepoda): Effects of Temperature and Salinity. *Mar. Pollut. Bull.* **2005**, *51*, 830–837. [CrossRef]
35. Paiva, F.; Pauli, N.; Briski, E. Are Juveniles as Tolerant to Salinity Stress as Adults? A Case Study of Northern European, Ponto-Caspian and North American Species. *Divers. Distrib.* **2020**, *26*, 1627–1641. [CrossRef]
36. Mauchline, J. The Biology of Calanoid Copepods. In *Advances in Marine Biology*; Mauchline, J., Ed.; Academic Press: Cambridge, MA, USA, 1998; Volume 33, pp. 71–81.
37. Cutts, C.J. Culture of Harpacticoid Copepods: Potential as Live Feed for Rearing Marine Fish. *Adv. Mar. Biol.* **2003**, *44*, 295–316. [CrossRef] [PubMed]
38. Fava, G.; Crotti, E. Effect of Crowding on Nauplii Production during Mating Time in *Tisbe clodiensis* and *T. holothuriae* (Copepoda, Harpacticoida). *Helgoländer Wiss. Meeresunters.* **1979**, *32*, 466–475. [CrossRef]
39. Polis, G.A. The Evolution and Dynamics of Intraspecific Predation. *Annu. Rev. Ecol. Syst.* **1981**, *12*, 225–251. [CrossRef]
40. Luppi, T.A.; Spivak, E.D.; Anger, K. Experimental Studies on Predation and Cannibalism of the Settlers of *Chasmagnathus granulate* and *Cyrtograpsus angulatus* (Brachyura: Grapsidae). *J. Exp. Mar. Biol. Ecol.* **2001**, *265*, 29–48. [CrossRef]
41. Moksnes, P.-O. The Relative Importance of Habitat-Specific Settlement, Predation and Juvenile Dispersal for Distribution and Abundance of Young Juvenile Shore Crabs *Carcinus maenas* L. *J. Exp. Mar. Biol. Ecol.* **2002**, *271*, 41–73. [CrossRef]
42. Landry, M.R. Switching between Herbivory and Carnivory by the Planktonic Marine Copepod *Calanus pacificus*. *Mar. Biol.* **1981**, *65*, 77–82. [CrossRef]
43. Conley, W.J.; Turner, J.T. Omnivory by the Coastal Marine Copepods *Centropages hamatus* and *Labidocera aestiva*. *Mar. Ecol. Prog. Ser.* **1985**, *21*, 113–120. [CrossRef]
44. Lazzaretto, I.; Salvato, B. Cannibalistic Behaviour in the Harpacticoid Copepod *Tigriopus fulvus*. *Mar. Biol.* **1992**, *113*, 579–582. [CrossRef]
45. Gallucci, F.; Ólafsson, E. Cannibalistic Behaviour of Rock-Pool Copepods: An Experimental Approach for Space, Food and Kinship. *J. Exp. Mar. Biol. Ecol.* **2007**, *342*, 325–331. [CrossRef]
46. Lewis, A.G.; Chatters, L.; Raudsepp, M. Feeding Structures and Their Functions in Adult and Preadult *Tigriopus californicus* (Copepoda: Harpacticoida). *J. Mar. Biol. Assoc. United Kingd.* **1998**, *78*, 451–466. [CrossRef]

47. Puello-Cruz, A.C.; Mezo-Villalobos, S.; González-Rodríguez, B.; Voltolina, D. Culture of the Calanoid Copepod *Pseudodiaptomus euryhalinus* (Johnson 1939) with Different Microalgal Diets. *Aquaculture* **2009**, *290*, 317–319. [CrossRef]
48. Gaudy, R.; Guerin, J.P.; Moraitou-Apostolopoulou, M. Effect of Temperature and Salinity on the Population Dynamics of *Tisbe holothuriae* Humes (Copepoda: Harpacticoida) Fed on Two Different Diets. *J. Exp. Mar. Biol. Ecol.* **1982**, *57*, 257–271. [CrossRef]
49. Edmands, S. Phylogeography of the Intertidal Copepod *Tigriopus californicus* Reveals Substantially Reduced Population Differentiation at Northern Latitudes. *Mol. Ecol.* **2001**, *10*, 1743–1750. [CrossRef] [PubMed]
50. Hassett, R.P.; Landry, M.R. Effects of Diet and Starvation on Digestive Enzyme Activity and Feeding Behavior of the Marine Copepod *Calanus pacificus*. *J. Plankton Res.* **1990**, *12*, 991–1010. [CrossRef]
51. Tsuda, A. Starvation Tolerance of a Planktonic Marine Copepod *Pseudocalanus newmani* Frost. *J. Exp. Mar. Biol. Ecol.* **1994**, *181*, 81–89. [CrossRef]
52. Norsker, N.-H.; Støttrup, J.G. The Importance of Dietary HUFAs for Fecundity and HUFA Content in the Harpacticoid, *Tisbe holothuriae* Humes. *Aquaculture* **1994**, *125*, 155–166. [CrossRef]
53. Miles, D.; Auchterlonie, N.; Cutts, C.; de Quero, C.M. *Rearing of the Harpacticoid Copepod Tisbe holothuriae and Its Application for the Hatchery Production of Atlantic Halibut Hippoglossus hippoglossus*; Link Project FIN 23; SEAFISH Development Aquaculture: Edinburgh, UK, 2001.
54. de Troch, M.; Chepurnov, V.; Gheerardyn, H.; Vanreusel, A.; Ólafsson, E. Is Diatom Size Selection by Harpacticoid Copepods Related to Grazer Body Size? *J. Exp. Mar. Biol. Ecol.* **2006**, *332*, 1–11. [CrossRef]
55. Kawecki, T.J.; Ebert, D. Conceptual Issues in Local Adaptation. *Ecol. Lett.* **2004**, *7*, 1225–1241. [CrossRef]

Journal of
*Marine Science
and Engineering*

MDPI

Article

A Preliminary Assessment of Microalgal Diets for Echinopluteus Larvae Culture of the Sea Urchin *Sphaerechinus granularis* (Lamarck, 1816) (Echinoidea: Toxopneustidae)

Ricardo Luís [1,*], Ricardo José [1], João Castro [1] and Carlos Andrade [2]

[1] Regional Agency for the Development of Research, Technology and Innovation (ARDITI), Madeira Tecnopolo, 9020-105 Funchal, Madeira, Portugal
[2] MARE—Marine and Environmental Sciences Centre/ARNET—Aquatic Research Network, Regional Agency for the Development of Research, Technology and Innovation (ARDITI), 9020-105 Funchal, Madeira, Portugal
* Correspondence: ricardoluis@arditi.pt

Abstract: Sea urchins play an important role in coastal marine ecosystems and are an economically valuable resource for their gonads (roe or uni). Increased demand by Asian and European countries caused overfishing of wild stocks resulting in a sharp decline of sea urchin populations. The increased efforts in developing cost-effective protocols are focused on larval development up to pre- and post-metamorphosis stages, as they are one of the main difficulties of echinoculture. This is a preliminary study to evaluate the influence of microalgal varying diet rationing on larval development, growth, and survivorship at pre-metamorphosis of *Sphaerechinus granularis* echinopluteus (30 days post-fertilization). Three microalgae diets based on *Dunaliella tertiolecta* and *Rhodomonas marina* and the combination of both species were tested. Each diet used three ration treatments: low ration (500–3000 cells·ml^{-1}); medium ration (1000–6000 cells·ml^{-1}); and high ration (4000–24,000 cells·ml^{-1}), which were cell-density adjusted to larval development. Assays used three triplicated treatments in 8 L cylindrical–conical glass incubators. Results showed that the survival of *S. granularis* echinopluteus was positively influenced by diets of lower cellular densities and a combined microalgae diet. Adoption of this type of diet regime is expected to improve larval production and provide the groundwork for future research on *S. granularis* in aquaculture.

Keywords: *Sphaerechinus granularis*; echinoculture; microalgae; survival; sustainability; marine aquaculture; diversification

Citation: Luís, R.; José, R.; Castro, J.; Andrade, C. A Preliminary Assessment of Microalgal Diets for Echinopluteus Larvae Culture of the Sea Urchin *Sphaerechinus granularis* (Lamarck, 1816) (Echinoidea: Toxopneustidae). *J. Mar. Sci. Eng.* **2023**, *11*, 1870. https://doi.org/10.3390/jmse11101870

Academic Editors: Daniel P. Costa and Ernesto Weil

Received: 1 August 2023
Revised: 5 September 2023
Accepted: 15 September 2023
Published: 26 September 2023

1. Introduction

Several species of sea urchins such as *Paracentrotus lividus*, *Psammechinus miliaris*, *Strongylocentrotus droebachiensis*, and *Sphaerechinus granularis* are exploited worldwide for their gonads (roe or uni) due to being highly regarded as a culinary delicacy. In regions such as Asia (with Japan as the principal consumer), Europe, Chile, and New Zealand, wild populations have been declining from over-exploitation [1–3].

Due to the unsustainable exploitation and crescent demand for seafood, aquaculture production is probably the best sustainable alternative to support the needs of global demand for uni [4,5]. Most aquaculture research in sea urchins focuses on the development and survival rates of larval planktonic stages up to post-metamorphosed juveniles, as they represent the major bottlenecks of echinoculture development [4,6–8].

The sea urchin *S. granularis* (Lamarck, 1816), commonly known as the purple or blunt sea urchin, is a common grazer and displays a typical covering behavior using shell fragments, pebbles, and algae [9]. They occur in the NE Atlantic, the Gulf of Guinea, and the Mediterranean Sea. This species can be found from the intertidal zone up to 130 m depth [3,9,10]. *S. granularis* has a life expectancy of 5 years and a high growth rate compared to other sea urchin species [11]. *S. granularis* urchins can attain 100 mm TD in

4 years, whereas *P. lividus* achieves 45–65 mm TD over the same 4-year duration in culture conditions [3,12,13].

Despite the increased importance of *S. granularis* as a fishery resource, the knowledge of its biology is still scarce [9]. Regarding aquaculture, research is needed to develop and optimize cost-effective methods to create protocols for rearing *S. granularis* under optimal conditions [13].

The establishment of an adequate diet to produce high-survival and good-quality larvae is a decisive step to support further research and development of aquaculture production methods for *S. granularis*. The algal species *D. tertiolecta* and *R. marina* have exhibited some of the highest survival rates in prior studies concerning the larval development of *P. lividus* echinopluteus [7,14]. Additionally, a connection has been established between the cellular size of microalgae and the mouth size of echinopluteus, as described in the literature [15]. For the echinopluteus larvae particles, 5 to 50 μm is noted to be the preferred size for feeding. This study focused on the evaluation of larval development, growth, and survivorship at pre-metamorphosis of *S. granularis* echinopluteus when fed with varying diet rations of microalgae.

2. Materials and Methods

2.1. Broodstock Collection and Rearing

Adult *S. granularis* broodstock (test size > 50 mm) were collected from local wild populations by snorkeling in the subtidal at east Madeira Island (Quinta-do-Lorde; 32°74′11.25″ N; 16°70′96.36″ W) in August 2021. The specimens were placed in 25 L containers with natural seawater and transported to Calheta Mariculture Center (CMC) where they were kept in two outdoor 750 L tanks with running ambient seawater and water flow of approximately 90 L per hour. The animals were fed *Zea mays*, at 0.7% of the biomass present in the rearing tanks three times a week, and before each feeding the uneaten food and feces were siphoned.

2.2. Spawning and Larval Rearing

Three assays were conducted. Each assay randomly selected three males and two females from the broodstock, using a total of 15 specimens with the following biometric measurements (average ± SD): humid body weight: 153.68 ± 23.92 g; body diameter: 68.36 ± 4.05 mm; and body height: 41.63 ± 2.53 mm. Sea urchins were cleaned with filtered and autoclaved seawater (20 μm; 121 °C, 15 min; hereafter referred to as FAS) and induced to spawn by agitation method [16]. Spawning urchins were placed individually with the oral side facing down in glass cubes with 1.5 L FAS for less than 30 min from the beginning of induction procedures. Eggs and sperm were pooled separately in 500 mL beakers, suspended in FAS. Egg and sperm densities were estimated with the Seggewick Rafter counting chamber and Malassez counting chamber, respectively, using a light microscope (Axioskop 2 plus, Carl Zeiss, Aalen, Germany). Egg fertilization was achieved by adding sperm volume that guaranteed the proportion of 500 spermatozoa: 1 oocyte [17]. Fertilization success (~95%) was calculated by counting the number of eggs featuring the fertilization envelope two hours after the fertilization procedure. The fertilized eggs were incubated in 6 L laboratory glass balloon bottles with FAS, at 16 ± 1 °C and medium aeration for 72 h [13].

The larval culture methods described were based on previous studies focused on development and survival in response to varying microalgae diet quantity and species [6,14,18]. For technical details, the larvae were transferred to nine 8 L cylindrical–conical glass incubators after reaching the two-arm (2 b) stage. The incubators were filled with 7 L filtered (20 μm) and ultraviolet sterilized seawater (hereafter referred to as FUVS) (AQUA—UV, De BARY, Frankfurt, Germany). Incubators were maintained partially submerged in a 200 L tank with a controlled temperature of 19 °C (water chiller HC 500A, Hailea, Chaozhou, China). The larvae were reared with a stocking density of 2 larvae·mL^{-1} in aerated, static FUVS and under fluorescent light for a constant photoperiod of 12 h light: 12 h dark.

Water quality parameters of oxygen (O_2) (multiparametric meter HandyPolaris, Oxigard®, Farum, Denmark), pH (pH Checker, HANNA, Villafranca Padovana, Italy), and salinity (refractometer, H_2Ocean, Stuart, FL, USA) were registered three times a week. Then, water changes (10% volume) were done in the incubators, and temperature was registered every hour (Logger HOBO MX2201, Onset, Cape Cod, MA, USA) for the duration of the assays. Ammonia concentration was monitored twice weekly (NH4/NH3 Test, SERA, Heinsberg, Germany) between water changes days.

2.3. Microalgae Culture

The microalgae *Dunaliella tertiolecta* (Chlamydomonales, CCAP 19/613) and *Rhodomonas marina* (Pyrenomonadales, CCAP 995/5), were grown in 5 L batch cultures [19] using a commercial culture medium (Nutribloom©, Necton, Olhão, Portugal). Fluorescent lights were used to keep a continuous photoperiod (24 h light) at 16 ± 1 °C and medium aeration. Microalgae were collected during the exponential growth phase and used in the assays.

2.4. Assays and Treatments

To study the influence of diet rationing, measured as algal cell density [18], three independent assays (D1, D2, and D3) were performed on the larval culture of *S. granularis*. The assays were performed sequentially between July and November 2022. Each assay used 126,000 larvae and a different microalga diet, i.e., assay D1 used a monospecific diet of *D. tertiolecta*, assay D2 used a monospecific diet of *R. marina*, and assay D3 used a combination of both algal species in the same proportion (*D. tertiolecta* + *R. marina* 1:1). Larval feeding was done three times a week on alternated days. Each assay D1, D2, and D3 used the following treatments: low ration (LR), medium ration (MR), and high ration (HR); the algal cell density treatments cells·ml^{-1} were adjusted accordingly with the larval stage (Table 1) following previous studies [14,18,20,21]. Three replicates for each treatment were used.

Table 1. Microalgae cell density (cells·mL^{-1}) used in the different treatments: low ration (LR); medium ration (MR); and high ration (HR) of assay D1 (*Dunaliella tertiolecta*), D2 (*Rhodomonas marina*), and D3 (*D. tertiolecta* + *R. marina* 1:1); and adjusted to the different larval stages: 2-arm pluteus (2 b); 4-arm pluteus (4 b); 6-arm pluteus (6 b); and 8-arm pluteus (8 b).

Larvae Stage	Low Ration (LR)	Medium Ration (MR)	High Ration (HR)
2 b and 4 b	500	1000	4000
6 b	1500	3000	12,000
8 b	3000	6000	24,000

2.5. Development, Morphology, and Survival

Larval development was assessed by identification of the larval stages. Identification was performed twice weekly based on the formation of new pairs of arms in the echinopluteus larvae (Figure 1). Stages identified in the present study were: two-arm pluteus (2 b); four-arm pluteus (4 b); six-arm pluteus (6 b); eight-arm pluteus (8 b); and competent/premetamorphic larvae (Figure 1). Developmental stages used to perform the shift in the algal cell density within the same treatment (see Table 1) were established when the ratio of larvae at the corresponding stages was at least 75% (Figure 2) [2,13,17]. Observations were done using a light microscope (Axioskop 2 plus, Carl Zeiss, Aalen, Germany).

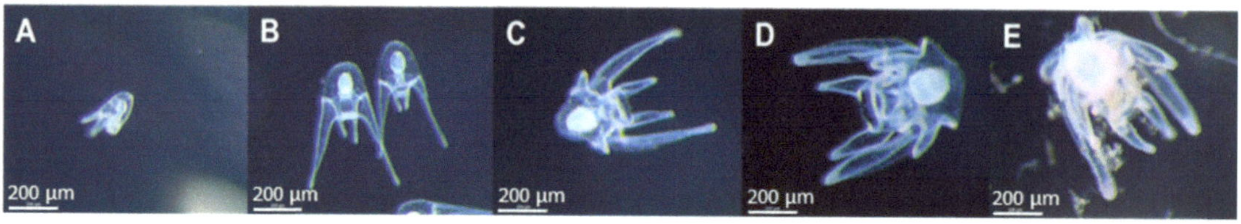

Figure 1. *Sphaerechinus granularis* larval development by larvae stage. (**A**): two-arm pluteus (2 b); (**B**): four-arm pluteus (4 b); (**C**): six-arm pluteus (6 b); (**D**): eight-arm pluteus (8 b); (**E**): competent larvae/pre-metamorphic.

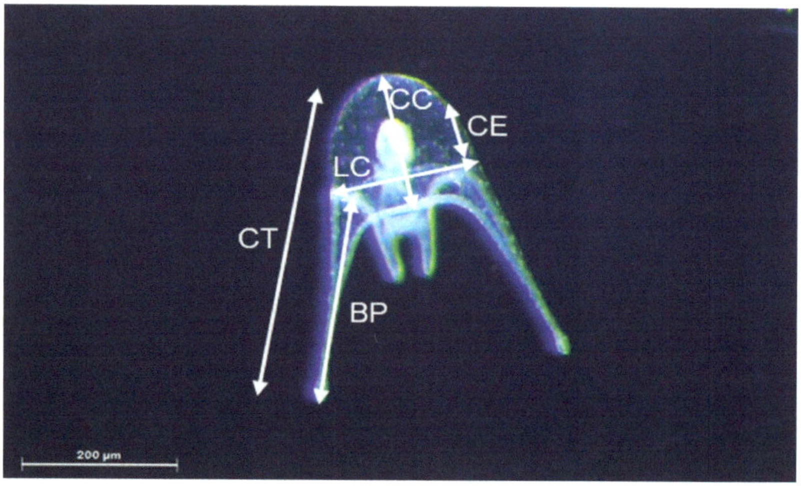

Figure 2. *Sphaerechinus granularis* four-arm pluteus (4 b). Larvae biometric parameters: CT: total length; CC: body length; LC: body width; BP: post-oral arm length; CE: stomach length.

In addition to weekly larval stage assessment, the larvae were sampled (10% of the volume in the incubators) in all assays (D1, D2, and D3) at 10, 25, and 30 days post-fertilization (DPF) [14] to measure the following biometric parameters: total length (CT); body length (CC); body width (LC); post-oral arm length (BP); and stomach length (CE) following the method described by Kelly et al. [18] (Figure 2). Biometric parameters were represented as larvae stages pooled together, reflecting larval development.

Measurements were performed using a stereo microscope (M165 C, LEICA, Germany) coupled with a digital camera (IC90 E, LEICA, Munich, Germany) and image analysis software (LAS X, LEICA, Munich, Germany). Survival (percentage values) was calculated by dividing the initial number of larvae transferred to the cylindrical–conical incubators by the total number of living larvae at 30 DPF.

2.6. Statistical Analyses

Assays were statistically analyzed separately. Larval development analyses were performed within each age (10, 25, and 30 DPF); the ratio of each larval stage was analyzed separately as a dependent variable using the ration treatments as factors. Larval morphology analyses were done within each age (10, 25, and 30 DPF); each biometric parameter was analyzed separately as a dependent variable using the ration treatments as factors. Survival was analyzed considering survival values as dependent variables and the treatments as factors.

Statistical analyses were conducted using IBM SPSS^{TM} Statistics 25 (IBM Corporation, AMONK, New York, NY, USA). All data were tested using the Shapiro–Wilk test for normality and Leven's test for homogeneity. Parametric data were analyzed using one-way ANOVA (F_{df} = value; significance level p) followed by the post hoc Bonferroni test. Nonparametric data were analyzed using the Kurskal–Wallis test (H_{df} = value; significance level p) followed by the post hoc Games–Howell test [22]. Results were expressed as mean ± standard deviation (SD), and in all cases, the null hypothesis was rejected when $p < 0.05$ for all statistical analyses.

3. Results

3.1. Water Quality

Considering all assays, we maintained consistent water quality parameters for the duration of the trials with T = 19.23 ± 0.30 °C (n = 4784), dissolved O_2 = 6.58 ± 0.22 mg·L^{-1} (n = 315), pH = 8.08 ± 0.03 (n = 315), salinity = 36 ± 0 (n = 315), and ammonia was 0 ± 0 (n = 189).

3.2. Larval Development

Assay D1: there were no significant differences in the larvae stage distribution among treatments ($H_1 = 0.000$; $p = 1.00$) at 10 DPF. At this point, 100% (n = 422) of sampled larvae were at the development stage 4 b (Figure 3A). At 25 DPF, significant statistical differences were detected in the larvae stage distribution in the HR treatment when compared with LR and MR treatments ($H_1 = 3.319$; $p < 0.05$) (Games–Howell tests; $p < 0.05$). In the LR treatment, 9.1% (n = 1) of the larvae were at the 2 b stage and 90.9% (n = 10) at the 4 b stage. In the MR treatment, 75% (n = 6) of the larvae were at the 4 b stage and 25% (n = 2) at the 6 b stage. The HR treatment showed that 100% (n = 2) of the sampled larvae were at the 8 b stage (Figure 3B). At 30 DPF, statistical analyses detected differences in the larvae stage distribution among treatments ($H_1 = 16.688$; $p < 0.05$), these differences were significant among all treatments (Games–Howell tests; $p < 0.05$). In the LR treatment, 95.1 % (n = 77) of the larvae were at the 4 b stage and 4.9 % (n = 4) at stage 6 b. In the MR treatment, 71.1% (n = 54) of the larvae were at stage 4 b, 21.1% (n = 16) at the 6 b stage, and 7.9% (n = 6) at the 8 b stage. In the HR treatment 27% (n = 10) of the larvae were at the 4 b stage, 56.8% (n = 21) at the 6 b stage, and 16.2% (n = 6) at the 8 b stage (Figure 3C).

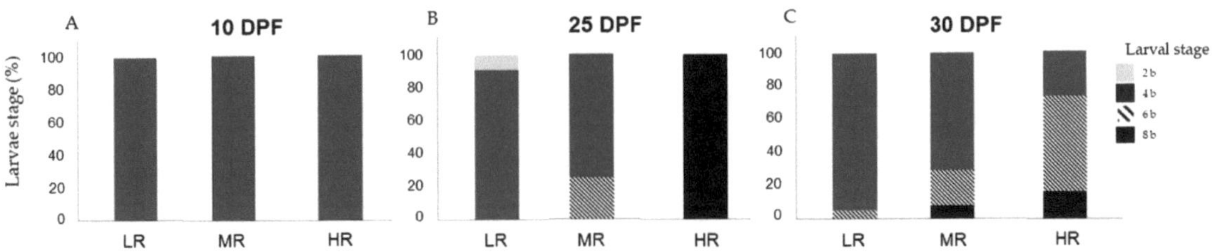

Figure 3. Percentage (%) of *Sphaerechinus granularis* larvae stage by treatment (treatment conditions are defined in Table 1) at days post-fertilization (DPF): 10 (**A**), 25 (**B**), and 30 (**C**), in assay D1 (*Dunaliella tertiolecta*). Differences among treatments were considered significant at $p < 0.05$.

In assay D2, at 10 DPF no statistical differences were detected in the larvae stage distribution among treatments ($H_1 = 1.000$; $p = 1.00$). One hundred percent of sampled larvae (n = 92) were at development stage 4 b (Figure 4A). At 25 DPF, statistical analyses found no statistical differences in the larvae stage distribution among treatments ($H_1 = 1.273$; $p = 0.259$). In the LR treatment, 66.7% (n = 8) of the larvae were at 8 b and 33.3% (n = 4) at the pre-metamorphic stage. In the MR treatment, 100% (n = 3) of the larvae were at the 8 b stage. In the HR treatment, 60% (n = 3) of the larvae were at the 8 b stage and 40% (n = 2) at the pre-metamorphic stage (Figure 4B). At 30 DPF, statistical differences were detected in the larvae stage distribution among treatments ($H_1 = 4.673$; $p < 0.05$). These differences were revealed to be significant when comparing the HR treatment with LR and MR treatments (Games–Howell tests; $p < 0.05$). In the LR treatment, 100% (n = 32) of the larvae were at the pre-metamorphic stage. In the MR treatment, 7.1% ($n = 1$) of the larvae were at the 6 b stage, 7.1% ($n = 1$) at the 8 b stage, and 85.7% ($n = 12$) at the pre-metamorphic stage. The HR treatment did not produce living larvae at this sampling point (Figure 4C).

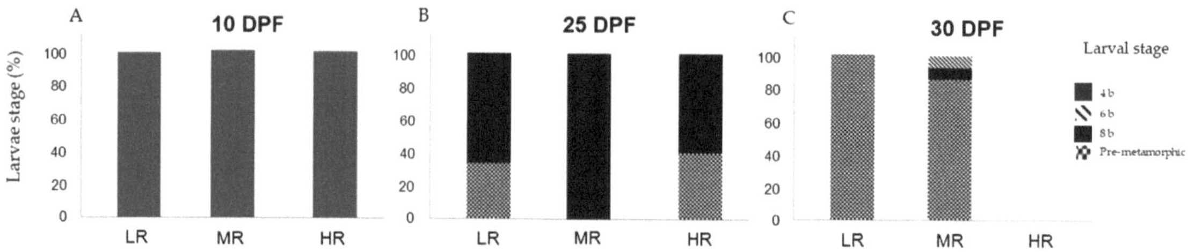

Figure 4. Percentage (%) of *Sphaerechinus granularis* larvae stage by treatment (treatment conditions are defined in Table 1) at days post-fertilization (DPF): 10 (**A**), 25 (**B**) and 30 (**C**), in assay D2 (*Rhodomonas marina*). Differences among treatments were considered significant at $p < 0.05$.

In assay D3, at 10 DPF no significant differences were detected in the larvae stage distribution among treatments ($H_1 = 0.000$; $p = 1.00$). One hundred percent of sampled larvae ($n = 270$) were at development stage 4 b (Figure 5A). At 25 DPF, statistical analyses found no statistical differences in the larvae stage distribution among treatments ($H_1 = 1.750$; $p = 0.186$). In the LR treatment, 100% ($n = 3$) of the sampled larvae were at the 8 b stage. In the MR treatment, 25% ($n = 1$) were at stage 4 b, 25% ($n = 1$) at the 6 b stage, and 50% at the 8 b stage (Figure 5B). At 30 DPF, larvae stage distribution among treatments was statistically different among treatments ($H_1 = 9.085$; $p < 0.05$). These differences were statistically significant when compared to the LR treatment with the MR treatment (Games–Howell tests; $p < 0.05$). In treatment LR 21.1% ($n = 4$) of the larvae were at the 4 b stage, 36.8% ($n = 7$) at the 6 b stage, and 42.1% ($n = 8$) at the 8 b stage. In the MR treatment, 3% ($n = 1$) of the larvae were at the 4 b stage, 15.2% ($n = 5$) at the 6 b stage, and 81.8% ($n = 27$) at the 8 b stage. In the HR treatment, 14.3% ($n = 1$) of the larvae were at the 4 b stage, 28.6% ($n = 2$) at the 6 b stage, and 57.1% ($n = 4$) at the 8 b stage (Figure 5C).

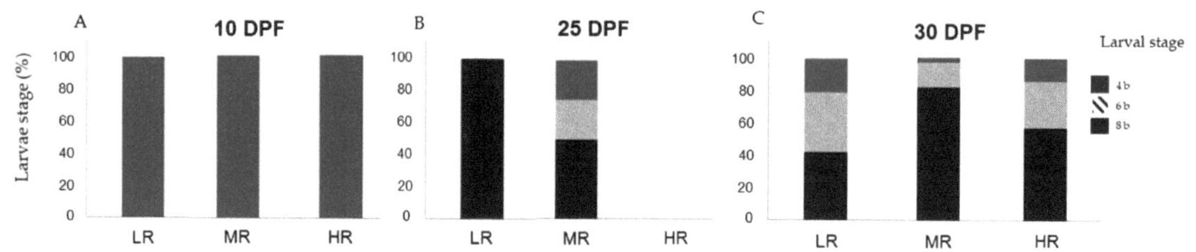

Figure 5. Percentage (%) of *Sphaerechinus granularis* larvae stage by treatment (treatment conditions are defined in Table 1) at days post-fertilization (DPF): 10 (**A**), 25 (**B**), and 30 (**C**), in assay D3 (*Dunaliella tertiolecta* and *Rhodomonas marina* at a 1:1 cellular ratio). Differences among treatments were considered significant at $p < 0.05$.

3.3. Larval Morphology

In assay D1, at 10 DPF statistical analyses detected differences in the distribution in larvae biometric parameters CT ($H_2 = 85.092$; $p < 0.05$); CC ($H_2 = 111.534$; $p < 0.05$); BP ($H_2 = 106.940$; $p < 0.05$); LC ($H_2 = 97.688$; $p < 0.05$); and CE ($H_2 = 105.053$; $p < 0.05$) among treatments (Figure 6). The distribution of CC (193.79 ± 30.39 μm); LC (173.38 ± 44.79 μm); and CE (106.97 ± 17.07 μm) were higher in the HR treatment and CT (501.57 ± 65.97 μm) and BP (352.57 ± 59.32 μm) were higher in the MR treatment (Games–Howell tests; $p < 0.05$) (Figure 6, column 10 DPF). At 25 DPF, differences were found in the distribution of larvae measurement LC ($H_2 = 9.270$; $p < 0.05$). Statistical analysis showed these differences were not significant for LC (362.08 ± 130.18 μm) when comparing among treatments (Games–Howell tests; $p < 0.05$) (Figure 6K). At 30 DPF, statistical differences were detected in the distribution of the biometric parameters CC (H_2 = 64.149; $p < 0.05$), LC (H_2 = 91.519; $p < 0.05$), and CE (H_2 = 58.490; $p < 0.05$). These differences were significant among treatments (LR, MR, HR) (Games–Howell tests; $p < 0.05$). The distribution of CC (246.17 ± 58.17 μm); LC (249.84 ± 77.47 μm); and CE (166.34 ± 44.41 μm) was higher in treatment HR (Figure 6F,M,P).

In assay D2, at 10 DPF differences were found in the distribution of larvae biometric parameters CC (H_2 = 10.187; $p < 0.05$); BP (H_2 = 1.114; $p < 0.05$); LC (H_2 = 15.070; $p < 0.05$); and CE (H_2 = 28.904; $p < 0.05$). The measurement CC (245.32 ± 36.42 μm) was significantly higher in the HR treatment and BP (233.28 ± 59.84 μm) showed to be higher in treatment MR as the biometric parameter LC (293.85 ± 55.29 μm) was significantly higher in treatment HR (Games–Howell tests; $p < 0.05$) (Figure 6D,G,J). The differences found in the distribution of the biometric parameter BP were not statistically significant among treatments (Games–Howell tests; $p < 0.05$) (Figure 6G). At 25 DPF, statistical analyses found differences in the distribution of larvae biometric parameter LC (H_2 = 6.501; $p < 0.05$), these differences were not significant when comparing among treatments (Games–Howell tests; $p < 0.05$) (Figure 6K). At 30 DPF, there were no differences in the distribution of the biometric parameters (CC; CT; BP; LC; CE) among treatments (LR, MR, HR) (Games–Howell tests; $p < 0.05$) (Figure 6, column 30 DPF).

In assay D3, at 10 DPF statistical analyses found differences in the distribution of parameters CT (H_2 = 74.940; $p < 0.05$); CC (H_2 = 175.583; $p < 0.05$); BP (H_2 = 223.119; $p < 0.05$); LC (H_2 = 189.158; $p < 0.05$); and CE (H_2 = 152.722; $p < 0.05$). The distribution of CT (523.04 ± 53.82 μm); CC (259.36 ± 25.33 μm); LC (273.07 ± 27.00 μm); and CE (105.71 ± 10.98 μm) was revealed to be significantly higher in the HR treatment (Games–Howell tests; $p < 0.05$) and the distribution of BP (304.88 ± 40.35 μm) showed to be significantly higher in treatment LR (Games–Howell tests; $p < 0.05$) (Figure 6, column 10 DPF). At 25 and 30 DPF, results showed no statistically significant differences in the distribution of all biometric parameters (CC; CT; BP; LC; CE) among treatments (LR; MR; HR) (Games–Howell tests; $p < 0.05$) (Figure 6, columns 25 and 30 DPF).

3.4. Larval Survival

In assay D1, statistical differences were detected in the survival distribution among treatments. Survival in the MR treatment (0.20 ± 0.05%) showed to be significantly lower when compared to the LR and HR treatments (0.25 ± 0.11 and 0.23 ± 0.05%, respectively) (H_2 = 36.356; $p < 0.05$) (Games–Howell tests; $p < 0.05$) (Figure 7). In assay D2, there were significant differences in the survival distribution among all treatments (H_2 = 37.780; $p < 0.05$), showing higher survival at LR (0.10 ± 0.02%). On the contrary, survival was negligible at HR (0.00 ± 0.00%) (Games–Howell tests; $p < 0.05$) (Figure 7). Assay D3 showed significantly higher survival in the LR and MR treatments (0.07 ± 0.03 and 0.07 ± 0.01%, respectively) when compared with the HR treatment (0.02 ± 0.01%) (H_2 = 19.841; $p < 0.05$) (Games–Howell tests; $p < 0.05$) (Figure 7).

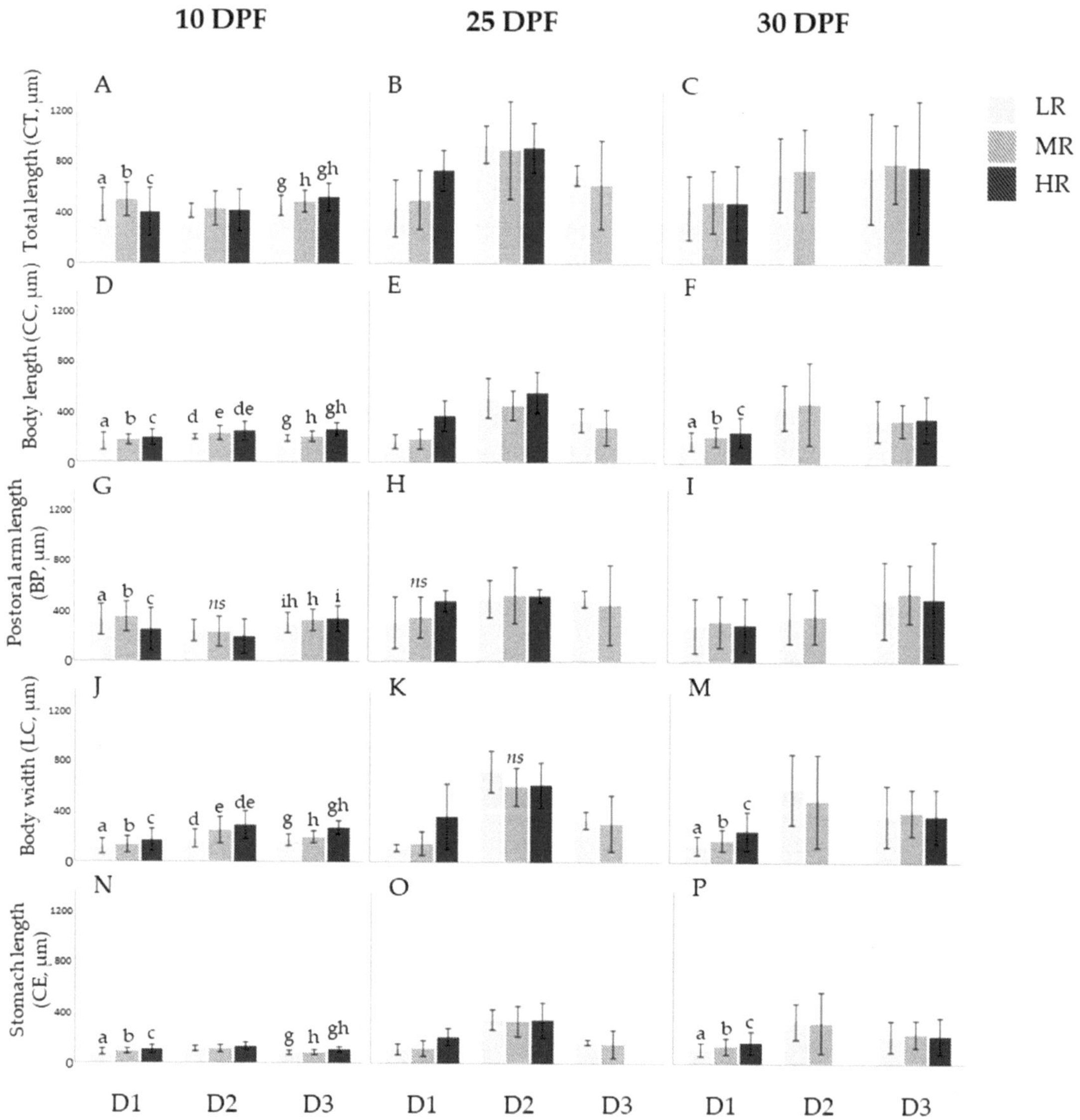

Figure 6. Mean (± DP) of the distribution of the biometric parameters: total length (CT), body length (CC), body width (LC), post-oral arm length (BP), and stomach length (CE) in microns (µm) registered in *S. granularis* larvae by treatment (LR, MR, HR) at 10, 25, and 30 days post-fertilization (DPF) for assays D1 (*D. tertiolecta*), D2 (*R. marina*), and D3 (*D. tertiolecta* and *R. marina* at a 1:1 cellular ratio) organized in subfigures (**A–P**). Significant differences ($p < 0.05$) in the distribution of biometric parameters among treatments are represented by superscript letters (a–i and non-significant differences ($p > 0.05$) are represented as *ns*. The absence of superscript letters indicates that there were no significant or non-significant differences in the distribution of biometric parameters among treatments.

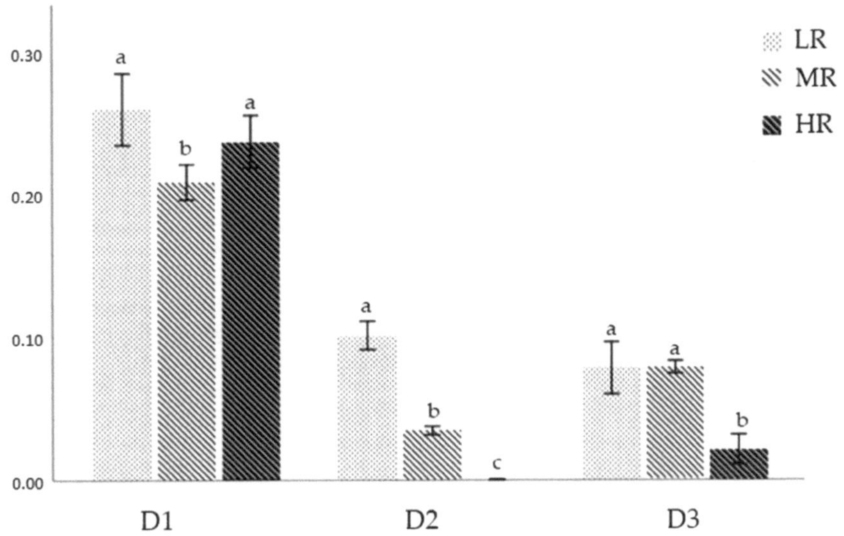

Figure 7. Survival of *S. granularis* larvae at 30 DPF for assays D1 (*D. tertiolecta*), D2 (*R. marina*), and D3 (*D. tertiolecta* and *R. marina* at a 1:1 cellular ratio). Results showed as mean ± SD. Significant differences ($p < 0.05$) among treatments are represented by different letters (a–c).

4. Discussion

Previous studies have provided insights into the impact of diet rations and microalgae species on the survival and morphology of echinoid larvae in cultured conditions [17,21].

All conducted assays indicate that diet rationing significantly influenced larval development. For instance, in the case of *S. granularis* larvae, a monospecific diet *of D. tertiolecta* with a high ration (HR) facilitated faster development compared to treatments with lower cell density (LR, MR). At 25 days post-fertilization (DPF), most larvae were in the eight-arm pluteus stage. Similarly, during assay D1, the HR treatment led to a distribution of more advanced stages (6 b and 8 b pluteus), while the LR and MR treatments exhibited less-developed larvae (four-arm and six-arm pluteus). Notably, the *D. tertiolecta* HR treatment at 10 DPF resulted in higher biometric development in parameters such as body length (CC), body width (LC), and stomach length (CE), while the MR treatment promoted greater post-oral arm length (BP) development. At 30 DPF, the larvae in assay D1 demonstrated a consistent biometric development pattern, aligning with higher values for post-oral arm length (BP) in the MR treatment, as seen in previous studies involving *E. esculentus*, *P. lividus*, and *P. miliaris*. Survival rates in assay D1 were highest for the LR treatment, in line with findings from other authors [6,18,21].

When it comes to a monospecific diet of *R. marina*, no significant differences in larval stage development were observed among various ration treatments. Results showed consistent developmental stage distributions across treatments throughout the assay. While the higher cell density treatment (HR) promoted better biometric development in parameters such as CC, LC, and BP, the lower cell density treatments (LR, MR) excelled in promoting post-oral arm length (BP) development. Survival rates at 30 DPF were notably higher for the LR and MR treatments compared to the HR treatment. However, it is worth noting that survival rate results in this assay (D2) might not be linked to inadequate diet rationing but instead to accelerated stage progression, resulting in larvae completing metamorphosis in the incubators. It is worth noting that water quality parameters remained consistent throughout the trials, with no detection of ammonia.

In assays involving a combined diet of *D. tertiolecta* and *R. marina* at a 1:1 cellular ratio, comparable larval development was evident when comparing LR and MR treatments up to 25 DPF. By 30 DPF, it became evident that the combined microalgae diet supported larval development up to the 8 b stage. All treatments exhibited similar distributions of advanced larval stages and survival rates. In this context, the HR treatment fostered higher biometric development in parameters such as CT, CC, LC, and CE, while the LR treatment led to greater BP development. Notably, LR and MR treatments demonstrated the highest survival rates.

Regarding the larval development of *S. granularis*, there is limited literature on the larval stage and biometric dimensions. However, it is established that the species follows a conventional development into planktotrophic and morphologically intricate larvae that require around 30 to 40 days in aquaculture conditions to reach competence [13,23]. This study underscores the influence of diet rationing and lower cell density diets on larval development and survival rates, aligning with findings from studies involving *P. lividus*, *P. miliaris*, and *S. droebachiensis* [2,14,17,18,24]. Notably, changes in biometric parameters during larval development, such as post-oral arm length (BP) and body width (LC), hold importance in assessing larval response to specific diet rations.

It has been demonstrated that microalgae diets with lower cellular density tend to yield better survival rates in *Echinus esculentus* [21], *P. lividus* [2], and *P. miliaris* [18] larvae. Research by Jimmy et al. [21] on *E. esculentus* suggests that higher food rations can be disadvantageous for developing echinopluteus larvae, leading to decreased metamorphosis numbers and survivorship. While apparent physiological impairments are not evident, other factors such as algal metabolites might impact larvae fed high-ration diets. Similar observations were made by Kelly et al. [18] in *P. miliaris* larvae fed a high-ration diet, resulting in shortened post-oral arms when fed microalgae *Pleurocrysis elongata*. This indicates that different species exhibit diverse development patterns despite similar survivorship outcomes.

Within the present study, a monospecific diet of *D. marina* fostered larval survivorship but slowed development, with no specimens reaching the pre-metamorphic stage by 30 DPF. This suggests that *D. marina* sustains *S. granularis* larvae but does not promote rapid stage progression and only supports larval development up to the 8 b stage in 30 DPF. Kelly et al. [18] highlight that *D. marina* lacks long-chain polyunsaturated fatty acids (PUFAs) such as 20:5n − 3 (eicosapentaenoic) and 22:6n − 3 (docosahexaenoic), and while these PUFAs might not be essential for *P. miliaris* development, the impact of these and other biochemical components on larval, juvenile, or broodstock diets remains unexplored.

Conversely, a monospecific diet of *R. marina* resulted in faster development up to the pre-metamorphic stage by 30 DPF, albeit with poorer larval survivorship. A combination of both microalgae species diets yielded improved survival compared to a monospecific *R. marina* diet, although no specimens reached the pre-metamorphic stage by 30 DPF. Although the combination of both microalgae species seems better suited to the echinopluteus development of *S. granularis*, it is important to acknowledge that larvae survival in all assays was less than 1%, suggesting that further studies using other microalgae species and cellular equivalence are needed.

Considering the imperative of a dependable source of high-quality juveniles for sustainable aquaculture development, enhancing larval development and survival rate up to competence under aquaculture conditions is paramount. Having established in this preliminary work that a combination of both microalgae species seems a more appropriate diet for *S. granularis* larvae, it is crucial for future studies to assess the impact of factors such as temperature, salinity, and water quality ranges, which significantly influence water quality maintenance and successful larval production. Additionally, investigations should delve into diet regimes, comparing various microalgae species and their biochemical components, given their substantial influence on larval development and survival.

Author Contributions: Conceptualization, R.L.; Formal analysis, R.L.; Funding acquisition, C.A.; Investigation, R.L.; Methodology, R.L., R.J. and J.C.; Project administration, C.A.; Supervision, C.A.; Writing—original draft, R.L.; Writing—review and editing, C.A. All authors have read and agreed to the published version of the manuscript.

Funding: This study was performed under the project ISLANDAP ADVANCED "R + D + i Towards Aquaponic Development in the UP islands and the Circular Economy. Interregional forward challenges (ISLANDAP ADVANCED, MAC2/1. 1ª/299)" founded by FEDER under the INTERREG MAC 2014-2020 Program.

Institutional Review Board Statement: The present work did not involve vertebrates or protected or endangered species. All experimental procedures on sea urchins were conducted in compliance with the Portuguese law and Directive 2010/63/EU.

Informed Consent Statement: Not applicable.

Data Availability Statement: The data that support the findings of this study are available from the corresponding author upon reasonable request.

Acknowledgments: The authors also would like to thank the Centro de Maricultura da Calheta technicians, especially António Abreu, for their assistance in the conception and assembly of the systems for broodstock maintenance and larval assays.

Conflicts of Interest: The authors declare no conflict of interest.

References

1. Andrew, N.; Agatsuma, Y.; Ballesteros, E.; Bazhin, A.; Creaser, E.; Barnes DK, A.; Botsford, L.W.; Bradbury, A.; Campbell, A.; Dixon, J.D.; et al. Status and management of world sea urchin fisheries. *Oceanogr. Mar. Biol.* **2002**, *40*, 343–425. [CrossRef]
2. Brundu, G.; Monleón, L.V.; Vallainc, D.; Carboni, S. Effects of larval diet and metamorphosis cue on survival and growth of sea urchin post-larvae (*Paracentrotus lividus*, Lamarck, 1816). *Aquaculture* **2016**, *465*, 265–271. [CrossRef]
3. Lourenço, S.; José, R.; Neves, P.; Gois, A.; Cordeiro, N.; Andrade, C.; Ribeiro, C. Population Density, Reproduction Cycle and Nutritional Value of *Sphaerechinus granularis* (Echinodermata: Echinoidea) in an Oceanic Insular Ecosystem. *Front. Mar. Sci.* **2022**, *8*, 699942. [CrossRef]
4. Brundu, G.; Vallainc, D.; Baroli, M.; Figus, A.M.; Pinna, A.; Carboni, S. Effects of on-demand feeding on sea urchin larvae (*Paracentrotus lividus*: Lamarck, 1816), development, survival and microalgae utilization. *Aquac. Res.* **2017**, *48*, 1550–1560. [CrossRef]
5. Martínez-Pita, I.; Sànchez-Espana, A.I.; García, F.J. Gonadal growth and reproduction in the sea urchin *Sphaerechinus granularis* (Lamarck 1816) (Echinodermata: Echinoidea) in southern Spain. *Sci. Mar.* **2008**, *72*, 603–611. [CrossRef]
6. Carboni, S.; Kelly, M.S.; Hughes, A.D.; Vignier, J.; Atack, T.; Migaud, H. Evaluation of flow through culture technique for commercial production of sea urchin (*Paracentrotus lividus*) larvae. *Aquac. Res.* **2012**, *1*, 5. [CrossRef]
7. Castilla-Gavilán, M.; Buzin, F.; Cognie, B.; Dumay, J.; Turpin, V.; Decottignies, P. Optimizing microalgae diets in sea urchin *Paracentrotus lividus* larviculture to promote aquaculture diversification. *Aquaculture* **2018**, *490*, 251–259. [CrossRef]
8. McEdward, L.R. Morphometric and metabolic analysis of the growth and form of an echinopluteus. *J. Exp. Mar. Biol. Ecol.* **1984**, *82*, 259–287. [CrossRef]
9. Vafidis, D.; Antoniadou, C.; Ioannidi, V. Population density, size structure, and reproductive cycle of the comestible sea urchin *Sphaerechinus granularis* (Echinodermata: Echinoidea) in the Pagasitikos gulf (Aegean Sea). *Animals* **2020**, *10*, 1506. [CrossRef] [PubMed]
10. Subasinghe, R.P.; Bondad-Reantaso, M.G.; McGladdery, S.E. Aquaculture development, health and wealth. In *Aquaculture in the Third Millennium. Technical Proceedings of the Conference on Aquaculture in the Third Millennium, Bangkok, Thailand, 20–25 February 2001*; Subasinghe, R.P., Bueno, P., Phillips, M.J., Hough, C., McGladdery, S.E., Arthur, J.R., Eds.; NACA: Bangkok, Thailand; FAO: Rome, Italy, 2001; pp. 167–191.
11. Guillou, M.; Lumingas, L.J.L. The reproductive cycle of the 'blunt' sea urchin. *Aquac. Int.* **1998**, *6*, 147–160. [CrossRef]
12. Boudouresque, C.F.; Verlaque, M. Ecology of *Paracentrotus lividus*. *Dev. Aquac. Fish. Sci.* **2007**, *37*, 243–285.
13. González-Suárez, D.; Hernández, J.C. Early developmental stages of the sea urchin *Sphaerechinus granularis* (Lamarck, 1816) (Echinoidea: Toxopneustidae). *Mar. Biol. Res.* **2022**, *18*, 266–277. [CrossRef]
14. Azad, A.K.; Pearce, C.M.; McKinley, R.S. Influence of microalgal species and dietary rations on larval development and survival of the purple sea urchin, *Strongylocentrotus purpuratus* (Stimpson, 1857). *Aquaculture* **2011**, *322*, 210–217. [CrossRef]
15. McEdward, L.R.; Miner, B.G. Echinoid larval ecology. *Dev. Aquac. Fish. Sci.* **2007**, *37*, 71–93. [CrossRef]
16. Gago, J.; Luís, O.J. Comparison of spawning induction techniques on *Paracentrotus lividus* (Echinodermata: Echinoidea) broodstock. *Aquac. Int.* **2011**, *19*, 181–191. [CrossRef]

17. Gomes, A.; Lourenço, S.; Santos, P.M.; Raposo, A.; Mendes, S.; Gonçalves, S.C.; Ferreira SM, F.; Pombo, A. Effects of single and mixed-diatom diets on growth condition, and survival of larvae of the sea urchin *Paracentrotus lividus* (Lamarcj, 1816). *Aquac. Int.* **2021**, *29*, 1069–1090. [CrossRef]
18. Kelly, M.S.; Hunter, A.J.; Scholfield, C.L.; McKenzie, J.D. Morphology and survivorship of larval *Psammechinus miliaris* (Gmelin) (Echinodermata: Echinoidea) in response to varying food quantity and quality. *Aquaculture* **2000**, *183*, 223–240. [CrossRef]
19. Ferreira, P.M.P. *Manual de Cultivo e Bioencapsulação da Cadeia Alimentar Para a Larvicultura de Peixes Marinhos;* Instituto Nacional de Recursos Biológicos IP: Lisbon, Portugal, 2009; p. 235. ISBN 978-972-9372-37-7.
20. Cárcamo, P.F. Massive production of larvae and seed of the sea urchin *Loxechinus albus.* In *Sea Urchins Fisheries and Ecologgy;* Lawrence, J.M., Guzmán, O., Eds.; DEStech Publications Inc.: Lancaster, PA, USA, 2004; pp. 299–306.
21. Jimmy, R.A.; Kelly, M.S.; Beumont, A.R. The effect of diet type and quantity on the development of common sea urchin larvae *Echinus esculentus. Aquaculture* **2003**, *220*, 261–275. [CrossRef]
22. Zar, J.H. *Biostatistical Analysis*, 5th ed.; Prentice-Hall/Pearson: Upper Saddle River, NJ, USA, 2010.
23. Guillou, M.; Michel, C. Reproduction and growth of *Sphaerechinus granularis* (echinodermata: Echinoidea) in southern Brittany. *J. Mar. Biol. Assoc. United Kingd.* **1993**, *73*, 179–192. [CrossRef]
24. Fenaux, L.; Cellario, C.; Etienne, M. Variations in the ingestion rate of algal cells with morphological development of larvae of *Paracentrotus lividus* (Echinodermata: Echinoidea). *Mar. Ecol. Prog. Ser.* **1985**, *24*, 161–165. [CrossRef]

Journal of
Marine Science and Engineering

MDPI

Article

Effect of Long-Term Day/Night Temperature Oscillations on the Overall Performance of Gilthead Seabream (*Sparus aurata*) Juveniles

Ana Catarina Matias [1,*,†], Ravi Luna Araújo [1,†], Laura Ribeiro [1], Narcisa Maria Bandarra [2], Amparo Gonçalves [2] and Pedro Pousão-Ferreira [1]

1 EPPO—Aquaculture Research Station, IPMA—Portuguese Institute for Sea and Atmosphere, Av. do Parque Natural da Ria Formosa s/n, 8700-194 Olhão, Portugal
2 DivAV—Aquaculture, Upgrading and Bioprospection Division, IPMA—Portuguese Institute for Sea and Atmosphere, Av. Alfredo Magalhães Ramalho 6, 1495-165 Algés, Portugal
* Correspondence: ana.matias@ipma.pt
† These authors contributed equally to this work.

Abstract: Water temperature variations affect fish growth and health, often leading to huge losses in fish production, especially during the cold season. To alleviate this constraint, fish farmers can use a water heating system driven by solar energy during daytime. This action will cause a water temperature drop during the night period, making it important to understand the physiological response of fish exposed to the resulting day/night temperature oscillations. To investigate this scenario, gilthead seabream juveniles (96.3 ± 1.0 g) were exposed to different thermal regimes for 67 days: Tconstant and Tdaily cycles. The latter group was exposed to daily water temperature oscillations between ~19 and 13 °C compared with a constant temperature of ~19 °C for the other experimental group. Temperature fluctuations compromised fish growth efficiency and reduced the proportion of fatty acids in several tissues, with implications for the whole proximate composition. Moreover, temperature oscillations influenced several blood parameters. These results favor the usage of a constant water temperature of ~19 °C for optimal gilthead seabream juvenile production instead of a day/night water temperature oscillating regime. Nevertheless, the type of energy used to warm the water will depend on the operational conditions and/or business strategy of fish farmers.

Keywords: gilthead seabream; temperature; physiology; production; solar energy; aquaculture

Citation: Matias, A.C.; Araújo, R.L.; Ribeiro, L.; Bandarra, N.M.; Gonçalves, A.; Pousão-Ferreira, P. Effect of Long-Term Day/Night Temperature Oscillations on the Overall Performance of Gilthead Seabream (*Sparus aurata*) Juveniles. *J. Mar. Sci. Eng.* **2023**, *11*, 1687. https://doi.org/10.3390/jmse11091687

Academic Editor: Ka Hou Chu

Received: 24 July 2023
Revised: 24 August 2023
Accepted: 24 August 2023
Published: 27 August 2023

1. Introduction

Water temperature is critical for ectothermic animals, such as fish, because they cannot regulate their body temperature. Consequently, all physiological processes are affected by this parameter. Numerous studies have shown that temperature has a huge impact on fish growth rate and food consumption, especially for farmed fish which cannot escape from unfavorable environmental conditions, such as low winter temperatures [1–3].

Gilthead seabream (*Sparus aurata*) is one of the most valuable farmed fish species in the Mediterranean Sea. Wild gilthead seabream is exposed to wide fluctuations in water temperature (11 to 30 °C), presenting a broad range of thermal tolerance [4]. Growth is compromised when the water temperature drops below 13 °C, which provokes fasting, growth arrest, general metabolic depression, and many other physiological alterations [5]. Winter syndrome is a pathology that may come from this complex biological response, often leading to acute mortality episodes and thus, considerable economic losses [6]. To overcome these issues in fish farms, rearing water temperature should be controlled, according to each species' optimal requirements, throughout the overall fish production period, independently of temperature weather variations. However, this scenario is economically inviable for fish farmers since it involves huge electricity and fuel costs, mainly in winter.

Photovoltaic technology may provide a sustainable and affordable energetic solution to overcome this obstacle by warming the rearing water only during daytime, causing a water temperature drop during the night period [7]. To test this hypothesis, we evaluated the effect of day/night water temperature fluctuations on the overall performance of gilthead seabream. In this trial, a group of fish was maintained in warm water during daytime to simulate water warming through solar energy and allow the normal feeding routine of the fish. Overnight, to simulate the natural temperature drop associated with the night period without the use of any water heating source, the water temperature was cooled down to ~13 °C, which corresponds to the average minimum temperature occurring over the year in the western Mediterranean Sea [8]. Although several studies have tested the physiological response of gilthead seabream exposed to water temperature oscillations [2,3,9,10], to our knowledge, this is the first study in which gilthead seabream is exposed to repeated day/night water temperature fluctuations where the cold period corresponds to the night period in which fish do not feed and overall fish metabolism is decreased. The decreased metabolism associated with cold temperatures was showed by [11].

Temperature variations can invoke biochemical and physiological changes at the organismal and cellular levels. At the organismal level, these changes are mediated by the neuroendocrine system and are generally characterized by changes in plasma metabolites, hematological features, and the concentration of circulating stress hormones, such as cortisol [12,13]. At the cellular level, thermal stress can induce the expression of heat shock proteins (Hsps) such as Hsp70, which is an abundant and well conserved protein [13,14]. When cells are exposed to a stressor, a rapid increase in Hsp70 levels occurs to protect the cells from the harmful effects of the stressor, such as temperature variations. Due to the greater need for molecular chaperones under stressful conditions (as the rate of cellular protein damage or refolding increases), the expression levels of Hsp70 in fish in both experimental groups were evaluated in the gills, a key tissue that is in direct contact with water temperature, and the liver, a central tissue in overall fish metabolism.

As metabolic functions are affected by temperature, the digestion process of fish can also be disturbed, partly due to alterations in the activity of digestive enzymes [15]. Since fish growth and health are dependent on the proper functioning of these enzymes and the consequent absorption of nutrients in the intestine, digestive enzymes constitute an important indicator to address the effect of long-term temperature fluctuations in fish. Similarly, the digestibility of lipids can be affected by temperature variations, resulting in changes in fatty acid composition [16] with possible effects on the fluidity of the cell membranes of fish. A reduction in temperature results in the accumulation of unsaturated fatty acids in the membranes during acclimation to low temperatures with concomitant decrease of membrane fluidity [17,18]. All these temperature-related effects have an impact on the quality of the final product, such as whole-body composition or flesh texture, as they reflect the metabolic changes that are occurring in the fish body [15,19,20].

Better control of fish rearing conditions such as temperature will lead to the improvement of aquaculture practices and the consequent optimization of production and associated costs. By precisely monitoring temperature, systems can be adjusted to ensure optimal growth conditions of aquatic organisms for various locations, be they warm or cold. Since daily temperature variations can induce changes in the physiological status of fish, we analyzed the effects of day/night temperature fluctuations on gilthead seabream juveniles' overall performance by monitoring growth and feed utilization parameters, blood and plasma biochemistry, digestive enzyme activity, cellular stress, tissue structure, and fatty acid composition.

2. Materials and Methods

2.1. Fish and Experimental Conditions

The fish trial was performed at the Aquaculture Research Station (EPPO) of the Portuguese Institute of Sea and Atmosphere (IPMA) (Olhão, Portugal), certified by the Direção Geral da Alimentação e Veterinária to execute animal experimentation under the authorization, 2018/12/17-025516. The experiment was directed by trained scientists (following category C FELASA recommendations) and conducted according to the European guidelines on the protection of animals used for scientific purposes (Directive 2010/63/UE of the European Parliament and the European Union Council) and the related Portuguese legislation guidelines (Decreto-Lei 113/2013) on animal experimentation and welfare.

Fish juveniles used in this trial (8 months old) derived from a single spawn of gilthead seabream broodstock maintained at the IPMA's Aquaculture Research Station (EPPO). A total of 720 gilthead seabream with a mean initial body weight of 96.3 ± 1.0 g were distributed in six 1.5 m^3 circular fiberglass tanks, resulting in a density of 7.7 ± 0.08 kg m^{-3}. Three tanks were subjected to water temperature variations, between ~13 °C (night) and ~19 °C (day), the Tdaily cycles group, and three tanks were kept at a constant temperature during the whole experimental period (~19 °C) for 67 days, the Tconstant group. The fish were fed three times a day (10.30 am, 2 pm and 4 pm) by hand to apparent satiety. Efforts were made to limit overfeeding, and a continuous record of feed intake was maintained during the entire trial. Rearing tanks were supplied with flow-through gravel-filtered, aerated seawater (salinity: 34 part per thousand; dissolved oxygen 65–85%) at 600 L h^{-1}. The photoperiod was set at 14 L:10 D, with lights on at 9 a.m. Ammonia and nitrite levels were kept around zero mg L^{-1}. At the end of the trial, the fish from each tank were weighed and biomass calculated. From each tank, 5 fish were used for blood collection, tissue sampling and histology, 5 more were used for texture analysis, and another 6 for proximal composition analysis.

2.2. Experimental Protocol—Day/Night Temperature Oscillations

The temperature variations for both groups, the Tconstant and Tdaily cycles, are represented in Figure 1. The water temperature of both experimental groups was kept warm during daytime for 7 h (warm period). The fish from the Tconstant group were kept at 18.8 ± 0.1 °C throughout the whole experimental period, while fish from the Tdaily cycles group were subjected to a temperature variation pattern during the night period. The water temperature of the Tdaily cycles group was programmed to decrease after 5 p.m. to initiate the cooling period. At this stage, the running warm water was replaced by cold water, and after 5 h, the water temperature reached 12.3 ± 0.2 °C. This was the time when the maximum difference between temperatures was attained (6.5 ± 0.1 °C). This cold temperature was maintained for 7 h, characterizing the cold period. At 5 a.m., the water temperature started to increase, taking approximately 5 h to reach the temperature of the warm water (~19 °C), signaling the end of the warming period. The water temperature oscillations were managed using a heat pump system, as described in [21].

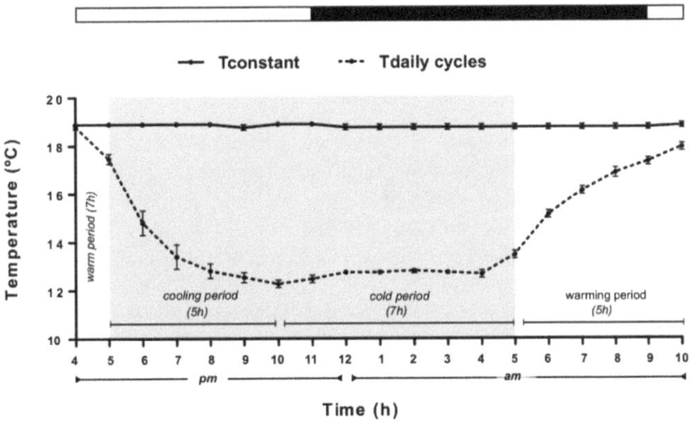

Figure 1. Representative temperature variations in the water tanks containing gilthead seabream juveniles during the experimental period (67 days). Fish were divided into two groups, with each group

exposed to different thermic treatments: Tconstant (solid line) and Tdaily cycles (dashed line). Vertical bars represent the mean ± SD. The shaded area represents the cold water period (5 p.m. to 5 a.m.). The temperature variation periods and their duration are indicated. White and black bars above the chart represent light and darkness, respectively.

2.3. Blood Collection and Tissue Sampling

Blood and tissue sampling occurred at 10 a.m. Blood collection was performed as described in [21]. After blood collection, the fish were immediately sacrificed, weighed, and measured and tissue samples (liver, gills, mid intestine, and white muscle) were collected. The dissection procedure of the tissues was performed on a cold platform. Samples were flash frozen in liquid nitrogen and stored at −80 °C until analysis. Liver samples for histology were collected from the upper part of the longer lobule. Liver and gill samples for histology were fixed with Bouin solution. Liver and visceral fat were weighed.

2.4. Liver Lipid Peroxidation

Lipid peroxidation is higher in well-irrigated tissues with lipid storage capacity, such as the liver, which is therefore considered a good marker of oxidative damage in fish. Lipid peroxidation levels were determined using a commercial kit (Bioxytech MDA-586, OXIS International, Portland, OR, USA, Cat. No. 21044), which is based on the reaction of a chromogenic reagent with MDA, a compound that arises from the oxidation of polyunsaturated fatty acids. Briefly, the liver samples were homogenized in 5 mL g^{-1} of cold PBS (137 mM NaCl, 2.5 mM KCl, 10 mM Na_2HPO_4, 1.8 mM KH_2PO_4, pH 7.4) using an Ultra-Turrax disperser. A standard curve was constructed with an MDA standard ranging from 0 to 4 µM. Tissue samples and standards were processed in parallel and at the end, 200 µL of each sample was added to a 96-well plate in triplicate. The formation of MDA was measured on a microplate reader at 586 nm (Multiskan GO, Thermo Scientific, Waltham, MA, USA). The results were presented as nmol MDA per mg of total protein. Protein concentrations were determined using the Bradford protein assay (5000006, BioRad, Hercules, CA, USA) [22].

2.5. Fatty Acid Profile Determination in Muscle and Liver Tissues

The fatty acid methyl esters, also known as FAME, were determined by acid-catalyzed transesterification in fish tissue samples, as described in [23]. The results were expressed as a percentage of its relative content.

2.6. Histology Assessment

Liver and gill samples were fixed with Bouin solution for 72 h, washed in tap water, and stored in ethanol 70% (*v*/*v*) until paraffin embedding. Sections of 6–7 µm, obtained using a microtome Leica RM-2155 (Leyca, Vienna, Austria), were stained with hematoxylin and eosin and examined under a light microscope. Microphotographs were examined under a binocular microscope, Nikon Eclipse Ci, and photographed using a Nikon DS-Fi2 camera. All sections were examined by the same researcher. Liver sections were evaluated for the integrity of the whole tissue and the hepatocyte cytoplasmic area was used to extrapolate the amount of energy reserves. The cytoplasmic area was determined in a defined image area (n = 3 per section; 15 fish per treatment), chosen to best represent the whole tissue section, considering the distinct dyed structures: dark stained nuclei and light stained cytoplasm and extracellular matrix. Cellular areas were determined using the ImageJ program v1.6.0. Gill sections were evaluated according to the length of the secondary lamellae. Three secondary lamellae were randomly selected from each fish section (15 fish per treatment). In total, 45 measurements per treatment were applied to calculate average values using ImageJ program v1.6.0.

2.7. Protein Extraction and Western Blot Analysis

Proteins were extracted based on the protocol adapted from [24,25]. Equivalent amounts of extracted proteins (75 µg) were separated on 12% (*w*/*v*) TGX stain-free polyacrylamide gels (BioRad, Hercules, CA, USA, Cat. No. 1610185) and transferred onto PVDF membranes (Amersham, UK, 0.45 μm) with prior exposure of stain-free gels to UV light for 1 min using the Chemidoc XRS+ imager (BioRad, Hercules, CA, USA, Cat No. 1708265). Immediately after transfer, the stain-free blots were again exposed to UV light for total protein band detection. Western blot analysis was executed as explained in [25] and the detected bands were quantified using Imagelab 6.0 software and normalized to total protein.

2.8. Hematology and Plasmatic Determinations

Before obtaining the plasma, a sub-sample of blood was removed from the initial blood sample drawn from each fish for hematocrit and hemoglobin analysis as described in [21,26]. The remaining blood was centrifuged at 2500× *g* for 10 min to obtain plasma, which was stored at −80 °C until analysis. Cortisol, glucose, total protein, and triglyceride levels were determined using commercial kits as described in [21]. Each sample underwent duplicate examination.

2.9. Digestive Enzymes

Mid intestine samples were collected and cut up on an ice surface according to [26]. The samples were then homogenized (IKA homogenizer) in 30 volumes (*w*/*v*) of ice-cold 2 mM Tris-HCl buffer at pH 7.0 containing 50 mM manitol for periods of 2 × 30 s. Two aliquots of homogenate were obtained. The first aliquot was used for the determination of digestive enzymes in the homogenate, which was centrifuged at 3300× *g* for 15 min at 4 °C, and the supernatant transferred to a clean tube. The second aliquot was used for brush border membrane purification, where 100 μL of 0.1 M calcium chloride was added to the homogenate, followed by centrifugation at 9000× g at 4 °C for 10 min. The supernatant was centrifuged again at 26,000× *g* at 4 °C for 40 min. The pellet was re-suspended in 5 mM Tris-Hepes buffer at pH 7.5 containing potassium chloride and 1 mM DTT, and was used for enzyme activity determination. Amylase activity was determined using starch as a substrate, dissolved in a pH 7.4 sodium phosphate buffer [27]. Alkaline phosphatase activity was assayed using 5 mM p-nitrophenyl-phosphate (pNPP) as a substrate in a 30 mM carbonate buffer at pH 9.8 [28]. Aminopeptidase activity was assayed using 0.1 M L-leucine p-nitroanilide as a substrate in an 80 mM solution of phosphate buffer at pH 7.0 [29]. Alkaline protease activity was determined according to the method described by [30], modified by [31]. Total soluble protein in enzyme extracts was determined in line with the Bradford technique, using bovine serum albumin as the standard protein [22]. Enzymatic activities were expressed as mU per mg of total protein.

2.10. Whole-Fish Proximate Composition Analysis

Proximate composition of the whole fish was analyzed to evaluate the impact of temperature variations on body carcasses. At the end of the experiment, 6 fish per tank were collected and frozen at −20 °C. The proximate composition of the whole fish was performed according to the reference AOAC methods [32]: dry matter by drying at 105 °C for 24 h; ash by combustion at 550 °C for 12 h; crude protein (Nx6.25) using a flash combustion technique, followed by gas chromatographic separation and thermal conductivity detection (LECO FP528); and crude fat after ethyl ether extraction using the Soxhlet method.

2.11. Whole-Fish Texture Analysis

Texture analysis was performed on the whole fish, without scaling, by compression, using a TA.XTPlus texture analyzer (Stable Micro Systems, Godalming, UK), equipped with a load cell of 30 kg, applying the single compression test. Whole-fish firmness (hardness) was determined using a 5 mm diameter spherical probe (P/5S) to simulate human finger (thumb) compression, which moved down 4 mm on fish at a constant speed of 1 mm s^{-1}. A total of 4 measurements was made on each fish in the zone between the operculum and the first ray of the dorsal fin (two readings on each side, right and left). Fifteen fish (5 fish from each replicate) from both treatments were analyzed.

2.12. Statistics

All data were tested for homogeneity using Levene's test, and the assumption of normal distribution was checked using the Shapiro–Wilk test. The Mann–Whitney U test was applied when the normality assumption was not verified. Values were reported as the mean ± SD, with the exception of Hsp70 expression, digestive enzymes, and hematologic parameters, which were expressed as mean ± SEM. The feed conversion ratio (FCR), specific growth rate (SGR), daily feed intake (DFI), and Hsp70 expression parameters were expressed as means of three replicates considering each tank as an experimental unit. The remaining parameters were expressed as means of 15 or 30 replicates, considering each fish an experimental unit per treatment. Student's *t*-test was used to check the differences between the Tconstant and the Tdaily cycle treatments. The results were considered significant at the 95% confidence level ($p < 0.05$). All statistical tests were performed using IBM-SPSS Statistics v25 software. Graphs were created using GraphPad Prism 5 software.

3. Results

3.1. Production Parameters

No fish mortality occurred throughout the experimental trial. The behavior of the fish varied in response to water temperature. The Tconstant group showed higher swimming activity and appetite than fish subjected to thermal fluctuations. The latter feature was supported by a higher daily feed intake (DFI) (Table 1). Fish final body weight (FBW) was greatest in fish maintained at constant temperature compared with fish exposed to temperature variations. Similarly, weight gain (WG), specific growth rate (SGR), and condition factor (CF) were greater in the Tconstant fish group. The feed conversion ratio (FCR) was lower for the Tconstant fish group, indicating significantly better feed utilization than fish exposed to water temperature fluctuations. The hepatosomatic index (HSI) and visceral fat somatic index (VFSI) were affected by water temperature variations, showing that the Tconstant group had higher values than the Tdaily cycles group for the latter parameter. The level of oxidized lipids was unaffected by water temperature variations (Table 1).

Table 1. Morphometric feed utilization parameters and liver oxidized lipid content of gilthead seabream from the Tconstant and Tdaily cycle groups.

Parameters	Tconstant	Tdaily Cycles	*p*-Value
IBW [1]	97.3 ± 1.9	96.1 ± 0.3	0.382
FBW [2]	173.2 ± 8.1	129.5 ± 1.4	0.001 *
WG [3]	8866 ± 789	4012 ± 200	<0.001 *
SGR [4]	0.86 ± 0.05	0.45 ± 0.02	<0.001 *
CF [5]	1.68 ± 0.13	1.59 ± 0.09	0.003 *
DFI [6]	1.36 ± 0.05	0.91 ± 0.01	<0.001 *
FCR [7]	1.65 ± 0.11	2.07 ± 0.08	0.005 *
HSI [8]	1.57 ± 0.21	1.90 ± 0.18	<0.001 *
VFSI [9]	2.06 ± 0.64	1.32 ± 0.47	0.001 *
Liver oxidized lipids [10]	164.9 ± 55.0	168.1 ± 61.2	0.884

Values represent the mean ± SD. * Indicates statistically significant differences between treatments (*t*-test, $p < 0.05$). The statistical value of *p* is indicated. [1] Initial body weight (IBW) (g): initial biomass (g)/ initial fish number. [2] Final body weight (FBW) (g): final biomass (g)/final fish number. [3] Weight gain (WG) (g): Final biomass (g)—Initial biomass (g). [4] Specific growth rate (SGR) (% day^{-1}): [(Ln final body weight (g)—Ln initial body weight (g))/Time (days)] × 100. [5] Condition factor (CF) (g cm^{-3}): body weight (g)/body length^3 (cm) × 100. [6] Daily feed intake (DFI) (% day^{-1}): 100 × total amount consumed feed/[Time days × (final weight + initial weight)/2]. [7] Feed conversion ratio (FCR): total amount of consumed feed (g)/weight gain (g). [8] Hepatosomatic index (HSI) (%): liver weight (g)/body weight (g) × 100. [9] Visceral fat somatic index (VFSI) (%): fat weight (g)/body weight (g) × 100. [10] Liver oxidized lipids (nmol MDA g^{-1} total protein).

3.2. Fatty Acid Composition

Table 2 shows the fatty acid composition of the liver and white muscle fillets of gilthead seabream exposed to different water temperature regimes. Total fatty acids were unaffected by temperature variations in both tissues ($p = 0.745$, liver; $p = 0.119$, muscle). However, the total saturated fatty acid (SFA) content was significantly affected in the group of fish exposed to daily temperature oscillations, with a reduction of C16:0 fatty acids in both tissues ($p < 0.001$, liver; $p < 0.001$, muscle) and C18:0 fatty acids only in muscle fillets ($p = 0.663$, liver; $p = 0.032$, muscle). The levels of monounsaturated fatty acids (MUFA, $p = 0.864$, liver; $p = 0.401$, muscle) and polyunsaturated fatty acids (PUFA, $p = 0.403$, liver; $p = 0.460$, muscle) were unaffected by water temperature changes. Total ω3 levels were unaffected in both tissues ($p = 0.461$, liver; $p = 0.227$, muscle); however, total ω6 levels increased in the muscle of fish exposed to water temperature oscillations ($p = 0.016$). Nevertheless, these temperature variations showed no effect on the ratio between these two PUFAs ($p = 0.632$, liver; $p = 0.635$, muscle). Eicosapentaenoic acid (EPA, $p = 0.473$, liver; $p = 0.465$, muscle) and docosahexaenoic acid (DHA, $p = 0.469$, liver; $p = 0.246$, muscle) levels were unchanged by water temperature oscillations in both fish tissues.

Table 2. Fatty acid composition of the liver and white muscle fillets of gilthead seabream exposed to different treatments, Tconstant and Tdaily cycles.

	Liver		Muscle	
% of Total Fatty Acids	**Tconstant**	**Tdaily Cycles**	**Tconstant**	**Tdaily Cycles**
C14:0	1.94 ± 0.16	1.88 ± 0.22	2.04 ± 0.26	2.14 ± 0.18
C16:0	17.31 ± 0.68 *	15.05 ± 0.73	17.11 ± 0.32 *	16.37 ± 0.29
C18:0	6.07 ± 0.86	6.19 ± 0.44	4.79 ± 0.39 *	4.52 ± 0.21
Other SFA	2.34 ± 0.53 *	3.08 ± 1.16	1.81 ± 0.11	1.81 ± 0.13
Total SFA	27.69 ± 1.32 *	26.19 ± 1.24	25.74 ± 0.43 *	24.81 ± 0.38
C16:1 (ω7 + ω9)	5.26 ± 1.23	5.71 ± 0.32	5.89 ± 0.61	6.16 ± 0.34
C18:1 (ω5 + ω7 + ω9)	31.19 ± 2.58	31.23 ± 2.86	29.49 ± 2.05	29.36 ± 1.11
C20:1 (ω7 + ω9)	1.47 ± 0.11 *	1.25 ± 0.26	1.50 ± 0.33	1.38 ± 0.12
Other MUFA	0.89 ± 0.10 *	0.81 ± 0.06	0.91 ± 0.10	0.87 ± 0.09
Total MUFA	38.80 ± 2.45	38.99 ± 3.17	37.79 ± 2.82	37.79 ± 1.49
C18:2 (ω6)	11.37 ± 0.87	11.36 ± 0.90	12.47 ± 0.41 *	12.84 ± 0.34
C18:3 (ω3)	0.84 ± 0.13	0.77 ± 0.12	0.99 ± 0.08	1.01 ± 0.05
C20:4 (ω6)	1.40 ± 0.27	1.61 ± 0.33	1.34 ± 0.30	1.32 ± 0.18
C20:5 (ω3) EPA	3.18 ± 0.49	3.31 ± 0.49	3.94 ± 0.31	4.01 ± 0.19
C22:6 (ω3) DHA	9.53 ± 2.38	10.16 ± 2.14	11.11 ± 2.30	11.27 ± 1.14
Other PUFA	4.30 ± 0.36	4.50 ± 0.30	4.72 ± 0.63	4.71 ± 0.20
Total PUFA	30.59 ± 3.74	31.77 ± 3.64	34.59 ± 2.50	35.19 ± 1.58
DHA/EPA	2.99 ± 0.48	3.03 ± 0.26	2.80 ± 0.38	2.81 ± 0.21
Total ω3	16.26 ± 3.04	17.08 ± 2.75	19.21 ± 2.47	19.39 ± 1.32
Total ω6	13.91 ± 0.91	14.21 ± 1.11	14.99 ± 0.26 *	15.34 ± 0.43
Ratio ω3/ω6	1.16 ± 0.17	1.19 ± 0.14	1.27 ± 0.16	1.26 ± 0.06
Total (SFA + MUFA + PUFA)	97.06 ± 1.09	96.96 ± 0.33	98.10 ± 0.66	97.81 ± 0.17
Total NI	2.94 ± 1.09	3.04 ± 0.33	1.90 ± 0.66	2.19 ± 0.17

Values are mean ± SD (n = 15). * Indicates statistically significant differences between each tissue from the experimental treatments (t-test, $p < 0.05$). SFA—Saturated fatty acids; MUFA—Monounsaturated fatty acids; PUFA—Polyunsaturated fatty acids; EPA—Eicosapentaenoic acid; DHA—Docosahexaenoic acid; Total NI—Non-identified fatty acids (100—Total SFA + MUFA + PUFA).

3.3. Liver and Gill Histology

No histomorphological changes were found in the liver or gill secondary lamellae structures of fish exposed to both experimental treatments (Figure 2). Moreover, there were no significant differences between secondary lamellae length of gills from fish exposed to water temperature fluctuations and constant temperature (Table 3). Similarly, thermal fluctuations had no significant impact on the energy reserves of fish hepatocytes, demonstrated by similar cytoplasmic areas.

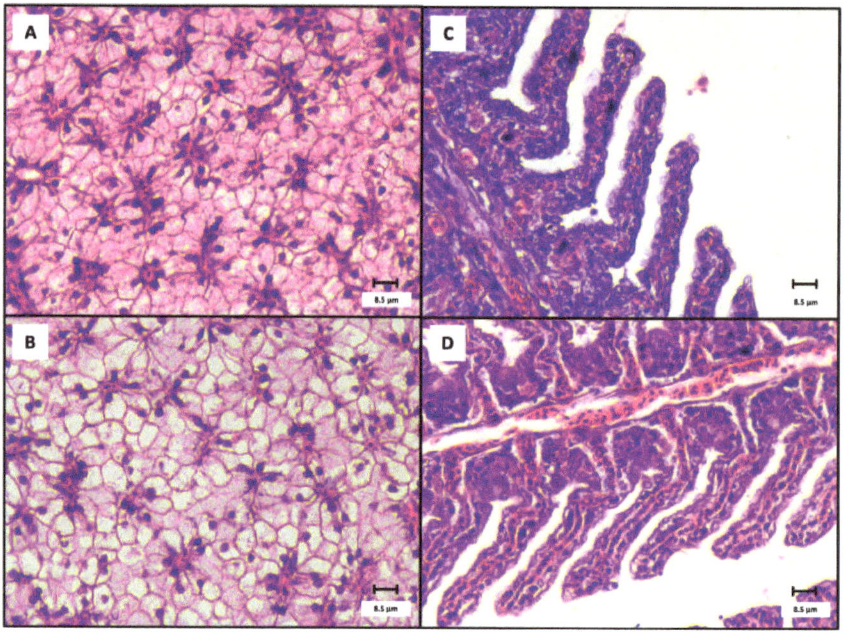

Figure 2. Light microphotographs of the liver (**A**,**B**) and gills (**C**,**D**) of gilthead seabream exposed to different water temperature treatments for 67 days, Tconstant (**A**,**C**) and Tdaily cycles (**B**,**D**). Scale bar: 8.5 µm; magnification 40×; hematoxylin-eosin staining.

Table 3. Measurements of gill and liver sections of gilthead seabream exposed to different water temperature treatments, Tconstant and Tdaily cycles.

Parameter	Tconstant	Tdaily Cycles	*p*-Value
HCA (%)	57.3 ± 4.81	60.6 ± 6.37	0.126
VL (µm g^{-1})	0.35 ± 0.05	0.34 ± 0.03	0.646

Values are means ± SD (n = 15). HCA—Hepatocyte cytoplasmic area: 100—nuclear area (%); VL—Secondary lamellae length (µm)/fish weight (g). The statistical value of p is indicated (t-test, $p < 0.05$).

3.4. Hsp70 Expression Levels

The relative expression of Hsp70 was monitored in the liver and gills of gilthead seabream (Figure 3). Either in the liver (p = 0.882) or in the gills (p = 0.159), Hsp70 expression levels were unaffected by water temperature variations.

3.5. Blood Parameters and Plasma Metabolites

Fish exposed to temperature variations exhibited greater hemoglobin values than those subjected to constant temperature during the whole experimental period (Table 4). No significant changes in hematocrit, plasmatic glucose and triglycerides were found between the different thermal treatments. Total protein levels were superior in fish exposed to temperature variations and cortisol levels rose in the group of fish exposed to constant temperature.

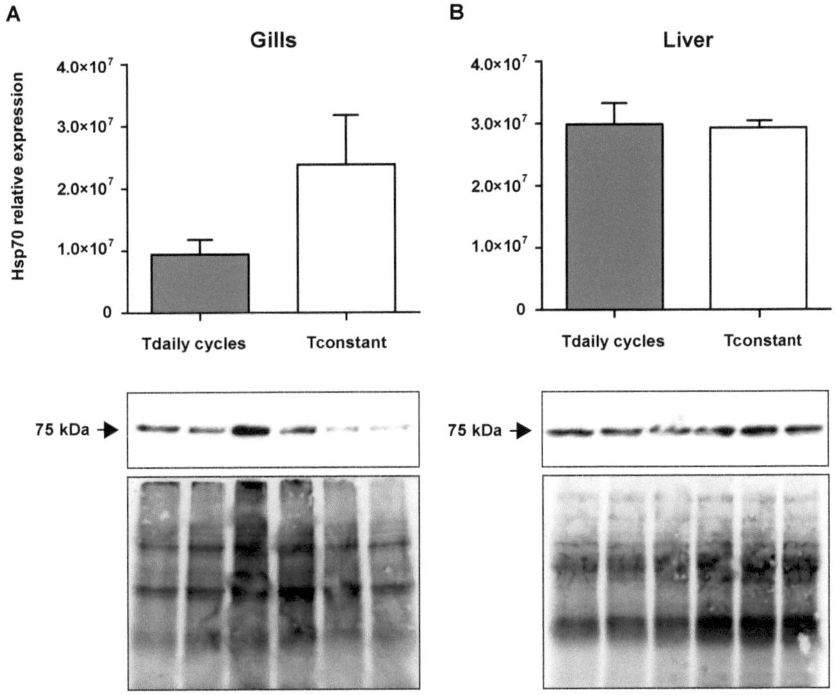

Figure 3. Relative expression of Hsp70 in the gills (**A**) and liver (**B**) of gilthead seabream experimental groups Tconstant and Tdaily cycles. Total protein was used as a normalization control for the western blots. Values are expressed as mean ± SEM of (n = 3) (t-test, $p < 0.05$).

Table 4. Blood parameters and plasma metabolites of gilthead seabream from Tconstant and Tdaily cycles groups.

Parameters	Tconstant	Tdaily Cycles	p-Value
Hemoglobin (g dL^{-1})	3.74 ± 0.46	6.43 ± 0.42	<0.001 *
Hematocrit (%)	32.9 ± 1.4	33.3 ± 0.9	0.841
Glucose (mg L^{-1})	54.3 ± 3.0	56.1 ± 3.6	0.714
Cortisol (μg dL^{-1})	35.9 ± 9.5	14.4 ± 4.4	0.031 *
Total proteins (g L^{-1})	30.9 ± 1.4	33.8 ± 1.4	0.026 *
Triglycerides (mg mL^{-1})	1.07 ± 0.04	1.15 ± 0.02	0.152

Values represent the mean ± SEM. * Indicates statistically significant differences between treatments (t-test, $p < 0.05$). The statistical value of p is indicated.

3.6. Activity of Intestinal Digestive Enzymes

Higher levels of alkaline phosphatase activity were found in the intestinal mucosa of gilthead seabream exposed to water temperature oscillations when compared with the group of fish subjected to constant temperature. The activity of aminopeptidase, amylase, and alkaline protease was unaffected by water temperature oscillations. Similar observations were registered for the analyzed brush border enzymes (aminopeptidase and alkaline phosphatase) (Table 5).

Table 5. Intestinal brush border (BB) and digestive enzymes (mU mg^{-1}) in gilthead seabream exposed to different water temperature regimes, Tconstant and Tdaily cycles.

Digestive Enzymes	Tconstant	Tdaily Cycles	*p*-Value
Aminopeptidase	42.0 ± 6.9	46.1 ± 7.3	0.689
Amylase	2500 ± 300	2300 ± 300	0.641
Alkaline Phosphatase	96.8 ± 18.7	248.3 ± 52.8	0.015 *
Alkaline proteases	5.6 ± 1.8	4.6 ± 1.3	0.867
Aminopeptidase (BB)	244.1 ± 70.5	235.5 ± 54.2	0.436
Alkaline Phosphatase (BB)	586 ± 136	1157 ± 263	0.098

Values represent the mean ± SEM (n = 15). * Indicates statistically significant differences between treatments (*t*-test, $p < 0.05$). The statistical value of p is indicated.

3.7. Whole-Fish Proximate Composition and Texture

Daily temperature variations significantly affected the dry matter, crude protein, and fat content of fish (Table 6). Dry matter and crude fat content decreased with temperature oscillations while crude protein content increased. In contrast, ash content was unaffected by this variable. Moreover, water temperature variations had no effect on fish texture (Table 6).

Table 6. Proximate composition analysis and texture of whole fish from Tconstant and Tdaily cycles gilthead seabream groups.

Parameters	Tconstant	Tdaily Cycles	*p*-Value
Crude protein (%)	45.8 ± 1.7	49.0 ± 1.9	<0.001 *
Crude fat (%)	37.8 ± 2.5	32.9 ± 2.5	<0.001 *
Dry matter (%)	36.1 ± 1.6	34.1 ± 1.4	<0.001 *
Ash (%)	17.5 ± 3.1	19.1 ± 2.5	0.099
Texture	0.99 ± 0.17	1.10 ± 0.15	0.088

Values represent the mean ± SD (n = 15). * Indicates statistically significant differences between treatments (*t*-test, $p < 0.05$). The statistical value of p is indicated.

4. Discussion

Repeated day/night water temperature oscillations induced several features associated with the winter syndrome, such as a reduction in growth, food ingestion, and increased HSI [6]. These observations may be due to the overall metabolic depression associated with the repeated cold periods [3]. Depressed metabolism may lead to alterations in the hepatic lipid metabolism of fish, causing an increase in liver mass due to the accumulation of lipids [33,34]. Despite a higher HSI, histological analysis showed no differences in the hepatocyte cytoplasmic area of fish exposed to both experimental treatments, suggesting similar energy reserves. In fact, gilthead seabream exposed to 5 cycles of thermal changes showed no lipid deposition in the liver [3]. Our results may indicate a moderate effect of thermal oscillations on the hepatic metabolism, perhaps caused by the continued feeding activity during daytime, which may compensate for any harsh hepatic damage caused by cold night periods. Indeed, liver fatty acid composition only differed in terms of total saturated fatty acid (SFA) content, which was significantly lower in the group of fish exposed to water temperature oscillations than in the Tconstant group, mainly due to the reduction of C16:0 fatty acid (palmitic acid). In muscle, thermal fluctuations significantly decreased the amount of SFA, primarily due to a reduction in C16:0 and C18:0 (stearic acid) fatty acids. This SFA decrease was accompanied by a significant increase in the levels of total ω6 polyunsaturated fatty acids (PUFA), mostly due to an increase in C18:2 fatty acid (linoleic acid). Low temperatures often lead to an increase in unsaturated fatty acids of phospholipids to maintain appropriate fluidity of cell membranes [34]. SFA in phospholipid tails is relatively straight. In contrast, unsaturated fatty acids contain some double bounds between carbon atoms, resulting in torsions of carbon chains. These twists in their tails push adjacent phospholipid molecules away, maintaining some space between the phospholipid molecules and resulting in a more fluid membrane [35]. For this reason, as the proportion of phospholipids with unsaturated fatty acids increases, the fluidity of membranes also increases [36]. The significant decrease in the proportion of SFA in the liver and muscle of gilthead seabream in response to thermal oscillations showed mild adaptation to the repeated cold periods. Our findings correspond to the response of steelhead trout to temperature drops, in which the concentration of unsaturated fatty acids of phospholipids increased in several tissues, including muscle [37]. Ibarz and colleagues reported no homeoviscous adaptation of liver membranes after exposing gilthead seabream to both short-term cold stress at two different rates and 5-cycle repetitive temperature fluctuations. However, they did report increased levels of unsaturated fatty acids in muscle tissue [20]. These contradictory effects on the liver of gilthead seabream in response to the cold may be due to the different experimental conditions, such as different temperature ranges, feeding regimes, and/or cold periods, causing different physiological responses in the fish.

Oxidative stress is an important component of the stress response in marine organisms, which are exposed to environmental stressors, such as temperature variations [38,39]. Cold water environments contain higher levels of dissolved oxygen, which may facilitate any oxidative injury. The higher proportion of unsaturated fatty acids found in the liver membranes of fish from the Tdaily cycles group can accentuate this scenario, since they are more susceptible to being attacked by oxygen [40]. However, these thermal oscillations caused no greater oxidative damage in the liver of fish compared with the group of fish exposed to constant temperature, as demonstrated by similar levels of the lipid peroxidation product, malondialdehyde (MDA). Studies on European sea bass demonstrated oxidative stress as the temperature moved away from thermal optimum [38]. Although daily temperature fluctuations can cause repetitive withdrawals when compared with the Tconstant group, fish are able to recover by remaining without significant oxidative stress injuries for some hours (7 h) of each cycle at ~19 °C. Indeed, the expression levels of Hsp70 showed no significant differences in the liver and gills of either experimental group of fish. Although cold temperatures can stimulate the expression of Hsp70 in the liver of gilthead seabream [41], the tested thermal oscillations did not induce any cellular thermal stress response, possibly because the levels of damaged or misfolded proteins were not high enough to trigger their action. Moreover, the thermal variations caused no histomorphological alterations to the secondary lamellae of the gills, with both groups exhibiting similar villi length. Although gills may adapt their structure to changes in the environment, such as temperature variations [42], our experimental thermal oscillations were not able to cause any change in their structure.

Previous observations of our group showed that 2 cycles of water temperature fluctuations decreased the metabolic activity of the gilthead seabream, with concomitant variation in the levels of several blood and plasmatic parameters, essentially observed close to the influence of the cold water period or during the water cooling period [21]. After a longer period of daily temperature fluctuations, gilthead seabream showed a significant increase in hemoglobin concentration, with no variation in hematocrit levels compared with the group of fish maintained at a constant temperature. The long-term repetitive water temperature oscillations might act as a stressor, forcing the fish to make daily adjustments in their oxygen carrying capacity. These adjustments may be induced either by higher swimming activity or lower dissolved oxygen levels during the warmer water periods or the opposite scenario during the colder phases [43]. Since blood sampling was performed at the beginning of the warm period (10 a.m.), the fish were coming from the warming period where they were adapting to the new warm temperature environment. Therefore, the higher hemoglobin values determined for the Tdaily cycles group of fish may be reflecting this shift in water temperature. Moreover, this group of fish demonstrated a reduction in cortisol levels, which could be associated with the different feeding activity exhibited by both experimental groups [21,44], mainly evidenced by the lower DFI in the Tdaily cycles group compared with the constant temperature group. Cortisol levels can influence glucose metabolism in fish [12]; however, under our experimental conditions, plasma glucose levels were unaltered in both groups. Triglyceride levels were also unchanged in both thermal regimes. This may indicate that these two blood-circulating metabolites are influenced by feeding since, by the time of sampling, both experimental groups were at similar post-prandial status. Drops in plasmatic total protein in fish are associated with fasting and low temperature scenarios [9]; however, in our study, we saw an increase in plasmatic total protein in the fish exposed to temperature fluctuations.

Temperature has a great effect on the structure and catalytic function of enzymes within metabolic pathways. For this reason, it is important to ensure an adequate rearing temperature level to guarantee that all metabolic pathways are working properly to produce high-quality, marketable-sized animals. Studies on the activity of digestive enzymes in fish can elucidate some aspects of the whole digestive capacity of fish. To evaluate the impact of day/night temperature oscillations on the digestive capacity of fish, the activity of pancreatic (amylase) and intestinal enzymes (aminopeptidase, alkaline phosphatase, and alkaline protease) was determined. Additionally, the activity of brush border membrane enzymes (aminopeptidase and alkaline phosphatase) was evaluated since these enzymes are responsible for the final stages of degradation and assimilation of nutrients by the intestinal cells. Overall, the long-term temperature fluctuations had no effect on the activity of brush border and digestive enzymes except for the cytosolic alkaline phosphatase, whose activity was higher in the fish exposed to temperature oscillations. A gradual decrease in water temperature from 18 to 2 °C caused a reduction in the intestinal alkaline phosphatase activity in *Carassius carassius* [45]. That study disagrees with our results perhaps because of the different experimental thermal oscillation conditions and the physiological response of the investigated species. In fact, the metabolism of the Tdaily cycles group exhibited higher plasmatic total protein levels and higher whole-fish protein content. Since the intestinal alkaline phosphatase dephosphorylates phosphorylated dietary moieties (e.g., proteins), the long-term day/night temperature variations could promote a higher protein metabolism to obtain energy to maintain physiological processes rather than growth. However, the fish reared at a constant temperature may have redirected energy from digested food to growth and fat storage, as directly evidenced by higher somatic growth (superior FBW, WG, SGR, CF), better feed efficiency (lower FCR), and increased visceral fat (VFSI) indicators [15,46]. Curiously, these differences in the whole-fish proximate composition analysis were not reflected in the fish texture. This parameter can be affected by several factors, such as lipid oxidation [47] and/or protein degradation [48]. Despite the differences in the whole-body fat and protein content between groups, the previously mentioned biochemical reactions had little or no effect on fish texture.

5. Conclusions

The overall aim of this study was to simulate the usage of a solar water heating system during daytime and investigate its effects on the whole performance of gilthead seabream juveniles. Long-term water temperature fluctuations reduced fish growth and affected several other physiological processes. Therefore, these findings favor the usage of a constant water temperature of ~19 °C for optimal gilthead seabream juvenile production instead of a day/night water temperature oscillating regime. Solar energy can be used to warm the rearing water 24/7. This decision can be done by fish farmers according to their operational conditions and/or business strategy.

Author Contributions: Conceptualization, L.R. and P.P.-F.; Data curation, A.C.M. and R.L.A.; Formal analysis, A.C.M., R.L.A., N.M.B. and A.G.; Funding acquisition, L.R. and P.P.-F.; Investigation, A.C.M. and R.L.A.; Methodology, A.C.M., R.L.A., L.R., N.M.B., A.G. and P.P.-F.; Project administration, L.R. and P.P.-F.; Supervision, L.R. and P.P.-F.; Visualization, A.C.M., R.L.A., N.M.B. and A.G.; Writing—original draft, A.C.M. and R.L.A.; Writing—review and editing, A.C.M., R.L.A., L.R., N.M.B. and A.G. All authors have read and agreed to the published version of the manuscript.

Funding: This work was supported by DIVERSIAQUA II project (MAR-02.01.01-FEAMP-0175) of the MAR2020 program.

Institutional Review Board Statement: This fish trial was performed at the Aquaculture Research Station (EPPO) of the Portuguese Institute of Sea and Atmosphere (IPMA) (Olhão, Portugal), certified by the Direção Geral da Alimentação e Veterinária to execute animal experimentation under the authorization 2018/12/17-025516. The experiment was directed by trained scientists (following category C FELASA recommendations) and conducted according to the European guidelines on the protection of animals used for scientific purposes (Directive 2010/63/UE of the European Parliament and the European Union Council) and related Portuguese legislation guidelines of (Decreto-Lei 113/2013) on animal experimentation and welfare. The experimental design respected all the procedures to protect animal welfare and ensure and extend the application of the 3Rs (reduce, refine, replace). The number of sacrificed animals was the minimum needed to obtain statistically significant results.

Informed Consent Statement: Not applicable.

Data Availability Statement: Data is contained within the manuscript.

Conflicts of Interest: The authors declare no conflict of interest.

Abbreviations

BB	Intestinal brush border
CF	Condition factor
DFI	Daily feed intake
DHA	Docosahexaenoic acid
EPA	Eicosapentaenoic acid
FAME	Fatty acid methyl esters
FBW	Final body weight
FCR	Feed conversion ratio
HCA	Hepatocyte cytoplasmic area
HSI	Hepatosomatic index
Hsp70	Heat Shock Proteins of 70 kDa
IBW	Initial body weight
MUFA	Monounsaturated fatty acid
PUFA	Polyunsaturated fatty acid
SFA	Saturated fatty acid
SGR	Specific growth rate
Total NI	Non-identified fatty acids
VFSI	Visceral fat somatic index
VL	Secondary lamellae length
WG	Weight gain

References

1. Tort, L.; Padrós, F.; Rotllant, J.; Crespo, S. Winter syndrome in the gilthead sea bream *Sparus aurata*. Immunological and histopathological features. *Fish Shellfish Immunol.* **1998**, *8*, 37–47. [CrossRef]
2. Tort, L.; Rotllant, J.; Liarte, C.; Acerete, L.; Hernández, A.; Ceulemans, S.; Coutteau, P.; Padros, F. Effects of temperature decrease on feeding rates, immune indicators and histopathological changes of gilthead sea bream *Sparus aurata* fed with an experimental diet. *Aquaculture* **2004**, *229*, 55–65. [CrossRef]
3. Ibarz, A.; Blasco, J.; Sala-Rabanal, M.; Gallardo, A.; Redondo, A.; Fernández-Borràs, J. Metabolic rate and tissue reserves in gilthead sea bream (*Sparus aurata*) under thermal fluctuations and fasting and their capacity for recovery. *Can. J. Fish. Aquat. Sci.* **2007**, *64*, 1034–1042. [CrossRef]
4. Fair-Fish Database. Available online: https://fair-fish-database.net/db/49/ (accessed on 20 July 2023).
5. Ibarz, A.; Fernández-Borràs, J.; Blasco, J.; Gallardo, M.A.; Sánchez, J. Oxygen consumption and feeding rates of gilthead sea bream (*Sparus aurata*) reveal lack of acclimation to cold. *Fish Physiol. Biochem.* **2003**, *29*, 313–321. [CrossRef]
6. Ibarz, A.; Padrós, F.; Gallardo, M.A.; Fernández-Borràs, J.; Blasco, J.; Tort, L. Low-temperature challenges to gilthead sea bream culture: Review of cold-induced alterations and 'Winter Syndrome'. *Rev. Fish Biol. Fish.* **2010**, *20*, 539–556. [CrossRef]
7. Columbus, C. Solar Power. In *The Potential for Renewable Energy Usage in Aquaculture*; Toner, D., Mathies, M., Eds.; Resource Development/Environment & Quality Section: Louth, Ireland; Dublin, Ireland, 2002; Chapter 6; pp. 34–38.

8. Villar-Torres, M.; Montero, F.E.; Raga, J.A.; Repullés-Albelda, A. The influence of water temperature on the life-cycle of *Sparicotyle chrysophrii* (Monogenea: Microcotylidae), a common parasite in gilthead seabream aquaculture. *Aquaculture* **2023**, *565*, 739103. [CrossRef]
9. Sala-Rabanal, M.; Sánchez, J.; Ibarz, A.; Fernandéz-Borràs, J.; Blasco, J.; Gallardo, M.A. Effects of low temperatures and fasting on hematology and plasma composition of gilthead sea bream (*Sparus aurata*). *Fish Physiol. Biochem.* **2003**, *29*, 105–115. [CrossRef]
10. Sánchez-Nuño, S.; Sanahuja, I.; Fernández-Alacid, L.; Ordóñez-Grande, B.; Fontanillas, R.; Fernández-Borràs, J.; Blasco, J.; Carbonell, T.; Ibarz, A. Redox challenge in a cultured temperate marine species during low temperature and temperature recovery. *Front. Physiol.* **2018**, *9*, 923. [CrossRef] [PubMed]
11. Ibarz, A.; Beltrán, M.; Fernández-Borràs, J.; Gallardo, M.A.; Sánchez, J.; Blasco, J. Alterations in lipid metabolism and use of energy depots of gilthead sea bream (*Sparus aurata*) at low temperatures. *Aquaculture* **2007**, *262*, 470–480. [CrossRef]
12. Mommsen, T.P.; Vijayan, M.M.; Moon, T.W. Cortisol in teleosts: Dynamics, mechanisms of action, and metabolic regulation. *Rev. Fish Biol. Fish.* **1999**, *9*, 211–268. [CrossRef]
13. Iwama, G.K.; Thomas, P.T.; Forsyth, R.B.; Vijayan, M.M. Heat shock protein expression in fish. *Rev. Fish Biol. Fish.* **1998**, *8*, 35–56. [CrossRef]
14. Yashamita, M.; Yabu, T.; Ojima, N. Stress protein Hsp70 in fish. *Aqua-BioSci. Monogr.* **2010**, *3*, 111–141. [CrossRef]
15. Volkoff, H.; Rønnestad, I. Effects of temperature on feeding and digestive process in fish. *Temperature* **2020**, *7*, 307–320. [CrossRef] [PubMed]
16. Araújo, B.C.; Miller, M.R.; Walker, S.P.; Symonds, J.E. The influence of temperature on performance, biological indices, composition, and nutrient retention of juvenile *Chinook salmon* (*Oncorhynchus tshawytscha*) reared in freshwater. *Comp. Biochem. Physiol. A Mol. Integr. Physiol.* **2023**, *280*, 111412. [CrossRef]
17. Christiansen, J.A. Changes in phospholipid classes and fatty acids and fatty acid desaturation and incorporation into phospholipids during temperature acclimation of green sunfish *Lepomis cyanellus* R. *Physiol. Zool.* **1984**, *57*, 481–492. [CrossRef]
18. Farkas, T.; Fodor, E.; Kitajka, K.; Halver, J.E. Response of fish membranes to environmental temperature. *Aquac. Res.* **2001**, *32*, 645–655. [CrossRef]
19. Johnston, I.A. Muscle development and growth: Potential implications for flesh quality in fish. *Aquaculture* **1999**, *177*, 99–115. [CrossRef]
20. Ibarz, A.; Blasco, J.; Beltrán, M.; Gallardo, M.A.; Sánchez, J.; Sala, R.; Fernández-Borràs, J. Cold-induced alterations on proximate composition and fatty acid profiles of several tissues in gilthead sea bream (*Sparus aurata*). *Aquaculture* **2005**, *249*, 477–486. [CrossRef]
21. Matias, A.C.; Ribeiro, L.; Araujo, R.L.; Pousão-Ferreira, P. Preliminary studies on hematological and plasmatic parameters in gilthead sea bream (*Sparus aurata*) held under day/night temperature variations. *Fish Physiol. Biochem.* **2018**, *44*, 273–282. [CrossRef]
22. Bradford, M.M. A rapid and sensitive method for the quantitation of microgram quantities of protein utilizing the principle of protein-dye binding. *Anal. Biochem.* **1976**, *7*, 248–254. [CrossRef]
23. Ferreira, I.; Gomes-Bispo, A.; Lourenço, H.; Matos, J.; Afonso, C.; Cardoso, C.; Castanheira, I.; Motta, C.; Prates, J.A.M.; Bandarra, N.M. The chemical composition and lipid profile of the chub mackerel (*Scomber colias*) show a strong seasonal dependence: Contribution to a nutritional evaluation. *Biochemie* **2020**, *178*, 181–189. [CrossRef]
24. Feidantsis, K.; Pörtner, H.O.; Lazou, A.; Kostoglou, B.; Michaelidis, B. Metabolic and molecular stress responses of the gilthead seabream *Sparus aurata* during long-term exposure to increasing temperatures. *Mar. Biol.* **2009**, *156*, 797–809. [CrossRef]
25. Matias, A.C.; Dias, J.; Barata, B.; Araujo, R.L.; Bragança, J.; Pousão-Ferreira, P. Taurine modulates protein turnover in several tissues of meagre juveniles. *Aquaculture* **2020**, *528*, 735478. [CrossRef]
26. Ribeiro, L.; Moura, J.; Santos, M.; Colen, R.; Rodrigues, V.; Bandarra, N.; Soares, F.; Ramalho, P.; Barata, M.; Moura, P.; et al. Effect of vegetable based diets on growth, intestinal morphology, activity of intestinal enzymes and haematological stress indicators in meagre (*Argyrosomus regius*). *Aquaculture* **2015**, *447*, 116–128. [CrossRef]
27. Métais, P.; Bieth, J. Détermination de l'a-amylase par une microtechnique. *Ann. Biol. Clin.* **1968**, *26*, 133–142.
28. Bessey, O.A.; Lowry, O.H.; Brock, M.J. A method for the rapid determination of alkaline phosphatase with five cubic millimeters of serum. *J. Biol. Chem.* **1946**, *164*, 321–329. [CrossRef] [PubMed]
29. Maroux, S.; Louvard, D.; Baratti, J. The aminopeptidase from hog intestinal brush border. *Biochim. Biophys. Acta* **1973**, *15*, 282–295. [CrossRef] [PubMed]
30. Kunitz, M. Crystalline soybean trypsin inhibitor II. General properties. *J. Gen. Physiol.* **1947**, *30*, 291–310. [CrossRef] [PubMed]
31. Walter, H.E. Proteinases: Methods with haemoglobin, casein and azocoll as substrates. In *Methods of Enzymatic Analysis*; Bergmeyer, H.U., Ed.; Verlag Chemie: Weinheim, Germany, 1984; pp. 270–277.
32. Horwitz, W.; Latimer, G.W. *Official Methods of Analysis of AOAC International*, 18th ed.; Association of Official Analytical Chemists International: Gaithersburg, MD, USA, 2006.
33. Gallardo, M.A.; Sala-Rabanal, M.; Ibarz, A.; Padrós, F.; Blasco, J.; Fernández, J.; Sánchez, J. Functional alterations associated with the 'winter syndrome' in gilthead sea bream (*Sparus aurata*). *Aquaculture* **2003**, *223*, 15–27. [CrossRef]
34. Hazel, J.R.; Prosser, C.L. Molecular mechanisms of temperature compensation in poikilotherms. *Physiol. Rev.* **1974**, *54*, 620–677. [CrossRef]

35. Hac-Wydro, K.; Wydro, P. The influence of fatty acids on model cholesterol/ phospholipid membranes. *Chem. Phys. Lipids* **2007**, *150*, 66–81. [CrossRef] [PubMed]
36. Cossins, A.R.; Prosser, C.L. Evolutionary adaptation of membranes to temperature. *Proc. Natl. Acad. Sci. USA* **1978**, *75*, 2040–2043. [CrossRef]
37. Liu, C.; Dong, S.; Zhou, Y.; Shi, K.; Pan, Z.; Sun, D.; Gao, Q. Temperature-dependent fatty acid composition change of phospholipid in steelhead trout (*Oncorhynchus mykiss*) tissues. *J. Ocean Univ. China* **2019**, *18*, 519–527. [CrossRef]
38. Vinagre, C.; Madeira, D.; Narciso, L.; Cabral, H.N.; Diniz, M. Effect of temperature on oxidative stress in fish: Lipid peroxidation and catalase activity in the muscle of juvenile seabass, *Dicentrarchus labrax*. *Ecol. Indic.* **2012**, *23*, 274–279. [CrossRef]
39. Ibarz, A.; Martín-Pérez, M.; Blasco, J.; Bellido, D.; Oliveira, d.E.; Fernández-Borràs, J. Gilthead sea bream liver proteome altered at low temperatures by oxidative stress. *Proteomics* **2010**, *10*, 963–975. [CrossRef]
40. Tocher, D.R. Metabolism and functions of lipids and fatty acids in teleost fish. *Rev. Fish. Sci.* **2003**, *11*, 107–184. [CrossRef]
41. Kyprianou, T.; Pörtner, H.O.; Anestis, A.; Kostoglou, B.; Feidantsis, K.; Michaelidis, B. Metabolic and molecular stress responses of gilthead sea bream Sparus aurata during exposure to low ambient temperature: An analysis of mechanisms underlying the winter syndrome. *J. Comp. Physiol. B* **2010**, *180*, 1005–1018. [CrossRef]
42. Sollid, J.; Nilsson, E. Plasticity of respiratory structures—Adaptative remodeling of fish gills induced by ambient oxygen and temperature. *Respir. Physiol. Neurobiol.* **2006**, *154*, 241–251. [CrossRef] [PubMed]
43. Olsen, R.E.; Sundell, K.; Ringø, E.; Myklebust, R.; Hemre, G.-I.; Hansen, T.; Karlsen, Ø. The acute stress response in fed and food deprived Atlantic cod, *Gadus morhua* L. *Aquaculture* **2008**, *280*, 232–241. [CrossRef]
44. Montoya, A.; López-Olmeda, J.F.; Garayzar, A.B.S.; Sánchez-Vázquez, F.J. Synchronization of daily rhythms of locomotor activity and plasma glucose, cortisol and thyroid hormones to feeding in Gilthead seabream (*Sparus aurata*) under a light-dark cycle. *Physiol. Behav.* **2010**, *101*, 101–107. [CrossRef] [PubMed]
45. Varis, J.; Haverinen, J.; Vornanen, M. Lowering temperature is the trigger for glycogen build-up and winter fasting in Crucian Carp (*Carassius carassius*). *Zool. Sci.* **2016**, *33*, 83–91. [CrossRef] [PubMed]
46. Hani, Y.M.I.; Marchand, A.; Turies, C.; Kerambrum, E.; Palluel, O.; Bado-Nilles, A.; Beaudouin, R.; Porcher, J.; Geffard, A.; Dedourge-Geffard, O. Digestive enzymes and gut morphometric parameters of threespine stickleback (*Gasterosteus aculeatus*): Influence of body size and temperature. *PLoS ONE* **2018**, *13*, e0194932. [CrossRef]
47. Huss, H.H. *Quality and Quality Changes in Fresh Fish*; FAO: Rome, Italy, 1995.
48. Bao, Y.; Ertbjerg, P. Effects of protein oxidation on the texture and water-holding of meat: A review. *Crit. Rev. Food Sci. Nutr.* **2019**, *59*, 3564–3578. [CrossRef] [PubMed]

Journal of
Marine Science and Engineering

MDPI

Article

Seasonal Variability in the Influence of Coastal Aquaculture Operation on Benthic–Pelagic Coupling Processes in Shallow Aquatic Ecosystems

Alexander Deen [1,2], Shu Kitajima [3], Waka Sato-Okoshi [3] and Toyonobu Fujii [2,*]

[1] Department of Forestry and Natural Resources, College of Agriculture, Purdue University, West Lafayette, IN 46202, USA; alexrdeen@gmail.com
[2] Laboratory of Marine & Coastal Ecosystem Science, Graduate School of Agricultural Science, Tohoku University, Onagawa 986-2248, Japan
[3] Laboratory of Biological Oceanography, Graduate School of Agricultural Science, Tohoku University, Sendai 980-8572, Japan
* Correspondence: toyonobu.fujii.a8@tohoku.ac.jp

Abstract: Coastal shellfish aquaculture can influence benthic–pelagic-coupled systems because cultured species consume phytoplankton in the water column and return the captured organic matter and nutrients to the environment as biodeposits, which fall to the seafloor, affecting local sediment characteristics and the benthic community. In 2023, we conducted monthly field surveys to characterize the relationships between shellfish aquaculture and the surrounding environment by examining a range of physical and biological variables along the benthic–pelagic gradient at multiple sampling locations in relation to their distances from the aquaculture facilities in Onagawa Bay, Japan. The abundances of benthic macrofauna were dominated by polychaetes (86.3%), followed by gastropods (4.7%), malacostracans (2.7%), ophiuroids (2.1%), and bivalves (1.5%). Both benthic biomass and biodiversity were markedly higher, but the chlorophyll-a concentration of the water column and the sediment organic matter content were significantly lower at the closest proximity to the aquaculture facilities. Although the physical presence of shellfish aquaculture may effectively enhance pelagic–benthic energy fluxes, such processes may also pose a new challenge under the influence of recent global warming, causing widespread hypoxic conditions due to increased stratification in the water column accompanied by excess organic inputs from the aquaculture.

Keywords: shellfish aquaculture; benthic–pelagic coupling; benthic macrofauna; phytoplankton; bio-deposit; global warming; hypoxia; socioecological system

Citation: Deen, A.; Kitajima, S.; Sato-Okoshi, W.; Fujii, T. Seasonal Variability in the Influence of Coastal Aquaculture Operation on Benthic–Pelagic Coupling Processes in Shallow Aquatic Ecosystems. *J. Mar. Sci. Eng.* **2024**, *12*, 1293. https://doi.org/10.3390/jmse12081293

Academic Editors: Ana Catarina Matias and Carlos Andrade

Received: 30 June 2024
Revised: 22 July 2024
Accepted: 29 July 2024
Published: 31 July 2024

1. Introduction

Aquaculture will continue to play a crucial role in meeting the increased demand for essential protein and nutrients for food globally, and while the global production of aquatic animals from both capture fisheries and aquaculture reached the highest record of 185 million tonnes in 2022, the production of animal species from aquaculture exceeded that from capture fisheries for the first time, with aquaculture accounting for 51 percent of the total production [1]. The outlook suggests that aquaculture production will increase greatly in the future [1], and the expected increase in aquaculture operations has raised concerns regarding the impacts of the associated activities on local environments [2–6]. Initially, attention was directed to the influence of finfish aquaculture on the environment because chemicals and excess nutrients from food and feces associated with finfish farming may disturb the surrounding area if the operations are intense [2,7,8]. Studies have also shown that aquaculture has a negative impact, to a varying degree, on the environment through changes in the physical, chemical, and biological attributes of the sediments below and the water column around aquaculture installations [2,9–11]. The culture of shellfish and other suspended filter-feeders, such as ascidians, is generally considered to have a smaller environmental impact than finfish aquaculture because shellfish are grown at comparatively lower intensities and no external feed is added [6,12–14]. However, shellfish farms typically cover a much greater area than finfish farms and may affect the surrounding environment in different ways [14–16].

Many towns and villages located along the northern Pacific coast of Japan are renowned for their commercial fishing and aquaculture industries. This region is known as the Sanriku Coast (Figure 1a), and major warm and cold ocean currents converge offshore of the region, providing some of the most productive fishing grounds on earth. Onagawa Bay accommodates one of the major fishing ports in this region, producing a variety of wild-caught and farmed seafood, including flatfish, Pacific cod, salmon, shellfish, and cultured ascidians. In Onagawa Bay alone, there are over 700 long-line systems for culturing scallops, oysters, and edible ascidians, as well as around 80 fish cages for farming salmon, which account for 84.5 and 15.5% of the total area occupied by the respective aquaculture facilities (Figure 1b). The long-line method is widely used to culture filter feeders (e.g., mussels, oysters, scallops, and ascidians), and this method may readily influence the sea bottom below and the surrounding water column through benthic–pelagic processes. This is because cultured shellfish and ascidians feed entirely on naturally occurring phytoplankton with other suspended particles in the water column, and the food and nutrients captured by them can be returned to the environment as undigested waste or feces in the form of biodeposits, which fall to the seafloor and can become important food sources for benthic-deposit feeders [12,13,15]. The suspended filter-feeder aquaculture may therefore enhance the organic material flux to the sediment layer, and populations of benthic invertebrates and microorganisms in turn decompose the biodeposits and thereby control the number of nutrients released to the sediment–water interface [15,17]. Further, an enhanced organic matter flux together with the increased remineralization processes on the bottom layer may lead to increased sediment oxygen uptake by the benthic community and thereby cause hypoxic conditions on the surrounding seafloor [16–18].

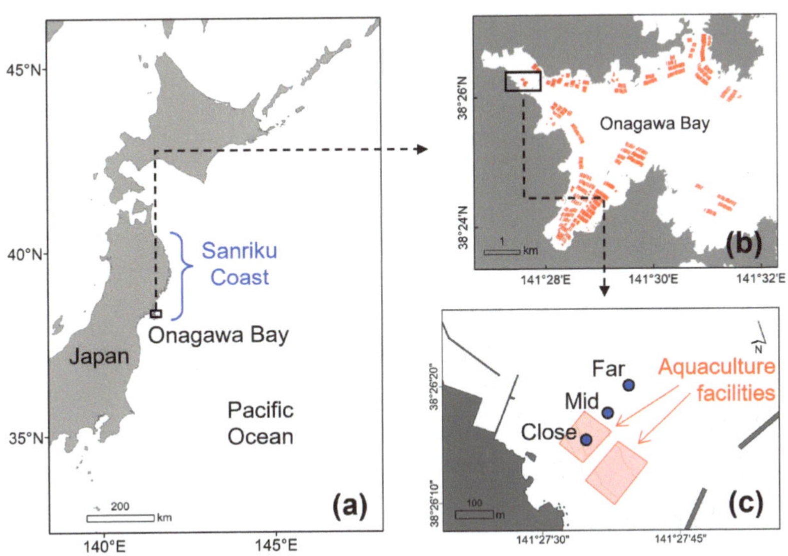

Figure 1. (**a**) Location of Onagawa Bay on the Sanriku Coast, Japan; (**b**) distribution of aquaculture facilities in Onagawa Bay (the red lines represent long-line aquaculture facilities); (**c**) locations of sampling stations in relation to the position of aquaculture facilities (the blue filled circles indicate the sampling stations).

Shellfish aquaculture facilities provide unique marine habitats in that they not only vertically stretch throughout the water column from the sea surface to the bottom but also occupy substantial areas horizontally in coastal shallow water environments, attracting various marine species ranging from benthic and demersal to pelagic organisms within the system [19,20]. The physical presence of shellfish aquaculture may therefore effectively mediate benthic–pelagic coupling processes through the enhanced biogeochemical cycle of organic matter from the surface to the bottom. However, such processes may also pose a challenge under the influence of recent global warming, causing widespread hypoxic conditions due to increased stratification of the water column accompanied by enhanced sediment oxygen uptake at the sea bottom triggered by the excess organic inputs generated from the pelagic aquaculture operations. The aim of this study was to characterize the relationships between shellfish aquaculture and the surrounding environment by examining a range of physical and biological variables along the benthic–pelagic gradient at multiple sampling locations in relation to their distances from the aquaculture facilities in Onagawa Bay, Japan. This was accomplished by investigating (1) the temporal variation in the community structures of benthic macrofauna observed in the sediment samples across the stations; (2) the temporal variation in the sediment properties observed in the sediment samples across the stations; (3) the temporal variation in the relative abundances of epibenthic megafauna observed in the underwater video data across stations; and (4) the temporal variation in the physical conditions of the water column observed in the CTD monitoring across stations. Based on the quantification of these variables, we discuss the implications of seasonal variability in the influence of coastal aquaculture operations on both benthic–pelagic coupling processes and subsequent shallow aquatic ecosystem dynamics.

2. Materials and Methods

2.1. Data Collection

Onagawa Bay is located on the Sanriku Coast of northern Japan (Figure 1a). The bay is characterized by a naturally sheltered deep-water harbor that is suitable for coastal aquaculture. Regarding farming methods, hanging culture using a long-line system is widely employed to farm organisms such as shellfish (oysters and scallops) and ascidians on vertical ropes, typically down to around 10 to 15 m water depths (Figure 1b). Three sampling stations, namely St. Far, St. Mid, and St. Close, were established in relation to their distances from the nearest aquaculture facilities to monitor a range of physical and biological variables along the benthic–pelagic gradient (Figure 1c, Table 1). At each station, we conducted field observations on board RV Kaisei (1) on a monthly basis between December 2022 and May 2023 (6 months) for sediment (benthic) sampling and (2) on a bi-weekly basis between December 2022 and December 2023 (13 months) for water column (pelagic) and underwater camera monitoring. We conducted sediment sampling for only the first 6 months in this study to monitor the benthic–pelagic coupling processes with particular reference to the influence of the springtime phytoplankton bloom on the benthic zone. We then continued to conduct water column and underwater camera observations to monitor the subsequent benthic–pelagic processes taking place over the rest of the study period.

Table 1. The geographic location of each sample station along with their mean water depth and relative distance to the nearest aquaculture facility.

Station	Position		Depth	Distance
	Lat	Lon	(m)	(m)
St.Far	38.4388	141.4606	24.0	165.7
St.Mid	38.4381	141.4602	19.8	79.8
St.Close	38.4376	141.4595	12.6	0.0

2.2. Sediment Sampling for Benthic Macrofauna and Sediment Properties (6 Months)

We sampled a total of 6 sediment samples at each station using the Ekman grab (sampling area of 0.02 m^2). For benthic macrofauna, we merged 3 of the full-grab samples and sieved them through a 1.0 mm mesh with filtered seawater on board. All organisms retained were preserved in 70% ethanol for subsequent sorting, species identification, enumeration, and biomass measurements. Species were identified to the lowest possible taxonomic levels, which was mostly the family level, using a stereomicroscope and a compound microscope where necessary, expressing abundance and biomass (wet weight) as ind./m^2 and g/m^2, respectively [19]. We also used separate sediment subsamples of the top 5 cm to determine both organic matter content and particle size composition, respectively. Organic matter content was measured as a loss on ignition over 8 h at 450 °C after drying the samples at 55 °C until constant weights were reached [21]. Particle size composition was determined using standard laser diffraction analysis (SALD-2300 Laser Diffraction Particle Size Analyzer, SHIMADZU, Kyoto, Japan) and categorized into 6 size classes defined here as "coarse sand" (<2 mm, >0.5 mm), "medium sand" (<0.5 mm, >0.25 mm), "fine sand" (<0.25 mm, >125 μm), "very fine sand" (<125 μm, >62.5 μm), "silt" (<62.5 μm, >3.9 μm), and "clay" (<3.9 μm).

2.3. Water Column and Underwater Camera Monitoring (13 Months)

At each sampling station, we recorded vertical profiles of the water temperature (°C), salinity, chlorophyll-a concentration (μg/L), dissolved oxygen (mg/L), sigma-t (water density: σt), and turbidity (FTU), using a CTD RINKO-Profiler (JFE Advantech Co., Ltd., Nishinomiya, Japan) at 0.5 m intervals from the surface to the bottom [19]. We also used an underwater video camera (FDR-X3000 4K Action Cam, SONY, Tokyo, Japan) with an underwater light to monitor the state of the seafloor by counting the number of epibenthic fauna or associated biological features encountered at each sampling station. During the observation, the video-capturing apparatus was lowered to the seafloor from the boat and kept approximately 60 cm above the bottom for 1 min to obtain continuous video data to count epibenthic fauna with any notable aquatic features or markings.

2.4. Data Analysis

We examined the biological metrics for the observed benthic macrofaunal communities obtained from the sediment samples across the seasons and stations (i.e., abundance, biomass, species richness, Shannon index H', and evenness index J'). We then conducted a multivariate community analysis in PRIMER v6 [22] to examine the spatial and temporal trends in the benthic macrofaunal community structure across the seasons and stations. Following the fourth-root transformation of macrofaunal counts to down-weight the influence of highly abundant species, we performed cluster analysis (group-average linkage) on a resemblance matrix of the transformed data based on the Bray–Curtis similarity index. Non-metric multidimensional scaling (nMDS) was used to illustrate the degree of similarity among the benthic community structures across the sampling stations and seasons. ANOSIM was performed to test for significant differences among/between the three sampling stations, namely St. Far, St. Mid, and St. Close. SIMPER (similarity percentage) analysis was then performed to identify the species most responsible for the similarity within the stations determined by cluster analysis. All the graphics were constructed using the "ggplot2" package [23] in R version 3.6.1 [24].

3. Results

3.1. Benthic Macrofaunal Community Structure

We collected 18 benthos samples (3 stations × 6 months) and identified a total of 729 individuals comprising 629 from class Polychaeta (86.3%), 34 from class Gastropoda (4.7%), 20 from class Malacostraca (2.7%), 15 from class Ophiuroidea (2.1%), 11 from class Bivalvia (1.5%), and 22 others (2.7%) (Tables 2 and S1). St. Far consistently showed the highest total abundance over the 6-month period, which was followed by St. Mid, with polychaetes being the most abundant at both stations (Table 2, Figure 2a). St. Close showed the lowest levels of abundance, yet not only polychaetes but also other taxonomic groups such as gastropods and malacostracans were markedly abundant at the station. With respect to polychaetes, St. Far was dominated by species belonging to the family Lumbrineridae, which was followed by other families such as Cirratulidae, Ampharetidae, Capitellidae, and Magelonidae (Figure 3). The abundances of similar polychaeta families also characterized St. Mid except for the family Ampharetidae, exhibiting markedly lower abundance when compared to St. Far (Figure 3). Although St. Close showed higher abundances of both lumbrinerid and cirratulid polychaetes, the site was more characterized by the higher abundance of maldanidae, unlike the other two stations (Figure 3).

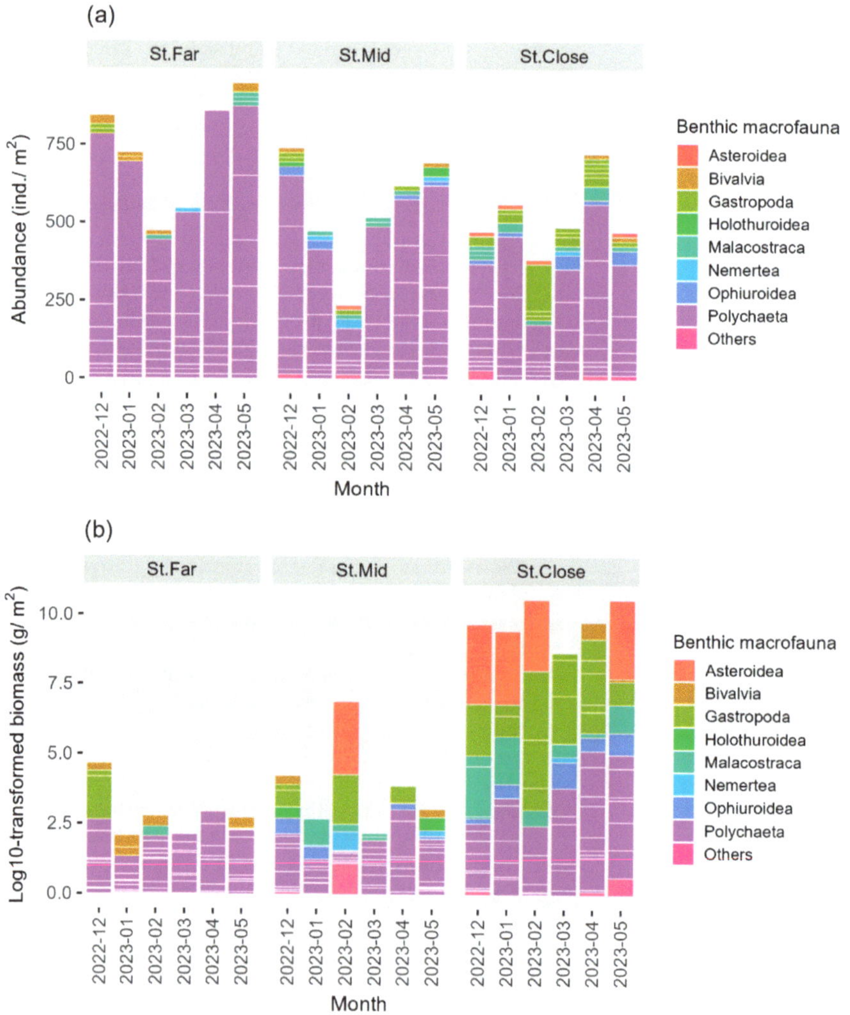

Figure 2. Temporal changes in (**a**) total abundance and (**b**) total biomass of the benthic macrofauna per taxonomic group (class) across sampling stations over the study period.

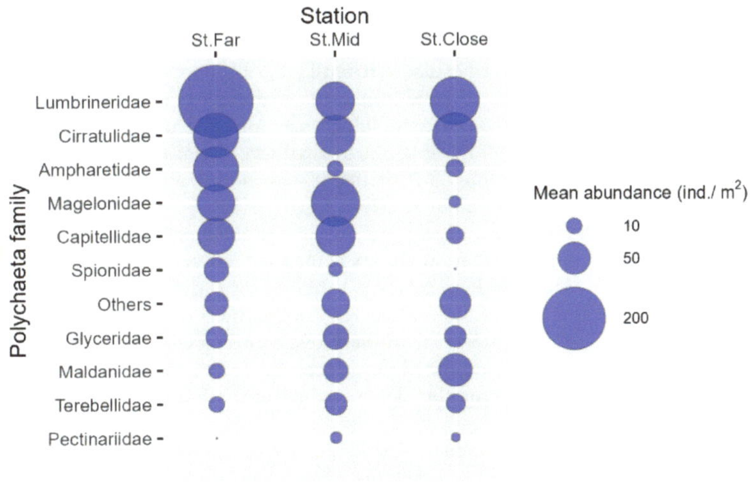

Figure 3. Mean abundance of polychaeta species collected per family across sampling stations.

Table 2. Summary data for the abundance of the benthic macrofauna in this study. N indicates the total number of benthic fauna per taxonomic group (ind.) collected over the entire study period.

Taxonomic Group	St.Far		St.Mid		St.Close		Total	
	N (ind.)	(%)	N (ind.)	(%)	N (ind.)	(%)	N (ind.)	(%)
Polychaeta	283	95.3	195	87.8	151	71.9	629	86.3
Gastropoda	2	0.7	4	1.8	28	13.3	34	4.7
Malacostraca	4	1.3	5	2.3	11	5.2	20	2.7
Ophiuroidea	-	-	6	2.7	9	4.3	15	2.1
Bivalvia	7	2.4	2	0.9	2	1.0	11	1.5
Nemertea	1	0.3	4	1.8	1	0.5	6	0.8
Asteroidea	-	-	1	0.5	4	1.9	5	0.7
Holothuroidea	-	-	3	1.4	-	-	3	0.4
Actinopterygii	-	-	1	0.5	1	0.5	2	0.3
Pycnogonida	-	-	-	-	2	1.0	2	0.3
Polycladida	-	-	1	0.5	-	-	1	0.1
Sipuncula	-	-	-	-	1	0.5	1	0.1
Total	297	100	222	100	210	100	729	100

In terms of biomass, class Asteroidea showed the highest value (70.3%), which was followed by Gastropoda (16.4%), Polychaeta (8.0%), Malacostraca (3.7%), Ophiuroidea (0.6%), and others (2.7%) (Table 3). Although St. Far maintained the highest macrofaunal abundance, the total biomass was the lowest among the three stations (Tables 2 and 3 and Figure 2b). St. Mid displayed intermediate amounts of biomass when compared to the other two stations (Table 3, Figure 2b). St. Close exhibited the highest levels of total biomass, which was characterized by the presence of various larger epibenthic megafauna such as asteroids, gastropods, and malacostracans (Table 3, Figure 2b).

The cluster analysis and the nMDS plot illustrate the clear separation of the benthic community structures systematically shifting from St. Close through St. Mid to St. Far (Figure 4a,b). The results of ANOSIM showed that the benthic community structures were significantly different among the three stations (ANOSIM: global R = 0.510, $p < 0.001$) (Table 4). Pairwise comparisons indicated that the community structure at St. Far was significantly different from those at both St. Mid (R = 0.380, $p < 0.01$) and St. Close (R = 0.872, $p < 0.01$), respectively, and the community structure at St. Mid was also significantly different from that at St. Close (R = 0.372, $p < 0.05$) (Table 4). SIMPER analysis revealed that the benthic community structures at St. Far, St. Mid, and St. Close showed an average similarity of 73.2, 59.3, and 58.7%, respectively (Table 5). The families of class Polycaeta such as Lumbrineridae, Cirratulidae, Ampharetidae, Capitellidae, Magelonidae, and Spionidae contributed predominantly to the average similarity at St. Far, whereas Magelonidae, Cirratulidae, Capitellidae, and Lumbrineridae were responsible for the similarity observed at St. Mid (Table 5). In addition to class Polychaeta, other taxonomic groups such as Ophiuroidea, Muricidae (Gastropoda), and Asterinidae (Asteroidea) contributed to the average similarity at St. Close (Table 5). St. Far tended to show the lowest overall values for taxonomic diversity indices, with the April sampling being the lowest for species richness and the Shannon index H', while the December sampling was the lowest for the evenness index J' (Figure 5a–c). While St. Mid displayed species richness values similar to those at St. Far, the values for the Shannon index H' were markedly higher than those at St. Far, showing the highest levels of the evenness index J' (Figure 5a–c). St. Close typically maintained the highest recorded values for both species richness and the Shannon index H', showing the highest values for both metrics in April (Figure 5a–c).

Table 3. Summary data for the biomass of the benthic macrofauna in this study. B indicates the total wet weight of benthic macrofauna per taxonomic group (g) collected over the entire study period.

Taxonomic Group	St.Far		St.Mid		St.Close		Total	
	B (g)	(%)	B (g)	(%)	B (g)	(%)	B (g)	(%)
Asteroidea	-	-	28.25	72.6	147.05	72.3	175.3	70.3
Gastropoda	2.29	31.9	4.30	11.1	34.37	16.9	41.0	16.4
Polychaeta	4.34	60.6	3.90	10.0	11.72	5.8	20.0	8.0
Malacostraca	0.09	1.3	0.63	1.6	8.52	4.2	9.24	3.7
Ophiuroidea	-	-	0.37	1.0	1.21	0.6	1.58	0.6
Actinopterygii	-	-	0.80	2.0	0.22	0.1	1.02	0.4
Bivalvia	0.44	6.2	0.14	0.4	0.23	0.1	0.81	0.3
Nemertea	0.004	0.1	0.30	0.8	0.04	0.02	0.34	0.1
Holothuroidea	-	-	0.23	0.6	-	-	0.23	0.1
Pycnogonida	-	-	-	-	0.03	0.01	0.03	0.01
Sipuncula	-	-	-	-	0.02	0.01	0.02	0.01
Polycladida	-	-	0.013	0.03	-	-	0.01	0.01
Total	7.2	100	38.9	100	203.4	100	249.5	100

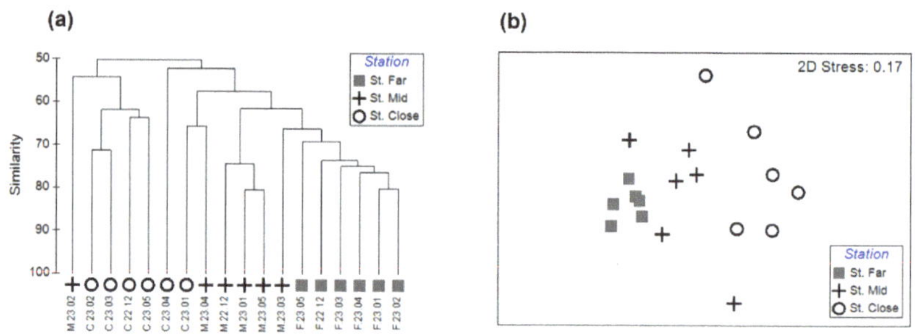

Figure 4. Results of the multivariate analysis of benthic macrofaunal community across sampling stations and months, showing (**a**) dendrogram and (**b**) non-metric multidimensional scaling (nMDS) ordination plot, based on fourth-root transformation and Bray–Curtis similarity and group average clustering.

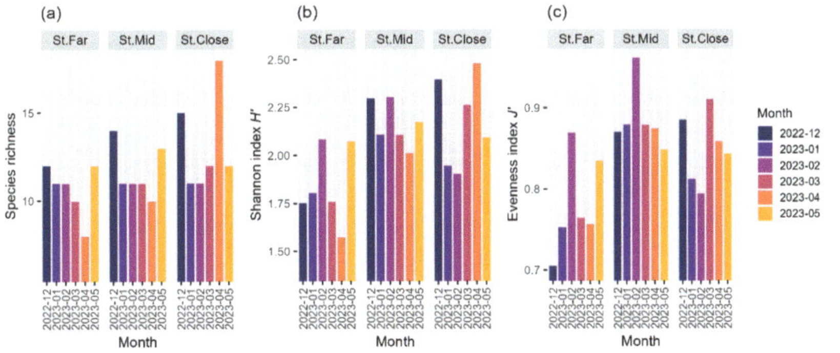

Figure 5. Temporal changes in (**a**) species richness, (**b**) Shannon index H', and (**c**) evenness index J', of the benthic macrofaunal community across sampling stations over the study period.

Table 4. ANOSIM results comparing variation in the structures of benthic macrofaunal communities between sampling stations.

Global Test	Pairwise Test	R	p
Global		0.510	<0.001
	St.Far × St.Mid	0.380	<0.01
	St.Far × St.Close	0.872	<0.01
	St.Mid × St.Close	0.372	<0.05

Table 5. Results of SIMPER analysis showing percentage contributions of major taxonomic groups that accounted for most of the similarity (>70%) within each station (Code: Av. Sim. = average similarity; Av. Abund. = average abundance; Contrib. = contribution; Cum. = cumulative contribution).

Station	Av. Sim. %	Taxonomic Group	Class	Av. Abund.	Contrib.%	Cum.%
St.Far	73.22	Lumbrineridae	Polychaeta	2.06	18.94	18.94
		Cirratulidae	Polychaeta	1.54	13.31	32.25
		Ampharetidae	Polychaeta	1.51	12.69	44.94
		Capitellidae	Polychaeta	1.41	12.51	57.46
		Magelonidae	Polychaeta	1.40	11.96	69.42
		Spionidae	Polychaeta	1.15	10.49	79.91
St.Mid	59.32	Magelonidae	Polychaeta	1.62	16.93	16.93
		Cirratulidae	Polychaeta	1.49	16.48	33.41
		Capitellidae	Polychaeta	1.48	15.84	49.25
		Lumbrineridae	Polychaeta	1.26	10.16	59.41
		Maldanidae	Polychaeta	0.99	8.39	67.8
		Other polychaeta	Polychaeta	1.03	7.97	75.77
St.Close	58.69	Cirratulidae	Polychaeta	1.57	16.3	16.3
		Lumbrineridae	Polychaeta	1.59	15.12	31.42
		Other polychaeta	Polychaeta	1.28	12.89	44.31
		Maldanidae	Polychaeta	1.31	12.85	57.16
		Ophiuroidea	Ophiuroidea	0.94	7.28	64.44
		Muricidae	Gastropoda	0.73	4.78	69.22
		Asterinidae	Asteroidea	0.67	4.73	73.96

3.2. Sediment Organic Matter Content and Particle Size Composition

The spatial patterns of sediment organic matter content decreased systematically from St. Far through St. Mid to St. Close as the distance to the nearest aquaculture facility decreased (Figure 6a, Table S2). The highest percentages of sediment organic matter were observed in May for the respective sites, although the increase was more pronounced at St.

Far and St. Mid (Figure 6a). In addition, the lowest percentage of sediment organic matter content was observed at the St. Close site in December (Figure 6a).

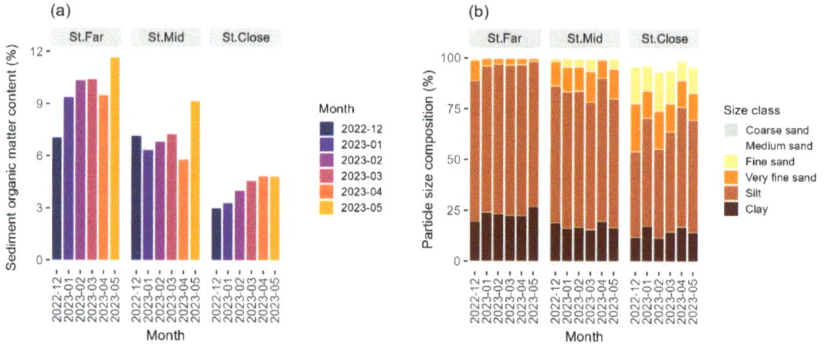

Figure 6. Temporal changes in (**a**) sediment organic matter content (%) and (**b**) particle size composition (%) of the seafloor sediment across sampling stations over the study period.

All three stations had silt as the dominant particle size class (Figure 6b, Table S3). However, sediment fractions of sand (i.e., coarse, medium, fine, and very fine sand) and mud (i.e., silt and clay) varied distinctively between the three stations, with St. Far and St. Close occupying opposite ends of the size composition spectrum (Figure 6b). The sediment particle size at St. Far consistently contained highly elevated proportions of silt, which was followed by clay, with a small fraction of grain sizes larger than fine sand present throughout the study period (Figure 6b). The sediment fractions of both silt and clay decreased, whereas that of larger sand increased systematically from St. Far through St. Mid to St. Close, exhibiting a clear gradient from muddy to sandy substrate as the proximity to the aquaculture facility became closer (Figure 6b).

3.3. Observation of Epibenthic Megafauna Based on the Video Data

The video data revealed that biogenic features such as tubes of polychaete worms, which belong to the family Maldanidae, and other faunal burrows dominated the bottom habitat at St. Far (Figure 7). St. Mid was characterized by more diverse epibenthic invertebrates such as asteroids, crinoids, and holothurians; however, faunal burrows and maldanid worm tubes still remained the conspicuous features recurring throughout the study period (Figure 7). St. Close differed drastically from both St. Far and St. Mid in that the seafloor at St. Close was characterized by a wider range of epibenthic megafauna including asteroids, crinoids, holothurians, echinoids, gastropods, and brachyuran decapods (Figure 7). The underwater camera also observed a variety of fish species in the class Actinopterygii, including gobies *Amblychaeturichthys sciistius*, *Pterogobius zacalles*, greenlings *Hexagrammos otakii*, and pufferfish *Takifugu snyderi*, most consistently present at St. Close but with the largest abundances occasionally observed at St. Mid (Figure 7). The abundances of the observed epibenthic megafauna and features tended to be the highest in cold months, particularly between February and April around the beginning of the study period across all the three stations. However, these temporal trends shifted to a rather steady decline in the warm months (e.g., between August and November 2023) observed across the three stations, with St. Far showing the most pronounced declining trend.

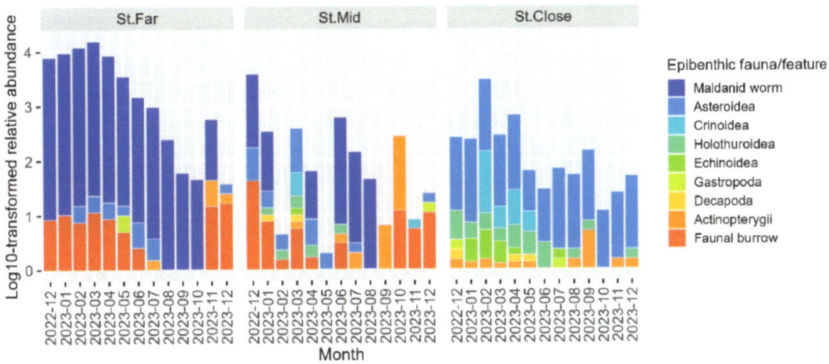

Figure 7. Temporal changes in the relative abundance of epibenthic fauna/features observed by underwater camera across sampling stations over the study period.

3.4. Observation of the Water Column Based on the CTD-Probe

Patterns of changes in temperature, salinity, and water density (Sigma-t) were almost identical between the three sampling stations, demonstrating clear seasonal patterns within the respective water depth layers (Figure 8a,c,e). For example, temporal changes in water temperature showed the lowest value around March and the highest value around September across all stations (Figure 8a). In terms of vertical trends, temperature values were almost the same among the depth layers during cold months between December and March; however, they gradually diverged, forming the largest stratification in the hottest month in September with higher temperatures always observed at shallower depth layers (Figure 8a). Spatiotemporal changes in density values showed precisely opposite patterns to those in temperature (Figure 8a,e), whereas salinity values exhibited a decreasing trend as the depth layers became shallower, with all three stations forming distinctive stratifications between the depth layers during warm months in the summer (Figure 8c).

In contrast, patterns of changes in the chlorophyll-a concentration, dissolved oxygen, and turbidity differed to a varying degree between the three stations (Figure 8b,d,f). Temporal changes in the chlorophyll-a concentration displayed similar peaks, once in spring around March at all depth layers and another in summer around August but only at shallower depths (0–10 m) across the three stations (Figure 8b). However, the values at 0–5 m and, particularly, 5–10 m depth layers declined systematically from St. Far through St. Mid to St. Close as the proximity to the nearest aquaculture facility became closer (Figures 1 and 8b). Dissolved oxygen (DO) showed essentially similar patterns between all three stations, exhibiting the highest and the lowest values observed across all depth layers around March and September, respectively. The DO values were also stratified around March and August, with the lowest values consistently observed at the bottom depths across all three stations (Figure 8d). Further, the values observed at depths deeper than 10 m decreased systematically from St. Far through St. Mid to St. Close as the proximity to the nearest aquaculture facility became closer (Figure 8d). Therefore, while the St. Far site exhibited minor anoxic conditions at a depth layer of 15–20 m in September, the St. Mid site displayed more adverse anoxic conditions at the same depth range, lasting a longer duration from August to October (Figure 8d). Finally, turbidity exhibited the lowest value around March and the highest value around August, with the values being very similar at all the depth layers except for the bottom layers, which showed the most turbid conditions across the three stations, with St. Close having the highest variation in turbidity (Figure 8f).

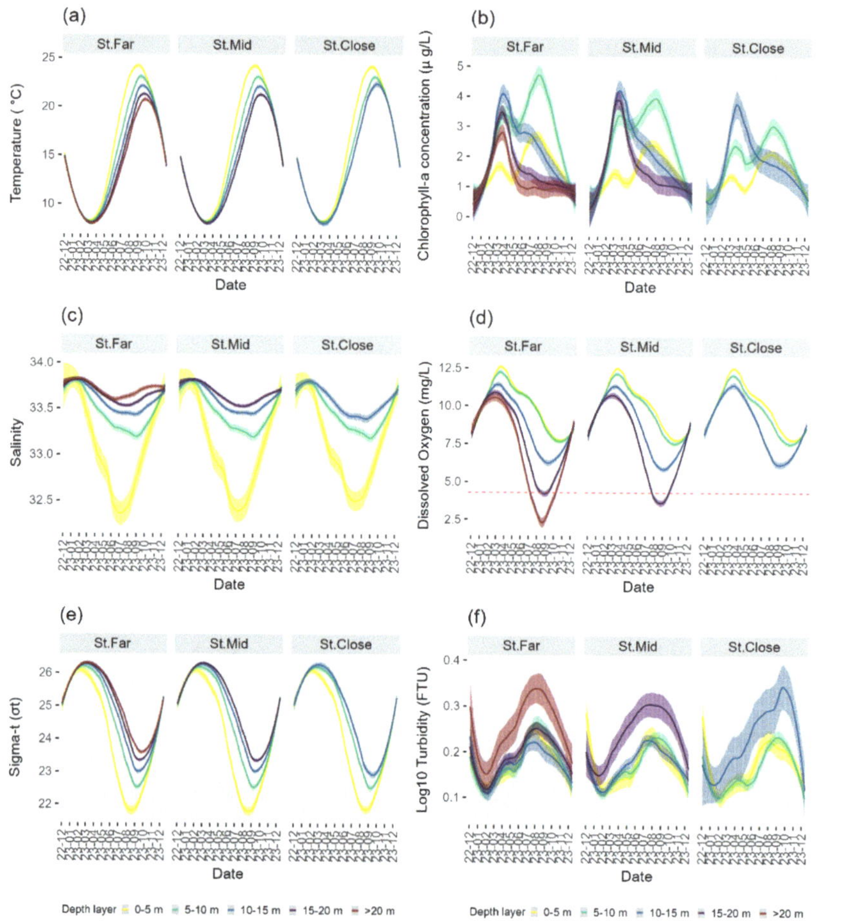

Figure 8. Temporal changes in environmental variables observed in the water column across sampling stations and depth layers (5 m intervals) for (**a**) water temperature; (**b**) chlorophyll-a concentration; (**c**) salinity; (**d**) dissolved oxygen; (**e**) sigma-t (seawater density); and (**f**) turbidity.

4. Discussion

This study revealed that the community structure of benthic macrofauna significantly changed in relation to the distance from the shellfish aquaculture facilities, and these changes were strongly associated with the variability in sediment properties of the respective seafloors in Onagawa Bay, Japan. While both benthic macrofaunal biodiversity and biomass became markedly higher, the sediment property of the seafloor shifted systematically from muddy to sandy substrate as the proximity to the nearest aquaculture facilities became closer, indicating that the physical presence of the shellfish aquaculture in the pelagic zone influences the bottom habitat below and thereby provides unique environmental gradients to which a variety of benthic fauna respond within a relatively small extent of space.

Polychaetes were the most abundant benthic macrofauna observed throughout the study site. The families Lumbrineridae, Magelonidae, and Cirratulidae showed the highest abundance in this study. Polychaetes are often bioturbators of the seafloor, playing essential roles in the biogeochemical processes by consuming and remineralizing organic matter supplied from the water column above through benthic–pelagic coupling interactions [25–27].

Lumbrinerid and magelonid polychaetes primarily inhabit sandy–muddy substrates in shallow coastal waters, actively participating in surface deposit feeding [28,29], whereas cirratulids tend to inhabit intertidal and nearshore soft-bottom seafloor with some living in crevices of hard substrates [30]. In addition, members of the families Capitellidae and Spionidae showed higher abundances at St. Far and St. Mid. They are often known as opportunistic species tolerant of stressful conditions and, hence, are used as bioindicators of an ecosystem's health [31–33]. The dominance of such opportunistic species at these stations may be indicative of the possible association of biological recolonization following a long-term severe hypoxia event in coastal ecosystems [34].

The organic carbon generated by shellfish aquaculture (i.e., biodeposits and detritus) provides substantial food and nutrient sources for these benthic and demersal biota [35]. However, the overall organic matter percentage was highest at St. Far rather than at St Close. A possible reason for this was that St. Close contained species that can readily utilize the organic matter material accumulated on the substrate and, as such, decrease its percentage within the benthic layer. While the video data showed that the tube constructions of polychaete Maldanidae were the most abundant at St. Far, the sediment grab samples revealed that the abundance of the maldanid polychaetes was the highest at St. Close. Bioturbation by maldanid polychaetes through their tube construction and feeding activities may influence sediment characteristics and thereby modify the benthic community structure of the seafloor [32,36]. Further, a certain maldanid species is capable of creating a porous substrate with high water content [37,38], which may have contributed to the formation of sandier sediment and the environmental improvement of the bottom habitat at St. Close.

The seafloor at St. Far was predominantly a muddier substrate and displayed the largest abundance of polychaetes yet lacked the species richness and diversity of other taxonomic groups. In contrast, the lowest abundance of polychaetes with a marked increase in the abundance of other epibenthic megafauna (e.g., starfish, gastropods, malacostracans, and ophiuroids), together with the occurrence of demersal fish, appeared to be responsible for the significant change in the benthic community structure observed at St. Close. In shallow aquatic ecosystems, seafloor conditions such as sediment granulometry critically influence the benthic community [39–42], and the physical presence of aquaculture facilities at St. Close may affect the seafloor environments through various pathways. For example, both aquaculture facilities (e.g., ropes, buoys, and anchors) and cultivated species (e.g., oysters and scallops) themselves provide unique three-dimensional hard substrates in the water column, which in turn attract a wide range of fouling and reef-associated fauna and flora including macroalgae, hydrozoans, crinoids, bivalves, barnacles, crustaceans, polychaetes, zooplankton, and fish [13]. These associated organisms can be dislodged and fall to the seafloor to form a structurally complex biogenic/shell mound immediately underneath the facilities [13]. Such shell mounds also offer unique sea-bottom habitats for many species, including reef-associated fish and epibenthic megafauna [13,43,44]. In addition, cultivated shellfish produce a substantial number of biodeposits such as undigested waste and feces, which increase the flux of organic matter and nutrients to the seafloor, altering local sediment characteristics and the benthic community composition [2,7,8,12,13,45,46], which may have facilitated the increase in the diversity of benthic infauna and epifauna at St. Close.

The formation of stratification in the water column accompanied by the severe hypoxic condition near the bottom occurred during the August–October sampling events that coincided with the substantial drop in the abundance of epibenthic megafauna observed in the video data across all sampling stations. The physical presence of shellfish facilities is known to impact local phytoplankton community structures [47]. For example, there was a significant decline in chlorophyll-a concentrations at 5–10 m depths at St. Close when compared to the other two stations. This was a clear indication of the enhanced consumption rate of phytoplankton caused by the shellfish aquaculture at the study site. Moreover, an unusually massive phytoplankton bloom occurred in summer, which coincided with the months of record-high temperatures and the development of severe hypoxic conditions in the benthic zone experienced throughout the study site, indicating that the combination of warmer waters and phytoplankton blooms may result in severe ecosystem-related consequences. Stratification of the water column by warm waters indicates that the water column does not mix well vertically, leading to stagnation and less opportunity for oxygen to be circulated to the lower depths. In addition, a higher amount of organic matter generated from the phytoplankton blooms can reach the bottom to fuel benthic fauna, facilitating the further depletion of oxygen and leading to the accelerated formation of anoxic zones [48–50]. In the case of our study, there was a strong coincidence between rapidly declining dissolved oxygen and the second phytoplankton bloom observed in August–September. At all three stations, dissolved oxygen levels drastically declined and the depth layers close to the bottom at St. Far and St. Mid reached severe anoxic conditions at the same time as the phytoplankton bloom in August–September. One interesting trend observed was the complete disappearance of any members of echinoids (e.g., sea urchins) beginning in June and lasting until the end of the study period. Echinoids are often important bioturbators that disturb the sediment to feed on organic matter and thereby affect sediment properties, providing complex biogeochemical interactions between the benthic and pelagic zones [51]. With the simultaneous disappearance of these bioturbators across all three stations, the development of the anoxic zone could well have been facilitated faster. The turbidity levels observed in this study also corroborate this linkage as at all three stations, extremely high levels of turbidity occurred during the second phytoplankton bloom, indicating a high amount of particulate matter in suspension (e.g., phytoplankton masses, algal growth, and substrate movement), which can further block sunlight for photosynthesis and other oxygen production-related processes. In a benthic–pelagic system where light cannot penetrate deep enough into the water column (e.g., high turbidity at depth), respiration can exceed the primary production at deeper depths, with low-light conditions leading to hypoxic or anoxic bottom water [52].

Overall, the physical presence of the shellfish aquaculture facilities may well effectively mediate pelagic–benthic energy fluxes. Currently, the benthic zones located close to or underneath the shellfish aquaculture facilities in Onagawa Bay appear to receive an enhanced amount of organic matter and detritus in the form of biodeposits generated from the farmed shellfish species, leading to an enhanced carrying capacity for accommodating larger populations of benthic infauna and megafauna along with demersal fish. However, our study demonstrated that such benthic–pelagic linkages may also pose a new challenge under the influence of recent global warming because it may cause shifts in the timing and magnitude of phytoplankton blooms, thereby causing widespread hypoxic conditions due to increased stratification of the water column accompanied by excess organic inputs from the surface primary production, as well as biodeposits generated from the filter-feeder aquaculture. Under such circumstances, one may well consider that aquaculture operation needs to be deeper at colder depths to avoid the negative impact of global warming. However, in this study, the dissolved oxygen levels observed at deeper layers decreased systematically as the proximity to the nearest aquaculture facility became closer. The possible expansion of aquaculture operation from shallow to deeper sites may therefore exacerbate the current problems caused by the warmer waters and hypoxic conditions. Although the present study only examined three stations situated in a local coastal area of Japan, the results clearly indicated complex interactions between the physical presence of the aquaculture facilities and the benthic–pelagic coupling processes. Given the large extent of marine aquaculture installed along coastal waters globally, there is a need to conduct more systematic and long-term monitoring at a number of locations covering wider geographical regions to identify the true ecological consequences of the expected increase in aquaculture operations in shallow aquatic ecosystems under the ever-increasing influence of global warming.

Supplementary Materials: The following supporting information can be downloaded at: https://www.mdpi.com/article/10.3390/jmse12081293/s1, Table S1: List of all the benthic macrofauna at class and family levels recorded across sampling stations over the study period; Table S2: Results of the sediment organic matter content analysis across sampling stations over the study period; Table S3: Results of the particle size composition analysis across sampling stations over the study period.

Author Contributions: Conceptualization, T.F.; methodology, T.F. and A.D.; validation, T.F. and W.S.-O.; formal analysis, T.F., A.D. and S.K.; investigation, T.F. and A.D.; resources, T.F. and W.S.-O.; data curation, T.F. and S.K.; writing—original draft preparation, T.F. and A.D.; writing—review and editing, A.D., S.K., W.S.-O. and T.F.; supervision, T.F.; project administration, T.F.; funding acquisition, A.D. and T.F. All authors have read and agreed to the published version of the manuscript.

Funding: This research was supported by the Tohoku University Junior Year Program (JYPE) in English 2022–2023 with the JASSO (Japan Student Services Organization) Scholarship. A.D. was supported by the Purdue College of Agriculture Scholarship and the Purdue Semester Abroad in Intercultural Learning (SAIL) Scholarship. T.F. was supported by the JST Grant Number JPMJPF2110 and KAKENHI-PROJECT-23H02230.

Institutional Review Board Statement: Not applicable.

Informed Consent Statement: Not applicable.

Data Availability Statement: Data are contained within the article.

Conflicts of Interest: The authors declare that the research was conducted in the absence of any commercial or financial relationships that could be construed as a potential conflict of interest.

References

1. FAO. *The State of World Fisheries and Aquaculture 2024*; FAO: Rome, Italy, 2024. [CrossRef]
2. Borja, Á.; Rodríguez, J.G.; Black, K.; Bodoy, A.; Emblow, C.; Fernandes, T.F.; Forte, J.; Karakassis, I.; Muxika, I.; Nickell, T.D.; et al. Assessing the suitability of a range of benthic indices in the evaluation of environmental impact of fin and shellfish aquaculture located in sites across Europe. *Aquaculture* **2009**, *293*, 231–240. [CrossRef]
3. Dumbauld, B.R.; Ruesink, J.L.; Rumrill, S.S. The ecological role of bivalve shellfish aquaculture in the estuarine environment: A review with application to oyster and clam culture in West Coast (USA) estuaries. *Aquaculture* **2009**, *290*, 196–223. [CrossRef]
4. Forrest, B.M.; Keeley, N.B.; Hopkins, G.A.; Webb, S.C.; Clement, D.M. Bivalve aquaculture in estuaries: Review and synthesis of oyster cultivation effects. *Aquaculture* **2009**, *298*, 1–15. [CrossRef]
5. McKindsey, C.W.; Archambault, P.; Callier, M.D.; Olivier, F. Influence of suspended and off-bottom mussel culture on the sea bottom and benthic habitats: A review. *Can. J. Zool.* **2011**, *89*, 622–646. [CrossRef]
6. Barrett, L.T.; Theuerkauf, S.J.; Rose, J.M.; Alleway, H.K.; Bricker, S.B.; Parker, M.; Petrolia, D.R.; Jones, R.C. Sustainable growth of non-fed aquaculture can generate valuable ecosystem benefits. *Ecosyst. Serv.* **2022**, *53*, 101396. [CrossRef]
7. Forchino, A.; Borja, A.; Brambilla, F.; Rodríguez, J.G.; Muxika, I.; Terova, G.; Saroglia, M. Evaluating the influence of offshore cage aquaculture on the benthic ecosystem in Alghero Bay (Sardinia, Italy) using AMBI and M-AMBI. *Ecol. Indic.* **2011**, *11*, 1112–1122. [CrossRef]
8. Tomassetti, P.; Gennaro, P.; Lattanzi, L.; Mercatali, I.; Persia, E.; Vani, D.; Porrello, S. Benthic community response to sediment organic enrichment by Mediterranean fish farms: Case studies. *Aquaculture* **2016**, *450*, 262–272. [CrossRef]
9. Tovar, A.; Moreno, C.; Manuel-Vez, M.P.; Garcia-Vargas, M. Environmental impacts of intensive aquaculture in marine waters. *Water Res.* **2000**, *34*, 334–342. [CrossRef]
10. Sarà, G.; Martire, M.L.; Sanfilippo, M.; Pulicanò, G.; Cortese, G.; Mazzola, A.; Manganaro, A.; Pusceddu, A. Impacts of marine aquaculture at large spatial scales: Evidences from N and P catchment loading and phytoplankton biomass. *Mar. Environ. Res.* **2011**, *71*, 317–324. [CrossRef]
11. Farmaki, E.; Thomaidis, N.; Pasias, I.; Baulard, C.; Papaharisis, L.; Efstathiou, C. Environmental impact of intensive aquaculture: Investigation on the accumulation of metals and nutrients in marine sediments of Greece. *Sci. Total Environ.* **2014**, *485*, 554–562. [CrossRef]
12. Shumway, S.E.; Davis, C.; Downey, R.; Karney, R.; Kraeuter, J.; Parsons, J.; Rheault, R.; Wikfors, G. Shellfish aquaculture–in praise of sustainable economies and environments. *World Aquac.* **2003**, *34*, 8–10.
13. McKindsey, C.W.; Anderson, M.R.; Barnes, P.; Courtenay, S.; Landry, T.; Skinner, M. *Effects of Shellfish Aquaculture on Fish Habitat*; Canadian Science Advisory Secretariat Research Document 2006/011; Fisheries and Oceans: Ottawa, ON, Canada, 2006; p. 84.
14. Weise, A.M.; Cromey, C.J.; Callier, M.D.; Archambault, P.; Chamberlain, J.; McKindsey, C.W. Shellfish-DEPOMOD: Modelling the biodeposition from suspended shellfish aquaculture and assessing benthic effects. *Aquaculture* **2009**, *288*, 239–253. [CrossRef]

15. Lee, J.S.; Kim, S.H.; Kim, Y.T.; Hong, S.J.; Han, J.H.; Hyun, J.H.; Shin, K.H. Influence of sea squirt (*Halocynthia roretzi*) aquaculture on benthic–pelagic coupling in coastal waters: A study of the South Sea in Korea. *Estuar. Coast. Shelf Sci.* **2012**, *99*, 10–20. [CrossRef]
16. Dimitriou, P.D.; Santi, I.; Moraitis, M.L.; Tsikopoulou, I.; Pitta, P.; Karakassis, I. Benthic–Pelagic Coupling in the Oligotrophic Eastern Mediterranean: A Synthesis of the HYPOXIA Project Results. *Front. Mar. Sci.* **2022**, *9*, 886335. [CrossRef]
17. Alonso-Pérez, F.; Ysebaert, T.; Castro, C.G. Effects of suspended mussel culture on benthic–pelagic coupling in a coastal upwelling system (Ría de Vigo, NW Iberian Peninsula). *J. Exp. Mar. Biol. Ecol.* **2010**, *382*, 96–107.
18. Christensen, P.B.; Glud, R.N.; Dalsgaard, T.; Gillespie, P. Impacts of longline mussel farming on oxygen and nitrogen dynamics and biological communities of coastal sediments. *Aquaculture* **2003**, *218*, 567–588. [CrossRef]
19. Fujii, T.; Kaneko, K.; Murata, H.; Yonezawa, C.; Katayama, A.; Kuraishi, M.; Nakamura, Y.; Takahashi, D.; Gomi, Y.; Abe, H.; et al. Spatio-Temporal Dynamics of Benthic Macrofaunal Communities in Relation to the Recovery of Coastal Aquaculture Operations Following the 2011 Great East Japan Earthquake and Tsunami. *Front. Mar. Sci.* **2019**, *5*, 535. [CrossRef]
20. Fujii, T.; Kaneko, K.; Nakamura, Y.; Murata, H.; Kuraishi, M.; Kijima, A. Assessment of coastal anthropo-ecological system dynamics in response to a tsunami catastrophe of an unprecedented magnitude encountered in Japan. *Sci. Total Environ.* **2021**, *783*, 146998. [CrossRef]
21. Pasternack, G.B.; Brush, G.S. Sedimentation cycles in a river-mouth tidal freshwater marsh. *Estuaries* **1998**, *21*, 407–415. [CrossRef]
22. Clarke, K.R.; Warwick, R.M. *Change in Marine Communities: An Approach to Statistical Analysis and Interpretation*; Primer-E: Plymouth, UK, 2001.
23. Wickham, H. *ggplot2: Elegant Graphics for Data Analysis*; Springer: New York, NY, USA, 2016.
24. R Development Core Team. *R: A Language and Environment for Statistical Computing*; R Foundation for Statistical Computing: Vienna, Austria, 2019.
25. Herman, P.M.J.; Middelburg, J.J.; Van De Koppel, J.; Heip, C.H.R. Ecology of estuarine macrobenthos. *Adv. Ecol. Res.* **1999**, *29*, 195–240.
26. Meysman, F.J.R.; Middelburg, J.J.; Heip, C.H.R. Bioturbation: A fresh look at Darwin's last idea. *Trends Ecol. Evol.* **2006**, *21*, 688–695. [CrossRef] [PubMed]
27. Ehrnsten, E.; Norkko, A.; Timmermann, K.; Gustafsson, B.G. Benthic-pelagic coupling in coastal seas–Modelling macrofaunal biomass and carbon processing in response to organic matter supply. *J. Mar. Syst.* **2019**, *196*, 36–47. [CrossRef]
28. Martins, R.; Carrera-Parra, L.F.; Quintino, V.; Rodrigues, A.M. Lumbrineridae (Polychaeta) from the Portuguese Continental Shelf (NE Atlantic) with the Description of Four New Species. *Zootaxa* **2012**, *3416*, 1–21. [CrossRef]
29. Mills, K.; Mortimer, K. Observations on the tubicolous annelid Magelona alleni (Magelonidae), with discussions on the relationship between morphology and behaviour of European magelonids. *J. Mar. Biol. Assoc. UK* **2019**, *99*, 715–727. [CrossRef]
30. Blake, J.A. Bitentaculate Cirratulidae (Annelida, Polychaeta) collected chiefly during cruises of the R/V Anton Bruun, USNS Eltanin, USCG Glacier, R/V Hero, RVIB Nathaniel B. Palmer, and R/V Polarstern from the Southern Ocean, Antarctica, and off Western South America. *Zootaxa* **2018**, *4537*, 1–130. [CrossRef]
31. Dean, H.K. The use of polychaetes (Annelida) as indicator species of marine pollution: A review. *Rev. Biol. Trop.* **2008**, *56*, 11–38.
32. Abe, H.; Kobayashi, G.; Sato-Okoshi, W. Impacts of the 2011 tsunami on the subtidal polychaete assemblage and the following recolonization in Onagawa Bay, northeastern Japan. *Mar. Environ. Res.* **2015**, *112*, 86–95.
33. Hilliard, J.; Hajduk, M.; Méndez, N.; Schulze, A. Capitella (Annelida: Capitellidae) species in the Gulf of Mexico: Delimitation, phylogeography and phylogeny. *Zool. Scr.* **2024**, *53*, 461–472. [CrossRef]
34. Steckbauer, A.; Duarte, C.M.; Carstensen, J.; Vaquer-Sunyer, R.; Conley, D.J. Ecosystem impacts of hypoxia: Thresholds of hypoxia and pathways to recovery. *Environ. Res. Lett.* **2011**, *6*, 025003. [CrossRef]
35. Theuerkauf, S.J.; Barrett, L.T.; Alleway, H.K.; Costa-Pierce, B.A.; St. Gelais, A.; Jones, R.C. Habitat value of bivalve shellfish and seaweed aquaculture for fish and invertebrates: Pathways, synthesis and next steps. *Rev. Aquac.* **2021**, *14*, 54–72. [CrossRef]
36. Levin, L.; Blair, N.; DeMaster, D.; Plaia, G.; Fornes, W.; Martin, C.; Thomas, C. Rapid subduction of organic matter by maldanid polychaetes on the North Carolina slope. *J. Mar. Res.* **1997**, *55*, 2237. [CrossRef]
37. Sanders, H.L.; Goudsmit, E.M.; Mills, E.L.; Hampson, G.E. A study of the intertidal fauna of Barnstable Harbor, Massachusetts. *Limnol. Oceanogr.* **1962**, *7*, 63–79. [CrossRef]
38. Mach, M.E.; Levings, C.D.; McDonald, P.S.; Chan, K.M. An Atlantic infaunal engineer is established in the northeast Pacific: *Clymenella torquata* (Polychaeta: Maldanidae) on the British Columbia and Washington Coasts. *Biol. Invasions* **2012**, *14*, 503–507. [CrossRef]
39. Constable, A.J. Ecology of benthic macro-invertebrates in soft-sediment environments: A review of progress towards quantitative models and predictions. *Aust. J. Ecol.* **1999**, *24*, 452–476. [CrossRef]
40. Thrush, S.F.; Hewitt, J.E.; Cummings, V.J.; Ellis, J.I.; Hatton, C.; Lohrer, A.; Norkko, A. Muddy waters: Elevating sediment input to coastal and estuarine habitats. *Front. Ecol. Environ.* **2004**, *2*, 299–306. [CrossRef]
41. Reiss, H.; Degraer, S.; Duineveld, G.C.A.; Kröncke, I.; Aldridge, J.; Craeymeersch, J.A.; Eggleton, J.D.; Hillewaert, H.; Lavaleye, M.S.S.; Moll, A.; et al. Spatial patterns of infauna, epifauna, and demersal fish communities in the North Sea. *ICES J. Mar. Sci.* **2010**, *67*, 278–293. [CrossRef]
42. Fujii, T. Climate change, sea-level rise and implications for coastal and estuarine shoreline management with particular reference to the ecology of intertidal benthic macrofauna in NW Europe. *Biology* **2012**, *1*, 597–616. [CrossRef]

43. Fujii, T. Potential influence of offshore oil and gas platforms on the feeding ecology of fish assemblages in the North Sea. *Mar. Ecol. Prog. Ser.* **2016**, *542*, 167–186. [CrossRef]
44. Meyer-Gutbrod, E.L.; Love, M.S.; Claisse, J.T.; Page, H.M.; Schroeder, D.M.; Miller, R.J. Decommissioning impacts on biotic assemblages associated with shell mounds beneath southern California offshore oil and gas platforms. *Bull. Mar. Sci.* **2019**, *95*, 683–701. [CrossRef]
45. Crawford, C.M.; Macleod, C.K.A.; Mitchell, I.M. Effects of shellfish farming on the benthic environment. *Aquaculture* **2003**, *224*, 117–140. [CrossRef]
46. Kalantzi, I.; Rico, A.; Mylona, K.; Pergantis, S.A.; Tsapakis, M. Fish farming, metals and antibiotics in the eastern Mediterranean Sea: Is there a threat to sediment wildlife? *Sci. Total Environ.* **2021**, *764*, 142843. [CrossRef]
47. Han, D.; Chen, Y.; Zhang, C.; Ren, Y.; Xue, Y.; Wan, R. Evaluating impacts of intensive shellfish aquaculture on a semi-closed marine ecosystem. *Ecol. Model.* **2017**, *359*, 193–200. [CrossRef]
48. Rabalais, N.N.; Turner, R.E.; JustiĆ, D.; Dortch, Q.; Wiseman, W.J.; SenGupta, B.K. Nutrient changes in the Mississippi River and system responses on the adjacent continental shelf. *Estuaries* **1994**, *19*, 386–407. [CrossRef]
49. Hagy, J.D.; Boynton, W.R.; Keefe, C.W.; Wood, K.V. Hypoxia in Chesapeake Bay, 1951–2001: Long-term change in relation to nutrient loading and river flow. *Estuaries* **2004**, *27*, 634–658. [CrossRef]
50. Ho, A.Y.; Xu, J.; Yin, K.; Yuan, X.; He, L.; Jiang, Y.; Lee, J.H.; Anderson, D.M.; Harrison, P.J. Seasonal and spatial dynamics of nutrients and phytoplankton biomass in Victoria Harbour and its vicinity before and after sewage abatement. *Mar. Pollut. Bull.* **2008**, *57*, 313–324. [CrossRef]
51. Lohrer, A.; Thrush, S.; Gibbs, M. Bioturbators enhance ecosystem function through complex biogeochemical interactions. *Nature* **2004**, *431*, 1092–1095. [CrossRef]
52. Baustian, M.M.; Hansen, G.J.A.; Kluijver, A.D.; Robinson, K.; Henry, E.N.; Knoll, L.B.; Rose, K.C.; Carey, C. Linking the bottom to the top in aquatic ecosystems: Mechanisms and stressors of benthic-pelagic coupling. In *Eco-DAS X Symposium Proceedings*; Kemp, P.F., Ed.; Association for the Sciences of Limnology and Oceanography: Washington, DC, USA, 2014; Volume 3, pp. 25–47. [CrossRef]

MDPI AG
Grosspeteranlage 5
4052 Basel
Switzerland
Tel.: +41 61 683 77 34

Journal of Marine Science and Engineering Editorial Office
E-mail: jmse@mdpi.com
www.mdpi.com/journal/jmse